New York, New York Columbus, Ohio Chicago, Illinois Woodland Hills, California

The McGraw-Hill Companies

Send all inquiries to:
Glencoe/McGraw-Hill
8787 Orion Place
Columbus, OH 43240-4027

ISBN: 978-0-07-891508-6 *(Student Edition)*
MHID: 0-07-891508-2 *(Student Edition)*
ISBN: 978-0-07-891509-3 *(Teacher Wraparound Edition)*
MHID: 0-07-891509-0 *(Teacher Wraparound Edition)*

Printed in the United States of America.

2 3 4 5 6 7 8 9 10 071 17 16 15 14 13 12 11 10 09

Handbook at a Glance

CONTENTS

Handbook Contents

PART TWO

3 Powers and Roots

4 Data, Statistics, and Probability

6 Algebra

6•5 Ratio and Proportion

6•6 Inequalities

6•7 Graphing on the Coordinate Plane

6•8 Slope and Intercept

6•9 Direct Variation

CONTENTS

CONTENTS

8 Measurement

9 Tools

Handbook Introduction

Why use this handbook?

You will use this handbook to refresh your memory of mathematics concepts and skills.

What are HotWords, and how do you find them?

HotWords are important mathematical terms. The **HotWords** section includes a glossary of terms, a collection of common or significant mathematical patterns, and lists of symbols and formulas in alphabetical order. Many entries in the glossary will refer you to chapters and topics in the **HotTopics** section for more detailed information.

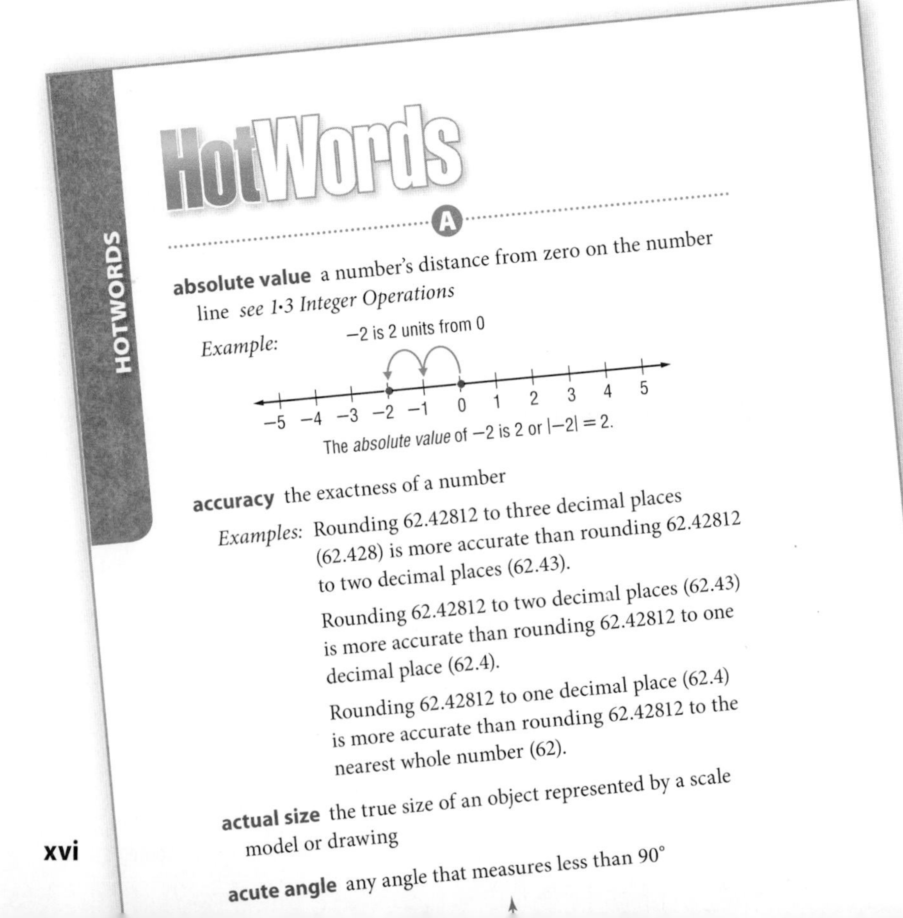

HOTWORDS

HotWords

A

absolute value a number's distance from zero on the number line *see 1•3 Integer Operations*

Example:

accuracy the exactness of a number

Examples: Rounding 62.42812 to three decimal places (62.428) is more accurate than rounding 62.42812 to two decimal places (62.43).

Rounding 62.42812 to two decimal places (62.43) is more accurate than rounding 62.42812 to one decimal place (62.4).

Rounding 62.42812 to one decimal place (62.4) is more accurate than rounding 62.42812 to the nearest whole number (62).

actual size the true size of an object represented by a scale model or drawing

acute angle any angle that measures less than 90°

What are HotTopics, and how do you use them?

HotTopics are key concepts that you need to know. The **HotTopics** section consists of nine chapters. Each chapter has several topics that give you to-the-point explanations of key mathematical concepts. Each topic includes one or more concepts. Each section includes Check It Out exercises, which you can use to check your understanding. At the end of each topic, there is an exercise set.

There are problems and a vocabulary list at the beginning and end of each chapter to help you preview what you know and review what you have learned.

What are HotSolutions?

The **HotSolutions** section gives you easy-to-locate answers to Check It Out and What Do You Know? problems. The **HotSolutions** section is at the back of the handbook.

HOTSOLUTIONS

HotSolutions

Chapter 1 Numbers and C

p. 72 **1.** $(4 + 7) \cdot 3 = 33$ **2.** $(30 + 15) \div$
3. no **4.** no **5.** yes **6.** no **7.** 2
9. $2 \cdot 5 \cdot 23$ **10.** 4 **11.** 5 **12.** 9
15. 90 **16.** 60

p. 73 **17.** 7, 7 **18.** 15, −15 **19.** 12, 12
21. >; −3 −2 −1 0 1 2 3
22. <; −8 −7 −6 −5 −4 −3 −2
23. >; −6 −5 −4 −3 −2 −1
24. >; −8 −7 −6 −5 −4 −3
25. 2 **26.** −4 **27.** −11 **28.** 1

FACTORS AND MULTIPLES 1.2

1•2 Factors and Multiples

Factors

Two numbers multiplied together to produce 12 … **factors** of 12. So, the factors of 12 are 1, 2, 3, 4, 6, …

To decide whether one number is a factor of anot… there is a remainder of 0, the number is a factor.

EXAMPLE **Finding the Factors of a Number**

What are the factors of 18?

$1 \cdot 18 = 18$
$2 \cdot 9 = 18$
$3 \cdot 6 = 18$

• Find all pairs of numbers t… to give the product.

1, 2, 3, 6, 9, 18

• List the factors in order, sta…

So, the factors of 18 are 1, 2, 3, 6, 9, and 18.

Check It Out

Find the factors of each number.

1. 8
2. 48

Common Factors

Factors that are the same for two or more numbers are **common factors**.

rotation
inverse
statistics
geomet

Part One 1

HotWords

The **HotWords** section includes a glossary of terms, lists of formulas and symbols, and a collection of common or significant mathematical patterns. Many entries in the glossary will refer to chapters and topics in the **HotTopics** section.

HotWords

A

absolute value a number's distance from zero on the number line *see 1·3 Integer Operations*

Example:

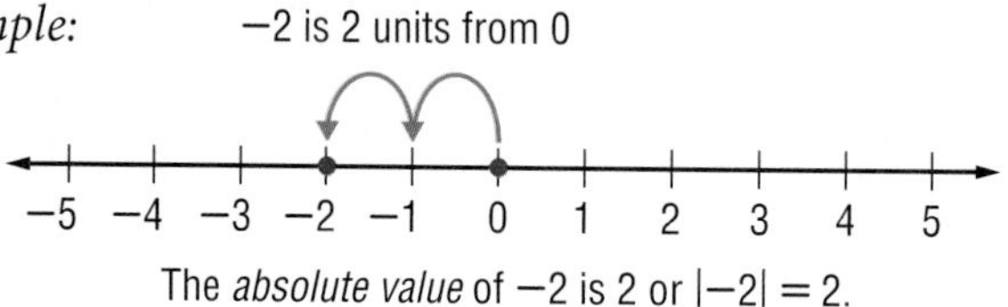

The *absolute value* of −2 is 2 or $|-2| = 2$.

accuracy the exactness of a number

Examples: Rounding 62.42812 to three decimal places (62.428) is more accurate than rounding 62.42812 to two decimal places (62.43).

Rounding 62.42812 to two decimal places (62.43) is more accurate than rounding 62.42812 to one decimal place (62.4).

Rounding 62.42812 to one decimal place (62.4) is more accurate than rounding 62.42812 to the nearest whole number (62).

actual size the true size of an object represented by a scale model or drawing

acute angle any angle that measures less than 90°

Example:

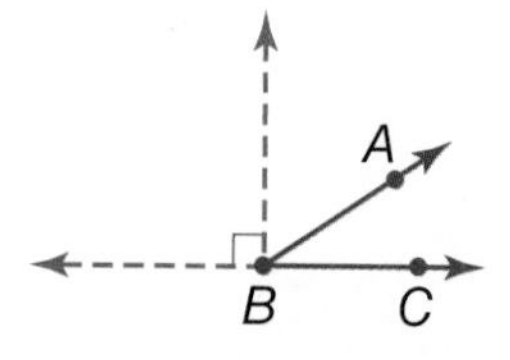

$\angle ABC$ is an *acute angle*.
$0° < m\angle ABC < 90°$

acute triangle a triangle in which all angles measure less than 90° *see 7•1 Classifying Angles and Triangles*

Example:

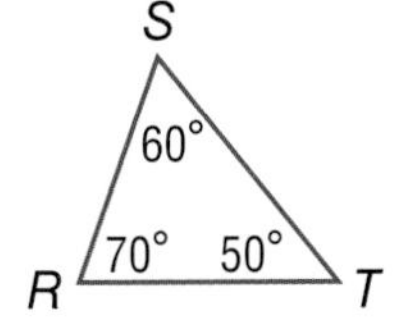

△*RST* is an *acute triangle.*

Addition Property of Equality the mathematical rule that states that if the same number is added to each side of an equation, the expressions remain equal *see 6•4 Solving Linear Equations*

Example: If $a = b$, then $a + c = b + c$.

additive inverse two integers that are opposite of each other; the sum of any number and its *additive inverse* is zero *see 6•4 Solving Linear Equations*

Example: $(+3) + (-3) = 0$
(-3) is the *additive inverse* of 3.

additive system a mathematical system in which the values of individual symbols are added together to determine the value of a sequence of symbols

Example: The Roman numeral system, which uses symbols such as I, V, D, and M, is a well-known *additive system.*

This is another example of an additive system:

▽▽□

If □ equals 1 and ▽ equals 7,
then ▽▽□ equals $7 + 7 + 1 = 15$.

algebra a branch of mathematics in which symbols are used to represent numbers and express mathematical relationships *see Chapter 6 Algebra*

algorithm a step-by-step procedure for a mathematical operation

alternate exterior angles in the figure below, transversal t intersects lines ℓ and m; $\angle 1$ and $\angle 7$, and $\angle 2$ and $\angle 8$ are alternate exterior angles; if lines ℓ and m are parallel, then these pairs of angles are congruent *see 7·1 Classifying Angles and Triangles*

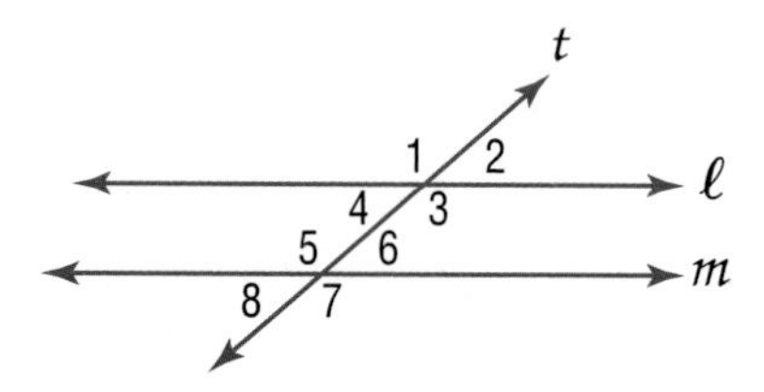

alternate interior angles in the figure below, transversal t intersects lines ℓ and m; $\angle 3$ and $\angle 5$, and $\angle 4$ and $\angle 6$ are alternate interior angles; if lines ℓ and m are parallel, then these pairs of angles are congruent *see 7·1 Classifying Angles and Triangles*

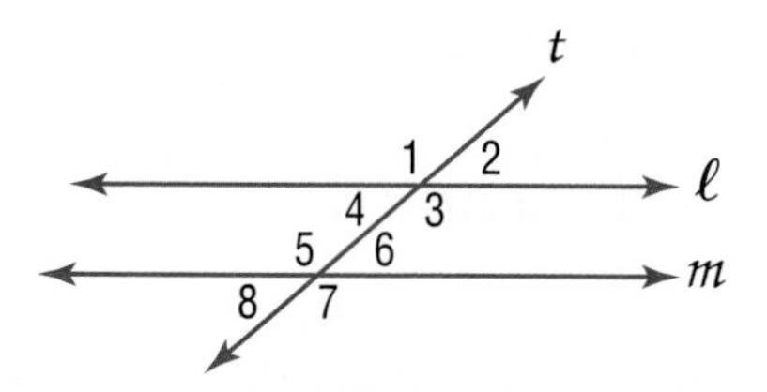

altitude the perpendicular distance from a vertex to the opposite side of a figure; *altitude* indicates the height of a figure

Example:

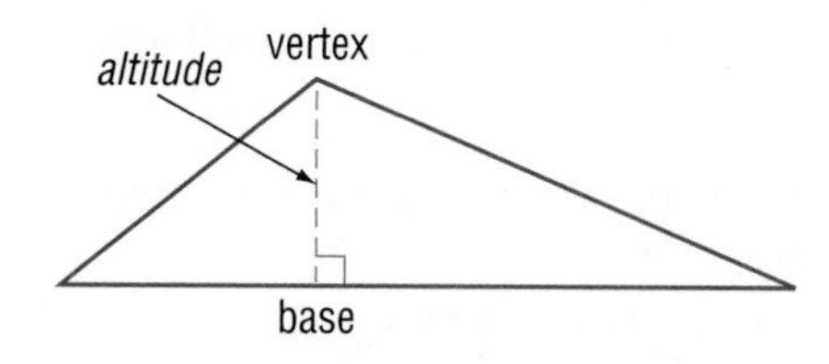

angle two rays that meet at a common endpoint

Example:

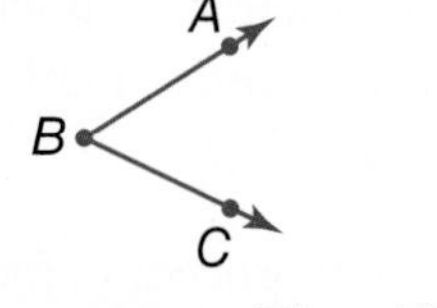

$\angle ABC$ is formed by $\overrightarrow{BA}$ and $\overrightarrow{BC}$.

angle of elevation the angle formed by a horizontal line and an upward line of sight

Example:

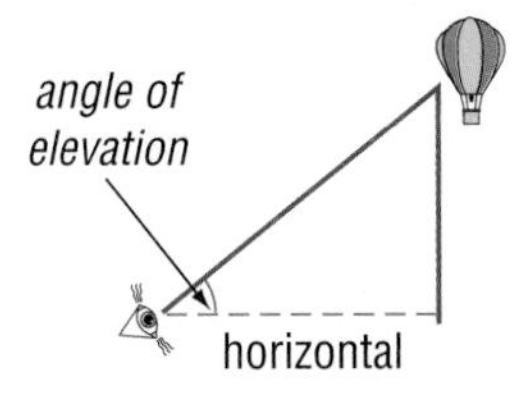

apothem a perpendicular line segment from the center of a regular polygon to one of its sides

Example:

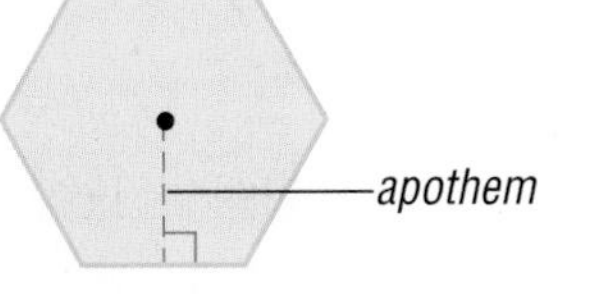

approximation an estimate of a mathematical value

Arabic numerals (or Hindu-Arabic numerals) the number symbols we presently use in our base-ten number system {0, 1, 2, 3, 4, 5, 6, 7, 8, 9}

arc a section of a circle *see 7•8 Circles*

Example:

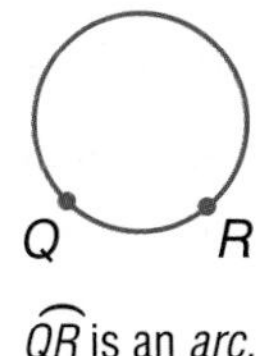

$\overset{\frown}{QR}$ is an *arc.*

area the measure of the interior region of a 2-dimensional figure or the surface of a 3-dimensional figure, expressed in square units *see Formulas page 64, 7•5 Area, 7•6 Surface Area, 7•8 Circles, 8•3 Area, Volume, and Capacity*

Example:

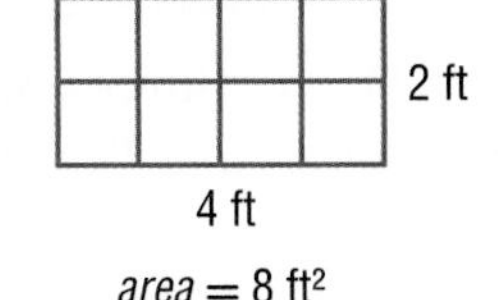

area = 8 ft^2

HOTWORDS

arithmetic expression a mathematical relationship expressed as a number, or two or more numbers with operation symbols *see expression*

arithmetic sequence *see Patterns page 67, 6•7 Graphing on the Coordinate Plane*

Associative Property the mathematical rule that states that the way in which numbers are grouped when they are added or multiplied does not change their sum or product
see 6•2 Simplifying Expressions

Examples: $(x + y) + z = x + (y + z)$
$x \cdot (y \cdot z) = (x \cdot y) \cdot z$

average the sum of a set of values divided by the number of values *see 4•4 Statistics*

Example: The *average* of 3, 4, 7, and 10 is
$(3 + 4 + 7 + 10) \div 4 = 6$.

average speed the average rate at which an object moves

axis (pl. *axes*) [1] a reference line by which a point on a coordinate graph may be located; [2] the imaginary line about which an object may be said to be symmetrical (*axis* of symmetry); [3] the line about which an object may revolve (*axis* of rotation) *see 6•7 Graphing on the Coordinate Plane*

B

bar graph a display of data that uses horizontal or vertical bars to compare quantities *see 4•2 Displaying Data*

base [1] the number used as the factor in exponential form; [2] two parallel congruent faces of a prism or the face opposite the apex of a pyramid or cone; [3] the side perpendicular to the height of a polygon; [4] the number of characters in a number system *see 3•1 Powers and Exponents, 7•5 Area, 7•7 Volume*

base-ten system the number system containing ten single-digit symbols {0, 1, 2, 3, 4, 5, 6, 7, 8, and 9} in which the numeral 10 represents the quantity ten

base-two system the number system containing two single-digit symbols {0 and 1} in which 10 represents the quantity two *see binary system*

benchmark a point of reference from which measurements and percents can be estimated *see 2·6 Percents*

best chance in a set of values, the event most likely to occur

biased sample a sample drawn in such a way that one or more parts of the population are favored over others
see 4·1 Collecting Data

bimodal distribution a statistical model that has two different peaks of frequency distribution *see 4·3 Analyzing Data*

binary system the base-two number system, in which combinations of the digits 1 and 0 represent different numbers or values

binomial an algebraic expression that has two terms

Examples: $x^2 + y$; $x + 1$; $a - 2b$

box plot a diagram that summarizes numerical data using the median, the upper and lower quartiles, and the maximum and minimum values *see 4·2 Displaying Data*

budget a spending plan based on an estimate of income and expenses

C

capacity the amount that can be held in a container

cell a small rectangle in a spreadsheet that stores information; each *cell* can store a label, number, or formula
see 9·3 Spreadsheets

center of the circle the point from which all points on a circle are equidistant *see 7•8 Circles*

chance the probability or likelihood of an occurrence, often expressed as a fraction, decimal, percentage, or ratio *see 4•6 Probability*

circle the set of all points in a plane that are equidistant from a fixed point called the center

Example:

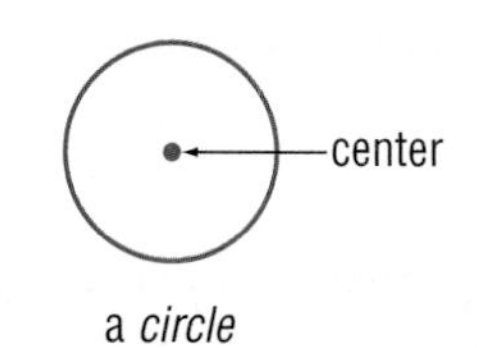

a *circle*

circle graph (pie chart) a display of statistical data that uses a circle divided into proportionally-sized "slices" *see 4•2 Displaying Data*

Example:

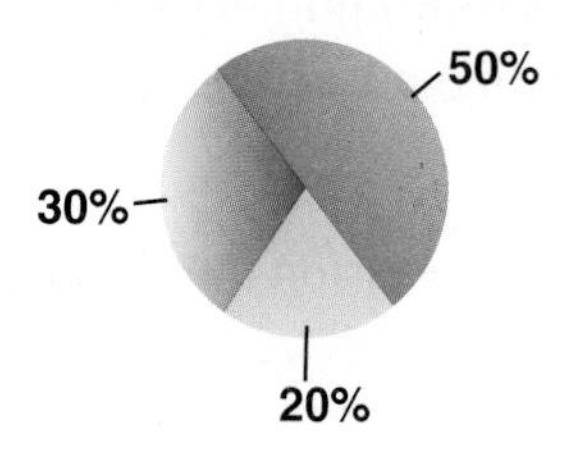

circumference the distance around (perimeter) a circle *see Formulas page 65, 7•8 Circles*

classification the grouping of elements into separate classes or sets

coefficient the numerical factor of a term that contains a variable *see 6•2 Simplifying Expressions*

collinear a set of points that lie on the same line

Example:

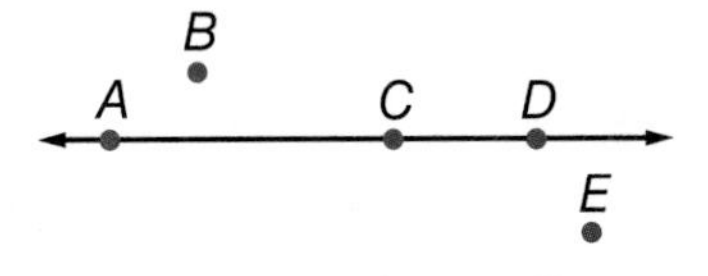

Points *A*, *C*, and *D* are *collinear.*

columns vertical lists of numbers or terms

combination a selection of elements from a larger set in which the order does not matter *see 4•5 Combinations and Permutations*

Example: 456, 564, and 654 are one *combination* of three digits from 4567.

common denominator a common multiple of the denominators of a group of fractions *see 2•2 Operations with Fractions*

Example: The fractions $\frac{3}{4}$ and $\frac{7}{8}$ have a *common denominator* of 8.

common difference the difference between any two consecutive terms in an arithmetic sequence

common factor a whole number that is a factor of each number in a set of numbers *see 1•2 Factors and Multiples*

Example: 5 is a *common factor* of 10, 15, 25, and 100.

common ratio the ratio of any term in a geometric sequence to the term that precedes it

Commutative Property the mathematical rule that states that the order in which numbers are added or multiplied does not change their sum or product *see 6•2 Simplifying Expressions*

Examples: $x + y = y + x$

$x \cdot y = y \cdot x$

compatible numbers two numbers that are easy to add, subtract, multiply, or divide mentally

complementary angles two angles are complementary if the sum of their measures is 90° *see 7·1 Classifying Angles and Triangles*

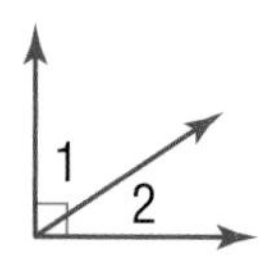

∠1 and ∠2 are *complementary angles.*

composite number a whole number greater than 1 having more than two factors *see 1·2 Factors and Multiples*

concave polygon a polygon that has an interior angle greater than 180°

Example:

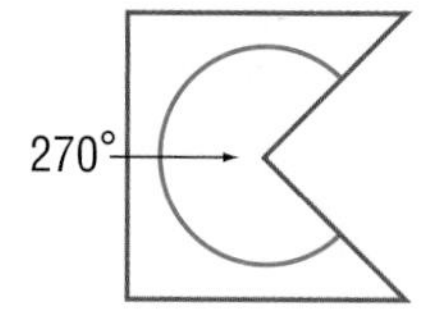

a *concave polygon*

conditional a statement that something is true or will be true provided that something else is also true *see 5·1 If/Then Statements*

Example: If a polygon has three sides, then it is a triangle.

cone a three-dimensional figure consisting of a circular base and one vertex

Example:

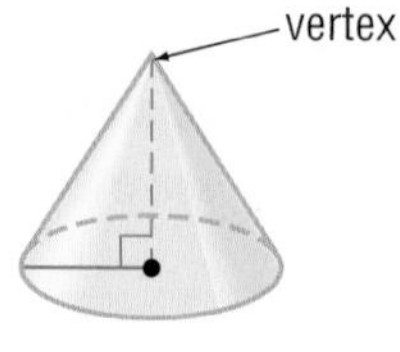

a *cone*

congruent having the same size and shape; the symbol $\cong$ is used to indicate congruence *see 7·1 Classifying Angles and Triangles*

Example:

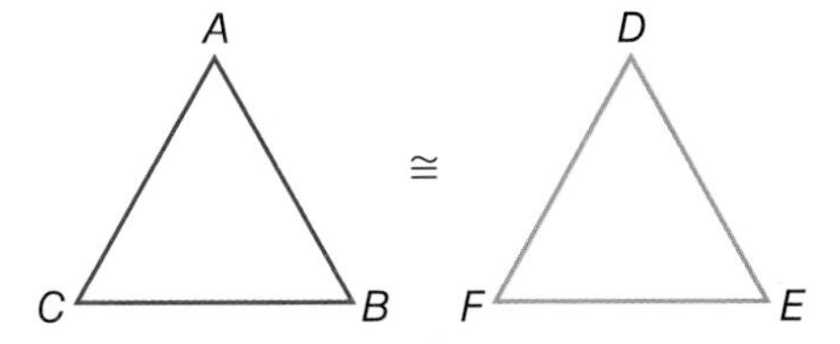

$\triangle ABC$ and $\triangle DEF$ are *congruent.*

congruent angles angles that have the same measure

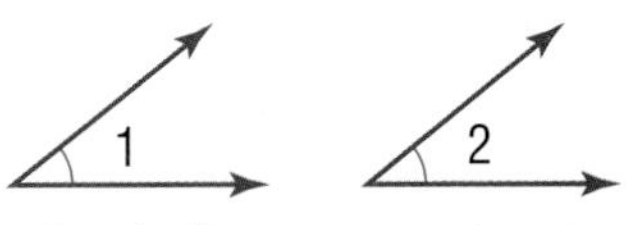

$\angle 1$ and $\angle 2$ are *congruent angles.*

conic section the curved shape that results when a conical surface is intersected by a plane

Example:

This ellipse is a *conic section.*

constant of variation a constant ratio in a direct variation *see 6·9 Direct Variation*

continuous data the complete range of values on the number line

Example: The possible sizes of apples are *continuous* data.

contrapositive a logical equivalent of a given conditional statement, often expressed in negative terms *see 5·1 If/Then Statements*

Example: "If *x*, then *y*" is a conditional statement; "if not *y*, then not *x*" is the *contrapositive* statement.

convenience sampling a sample obtained by surveying people who are easiest to reach; *convenience sampling* does not represent the entire population, therefore it is considered biased

converse a conditional statement in which terms are expressed in reverse order *see 5•1 If/Then Statements*

Example: "If x, then y" is a conditional statement; "if y, then x" is the *converse* statement.

convex polygon a polygon with all interior angles measuring less than 180°

Example:

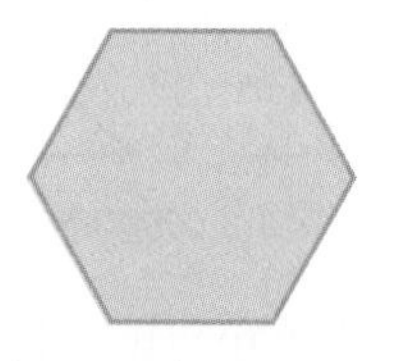

A regular hexagon is a *convex polygon.*

coordinate any number within a set of numbers that is used to define a point's location on a line, on a surface, or in space *see 1•3 Integer Operations*

coordinate plane a plane in which a horizontal number line and a vertical number line intersect at their zero points *see 6•7 Graphing on the Coordinate Plane*

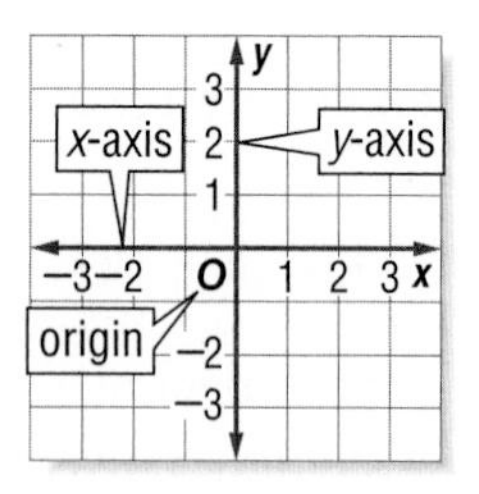

coplanar points or lines lying in the same plane

correlation the way in which a change in one variable corresponds to a change in another *see 4•3 Analyzing Data*

corresponding angles in the figure below, transversal t intersects lines ℓ and m; $\angle 1$ and $\angle 5$, $\angle 2$ and $\angle 6$, $\angle 4$ and $\angle 8$, and $\angle 3$ and $\angle 7$ are *corresponding angles;* if lines ℓ and m are parallel, then these pairs of angles are congruent
see 7·1 Classifying Angles and Triangles

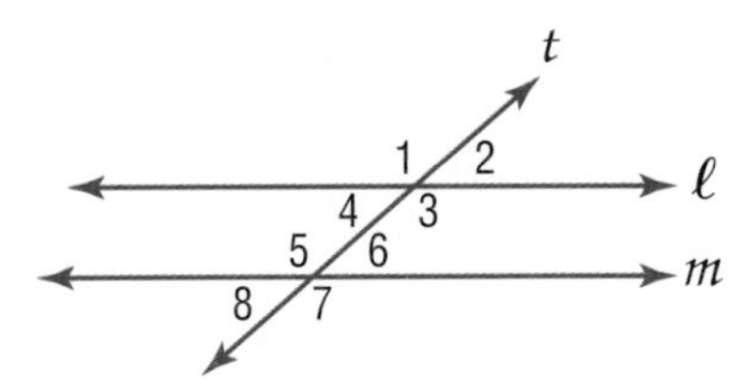

cost an amount paid or required in payment

cost estimate an approximate amount to be paid or to be required in payment

counterexample a statement or example that disproves a conjecture *see 5·2 Counterexamples*

counting numbers the set of positive whole numbers {1, 2, 3, 4 . . .} *see positive integers*

cross product a method used to solve proportions and test whether ratios are equal *see 2·1 Fractions, 6·5 Ratio and Proportion*

Example: $\frac{a}{b} = \frac{c}{d}$ if $a \cdot d = b \cdot c$

cross section the figure formed by the intersection of a solid and a plane

Example:

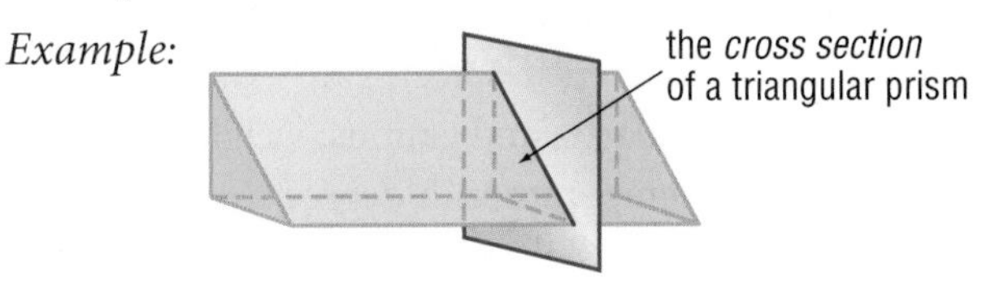

cube [1] a solid figure with six congruent square faces *see 7•2 Naming and Classifying Polygons and Polyhedrons*
[2] the product of three equal terms *see 3•1 Powers and Exponents*

Examples: [1]

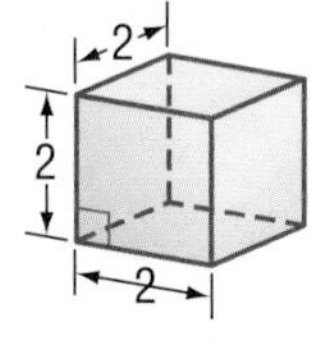

a *cube*

[2] $2^3 = 2 \cdot 2 \cdot 2 = 8$

cube root a number that when raised to the third power equals a given number *see 3•2 Square and Cube Roots*

Example: $\sqrt[3]{8} = 2$
2 is the *cube root* of 8

cubic centimeter the volume of a cube with edges that are 1 centimeter in length

cubic foot the volume of a cube with edges that are 1 foot in length

cubic inch the volume of a cube with edges that are 1 inch in length *see 7•7 Volume*

cubic meter the volume of a cube with edges that are 1 meter in length *see 7•7 Volume*

customary system units of measurement used in the United States to measure length in inches, feet, yards, and miles; capacity in cups, pints, quarts, and gallons; weight in ounces, pounds, and tons; and temperature in degrees Fahrenheit *see English system, 8•1 Systems of Measurement*

cylinder a solid shape with parallel circular bases

Example:

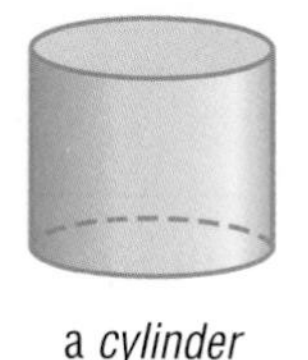

a *cylinder*

D

decagon a polygon with ten angles and ten sides

decimal system the most commonly used number system, in which whole numbers and fractions are represented using base ten

Example: Decimal numbers include 1230, 1.23, 0.23, and −13.

degree [1] (algebraic) the exponent of a single variable in a simple algebraic term; [2] (algebraic) the sum of the exponents of all the variables in a more complex algebraic term; [3] (algebraic) the highest degree of any term in a polynomial; [4] (geometric) a unit of measurement of an angle or arc, represented by the symbol °

Examples: [1] In the term $2x^4y^3z^2$, x has a *degree* of 4, y has a *degree* of 3, and z has a *degree* of 2.

[2] The term $2x^4y^3z^2$ as a whole has a *degree* of $4 + 3 + 2 = 9$.

[3] The equation $x^3 = 3x^2 + x$ is an equation of the third *degree.*

[4] An acute angle is an angle that measures less than 90°.

denominator the bottom number in a fraction representing the total number of equal parts in the whole *see 2·1 Fractions*

Example: In the fraction $\frac{a}{b}$, b is the *denominator.*

dependent events two events in which the outcome of one event is affected by the outcome of another event
see 4·6 Probability

HOTWORDS

diagonal a line segment connecting two non-adjacent vertices of a polygon *see 7•2 Naming and Classifying Polygons and Polyhedrons*

Example:

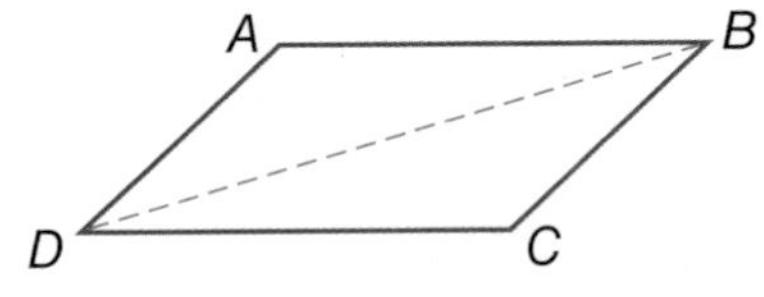

$\overline{BD}$ is a *diagonal* of parallelogram *ABCD*.

diameter a line segment connecting the center of a circle with two points on its perimeter *see 7•8 Circles*

Example:

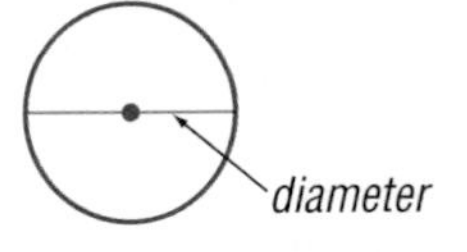

difference the result obtained when one number is subtracted from another *see 6•1 Writing Expressions and Equations*

dimension the number of measures needed to describe a figure geometrically

Examples: A point has 0 *dimensions.*
A line or curve has 1 *dimension.*
A plane figure has 2 *dimensions.*
A solid figure has 3 *dimensions.*

direct correlation the relationship between two or more elements that increase and decrease together *see 4•3 Analyzing Data*

Example: At an hourly pay rate, an increase in the number of hours worked means an increase in the amount paid, while a decrease in the number of hours worked means a decrease in the amount paid.

direct variation a relationship between two variable quantities with a constant ratio *see 6•9 Direct Variation*

discount a deduction made from the regular price of a product or service *see 2•7 Using and Finding Percents*

discrete data only a finite number of values is possible

Example: The number of parts damaged in a shipment is *discrete data.*

distance the length of the shortest line segment between two points, lines, planes, and so forth

distribution the frequency pattern for a set of data *see 4•3 Analyzing Data*

Distributive Property the mathematical rule that states that multiplying a sum by a number gives the same result as multiplying each addend by the number and then adding the products *see 6•2 Simplifying Expressions*

Example: $a(b + c) = a \cdot b + a \cdot c$

divisible a number is *divisible* by another number if their quotient has no remainder *see 1•2 Factors and Multiples*

division the operation in which a dividend is divided by a divisor to obtain a quotient

Example: $12 \div 3 = 4$

dividend | divisor | quotient

Division Property of Equality the mathematical rule that states that if each side of an equation is divided by the same nonzero number, the two sides remain equal *see 6•4 Solving Linear Equations*

Example: If $a = b$, then $\frac{a}{c} = \frac{b}{c}$.

domain the set of input values in a function *see 6•7 Graphing on the Coordinate Plane*

double-bar graph a display of data that uses paired horizontal or vertical bars to compare quantities *see 4•2 Displaying Data*

Example:

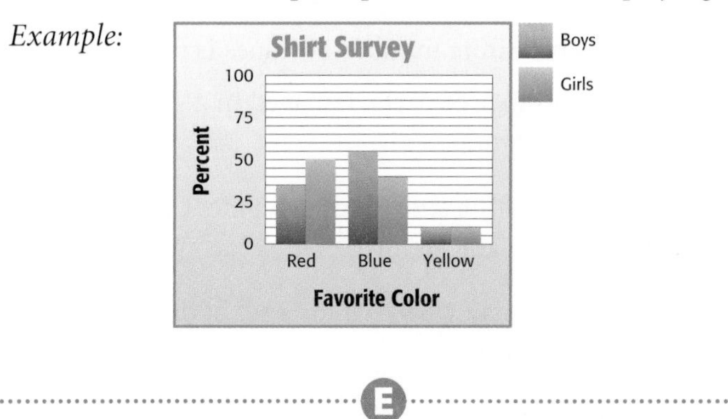

E

edge a line segment joining two planes of a polyhedron

English system units of measurement used in the United States that measure length in inches, feet, yards, and miles; capacity in cups, pints, quarts, and gallons; weight in ounces, pounds, and tons; and temperature in degrees Fahrenheit *see customary system*

equal angles angles that measure the same number of degrees

equally likely describes outcomes or events that have the same chance of occurring

equally unlikely describes outcomes or events that have the same chance of not occurring

equation a mathematical sentence stating that two expressions are equal *see 6•1 Writing Expressions and Equations, 6•8 Slope and Intercept*

Example: $3 \cdot (7 + 8) = 9 \cdot 5$

equiangular the property of a polygon in which all angles are congruent

equiangular triangle a triangle in which each angle is 60°

Example:

A

$m\angle A = m\angle B = m\angle C = 60°$
$\triangle ABC$ is *equiangular.*

C B

equilateral the property of a polygon in which all sides are congruent

equilateral triangle a triangle in which all sides are congruent

Example:

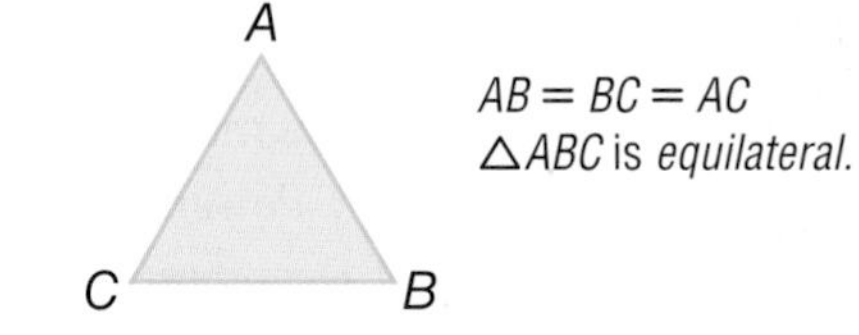

$AB = BC = AC$
$\triangle ABC$ is *equilateral.*

equivalent equal in value *see 6•1 Writing Expressions and Equations*

equivalent expressions expressions that always result in the same number, or have the same mathematical meaning for all replacement values of their variables *see 6•2 Simplifying Expressions*

Examples: $\frac{9}{3} + 2 = 10 - 5$
$2x + 3x = 5x$

equivalent fractions fractions that represent the same quotient but have different numerators and denominators *see 2•1 Fractions*

Example: $\frac{5}{6} = \frac{15}{18}$

equivalent ratios ratios that are equal

Example: $\frac{5}{4} = \frac{10}{8}$; 5:4 = 10:8

estimate an approximation or rough calculation *see 2•6 Percents*

even number any whole number that is a multiple of 2 {0, 2, 4, 6, 8, 10, 12 . . .}

HOTWORDS

event any happening to which probabilities can be assigned *see 4•5 Combinations and Permutations*

expanded form a method of writing a number that highlights the value of each digit

Example: $867 = (8 \cdot 100) + (6 \cdot 10) + (7 \cdot 1)$

expense an amount of money paid; cost

experimental probability the ratio of the total number of times the favorable outcome occurs to the total number of times the experiment is completed *see 4•6 Probability*

exponent a numeral that indicates how many times a number or variable is used as a factor *see 3•1 Powers and Exponents, 3•3 Scientific Notation, 3•4 Laws of Exponents*

Example: In the equation $2^3 = 8$, the *exponent* is 3.

expression a mathematical combination of numbers, variables, and operations *see 6•1 Writing Expressions and Equations, 6•2 Simplifying Expressions, 6•3 Evaluating Expressions and Formulas*

Example: $6x + y^2$

F

face a two-dimensional side of a three-dimensional figure *see 7•2 Naming and Classifying Polygons and Polyhedrons, 7•6 Surface Area*

factor a number or expression that is multiplied by another to yield a product *see 1•2 Factors and Multiples, 2•2 Operations with Fractions, 3•1 Powers and Exponents*

Example: 3 and 11 are *factors* of 33

factorial represented by the symbol !, the product of all the whole numbers between 1 and a given positive whole number *see 4•5 Combinations and Permutations*

Example: $5! = 1 \cdot 2 \cdot 3 \cdot 4 \cdot 5 = 120$

factor pair two unique numbers multiplied together to yield a product

fair describes a situation in which the theoretical probability of each outcome is equal

Fibonacci numbers *see Patterns page 67*

flat distribution a frequency graph that shows little difference between responses *see 4•3 Analyzing Data*

Example:

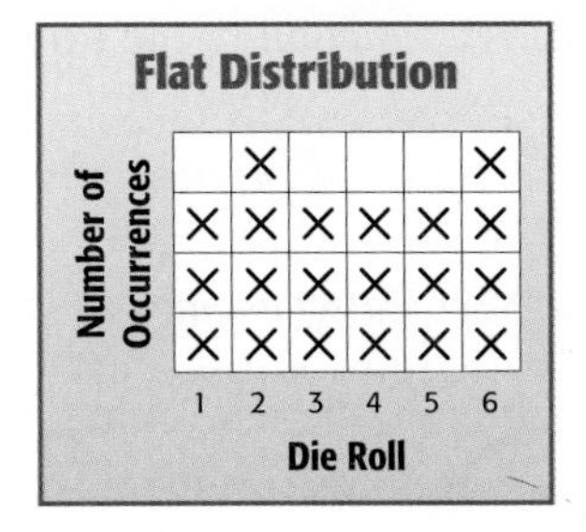

flip a transformation that produces the mirror image of a figure *see 7•3 Symmetry and Transformations*

Example:

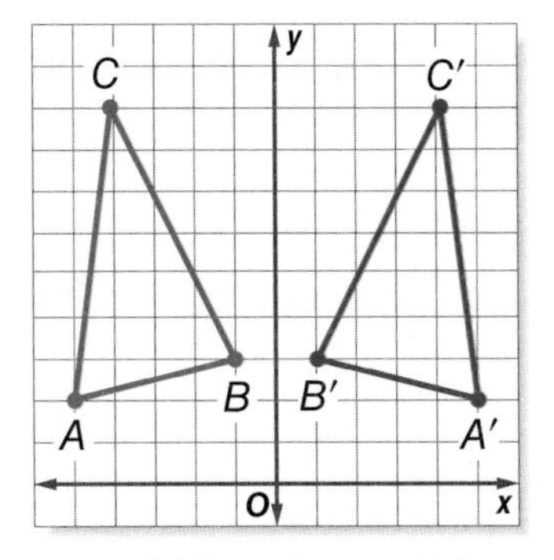

$\triangle A'B'C'$ is a *flip* of $\triangle ABC$.

HOTWORDS

formula an equation that shows the relationship between two or more quantities; a calculation performed by a spreadsheet *see Formulas pages 64–65, 9•3 Spreadsheets*

Example: $A = \pi r^2$ is the *formula* for calculating the area of a circle; A2 * B2 is a spreadsheet *formula.*

fraction a number representing part of a whole; a quotient in the form $\frac{a}{b}$ *see 2•1 Fractions*

function the assignment of exactly one output value to each input value *see 6•7 Graphing on the Coordinate Plane*

Example: You are driving at 50 miles per hour. There is a relationship between the amount of time you drive and the distance you will travel. You say that the distance is a *function* of the time.

G

geometric sequence *see Patterns page 67*

geometry the branch of mathematics that investigates the relations, properties, and measurements of solids, surfaces, lines, and angles *see Chapter 7 Geometry, 9•2 Geometry Tools*

gram a metric unit of mass *see 8•1 Systems of Measurement*

greatest common factor (GCF) the greatest number that is a factor of two or more numbers *see 1•2 Factors and Multiples, 2•1 Fractions*

Example: 30, 60, 75
The *greatest common factor* is 15.

H

harmonic sequence *see Patterns page 67*

height the perpendicular distance from a vertex to the opposite side of a figure

heptagon a polygon with seven angles and seven sides

Example:

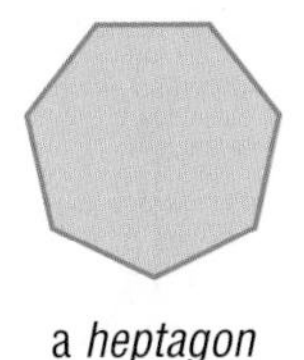

a *heptagon*

hexagon a polygon with six angles and six sides

Example:

a *hexagon*

hexagonal prism a prism that has two hexagonal bases and six rectangular sides

Example:

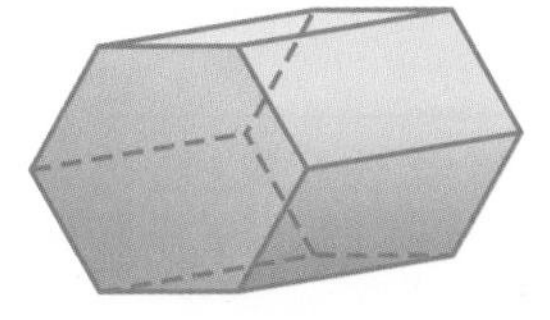

a *hexagonal prism*

hexahedron a polyhedron that has six faces

Example:

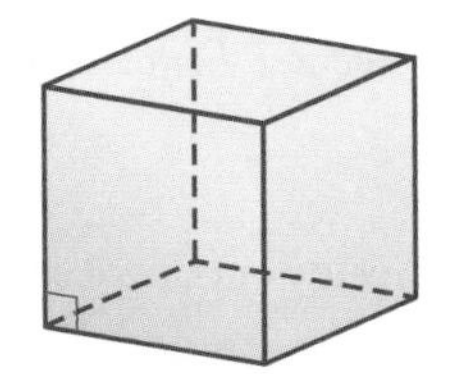

A cube is a *hexahedron.*

histogram a special kind of bar graph that displays the frequency of data that has been organized into equal intervals
see 4•2 Displaying Data

horizontal parallel to or in the plane of the horizon
see 6•7 Graphing on the Coordinate Plane

hyperbola the curve of an inverse variation function, such as $y = \frac{1}{x}$, is a *hyperbola*

Example:

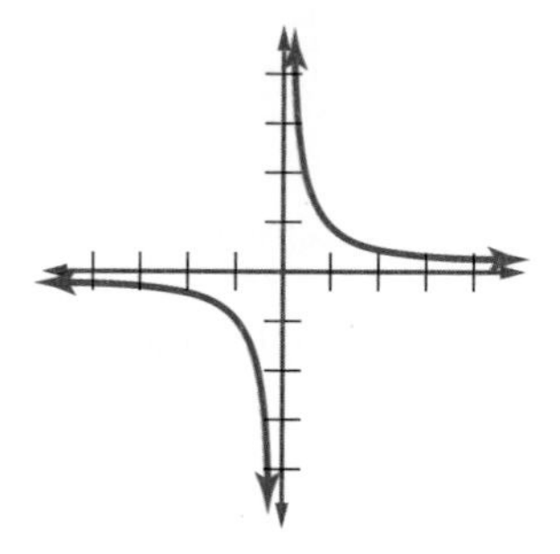

hypotenuse the side opposite the right angle in a right triangle *see 7·9 Pythagorean Theorem*

Example:

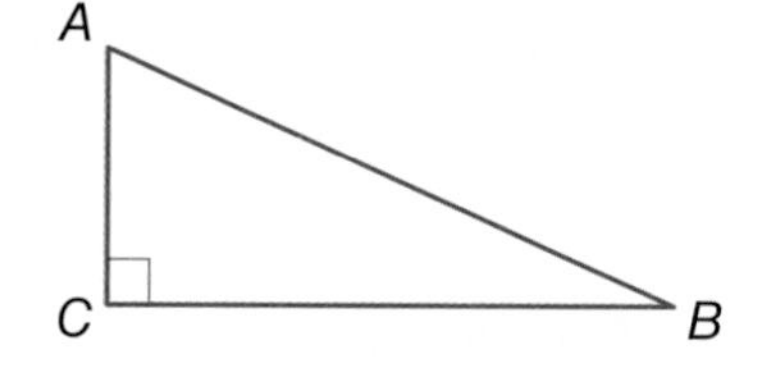

Side $\overline{AB}$ is the *hypotenuse* of this right triangle.

I

improper fraction a fraction in which the numerator is greater than the denominator *see 2·1 Fractions*

Examples: $\frac{21}{4}, \frac{4}{3}, \frac{2}{1}$

income the amount of money received for labor, services, or the sale of goods or property

independent events two events in which the outcome of one event is not affected by the outcome of another event *see 4·6 Probability*

inequality a statement that uses the symbols > (greater than), < (less than), ≥ (greater than or equal to), and ≤ (less than or equal to) to compare quantities *see 6·6 Inequalities*

Examples: $5 > 3$; $\frac{4}{5} < \frac{5}{4}$; $2(5 - x) > 3 + 1$

infinite, nonrepeating decimal irrational numbers, such as π and $\sqrt{2}$, that are decimals with digits that continue indefinitely but do not repeat

inscribed figure a figure that is enclosed by another figure as shown below

Examples:

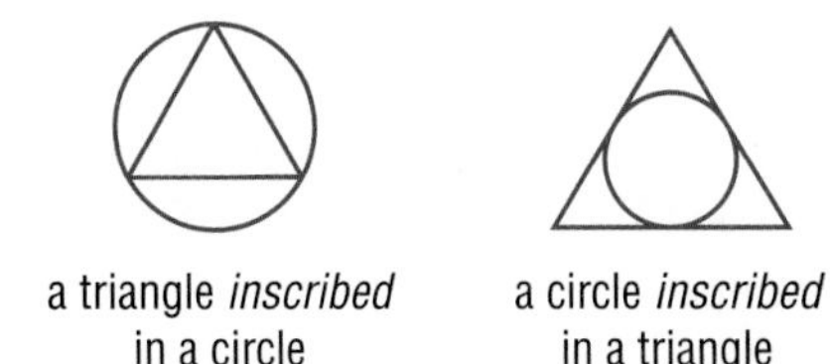

a triangle *inscribed* in a circle

a circle *inscribed* in a triangle

interquartile range the range of the middle half of a set of data; it is the difference between the upper and the lower quartile *see 4•4 Statistics*

integers the set of all whole numbers and their additive inverses {. . . , −5, −4, −3, −2, −1, 0, 1, 2, 3, 4, 5, . . .}

intercept [1] the cutting of a line, curve, or surface by another line, curve, or surface; [2] the point at which a line or curve cuts across a coordinate axis *see 6•8 Slope and Intercept*

intersection the set of elements common to two or more sets *see 5•3 Sets*

Example:

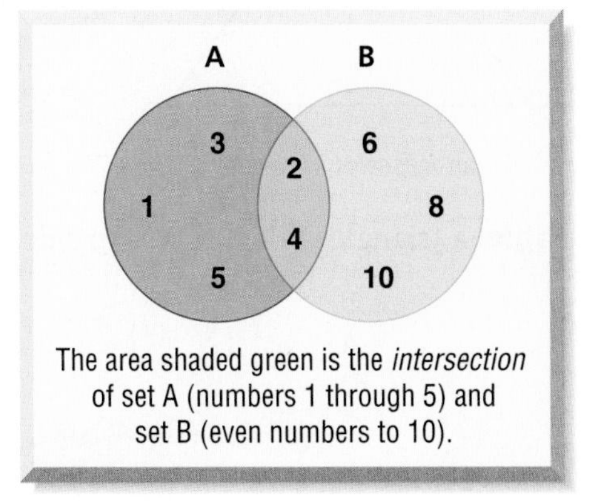

The area shaded green is the *intersection* of set A (numbers 1 through 5) and set B (even numbers to 10).

inverse negation of the *if* idea and the *then* idea of a conditional statement *see 5•1 If/Then Statements*

Example: "if *x*, then *y*" is a conditional statement;
"if not *x*, then not *y*" is the inverse statement

inverse operation the operation that reverses the effect of another operation

Examples: Subtraction is the *inverse operation* of addition.
Division is the *inverse operation* of multiplication.

irrational numbers the set of all numbers that cannot be expressed as finite or repeating decimals *see 2•5 The Real Number System*

Example: $\sqrt{2}$ (1.414214 . . .) and π (3.141592 . . .) are *irrational numbers.*

isometric drawing a two-dimensional representation of a three-dimensional object in which parallel edges are drawn as parallel lines

Example:

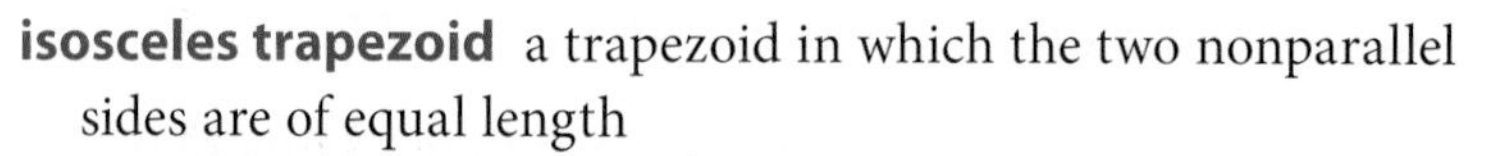

isosceles trapezoid a trapezoid in which the two nonparallel sides are of equal length

Example:

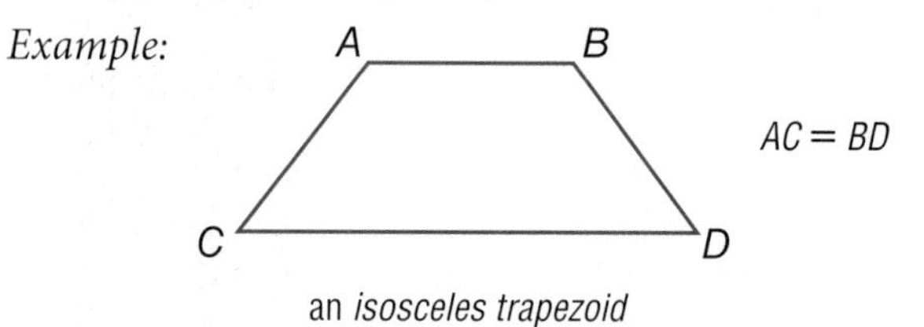

an *isosceles trapezoid*

isosceles triangle a triangle with at least two sides of equal length

Example:

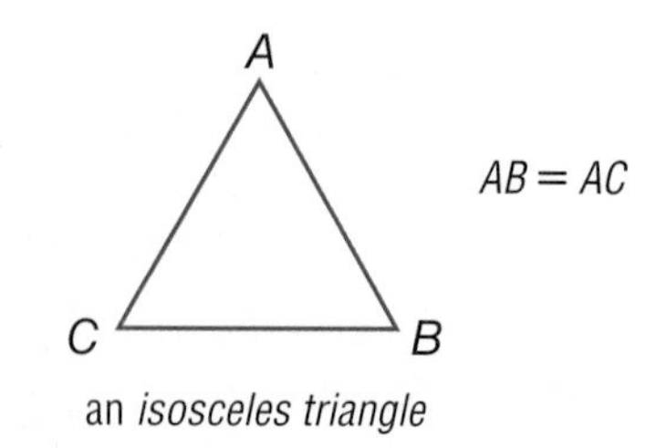

an *isosceles triangle*

leaf the unit digit of an item of numerical data between 1 and 99 *see stem-and-leaf plot, 4•2 Displaying Data*

least common denominator (LCD) the least common multiple of the denominators of two or more fractions

Example: The *least common denominator* of $\frac{1}{3}$, $\frac{2}{4}$, and $\frac{3}{6}$ is 12.

least common multiple (LCM) the smallest nonzero whole number that is a multiple of two or more whole numbers *see 1•2 Factors and Multiples*

Example: The *least common multiple* of 3, 9, and 12 is 36.

legs of a triangle the sides adjacent to the right angle of a right triangle

Example:

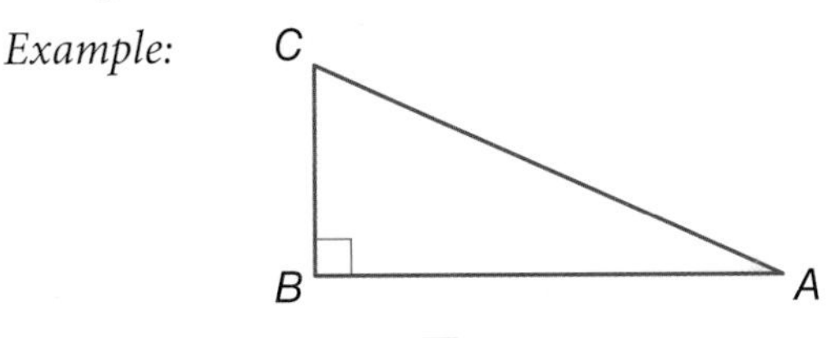

$\overline{AB}$ and $\overline{BC}$ are the *legs of* $\triangle ABC$.

length a measure of the distance of an object from end to end

likelihood the chance of a particular outcome occurring

like terms terms that include the same variables raised to the same powers; *like terms* can be combined *see 6•2 Simplifying Expressions*

Example: $5x^2$ and $6x^2$ are *like terms;* $3xy$ and $3zy$ are not like terms.

line a connected set of points extending forever in both directions

linear equation an equation with two variables (x and y) that takes the general form $y = mx + b$, where m is the slope of the line and b is the y-intercept

linear measure the measure of the distance between two points on a line

line graph a display of data that shows change over time
see 4•2 Displaying Data

Example:

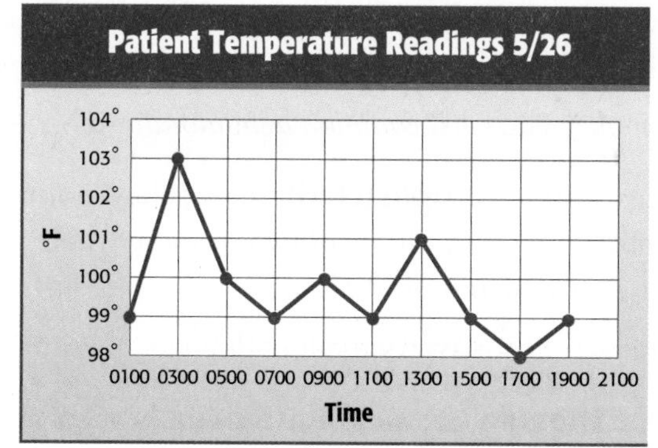

line of best fit on a scatter plot, a line drawn as near as possible to the various points so as to best represent the trend in data
see 4•3 Analyzing Data

Example:

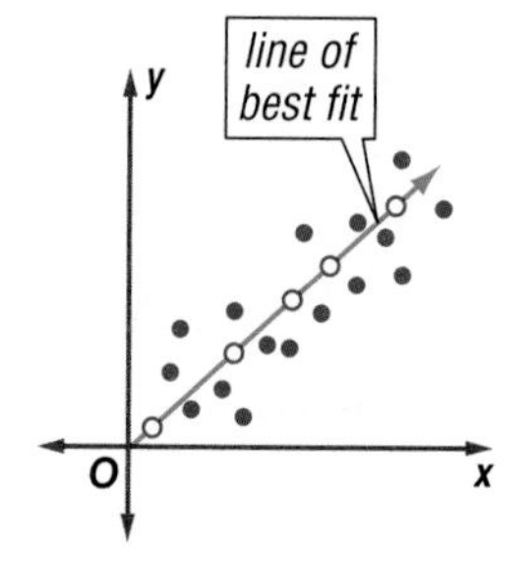

line of symmetry a line along which a figure can be folded so that the two resulting halves match

Example:

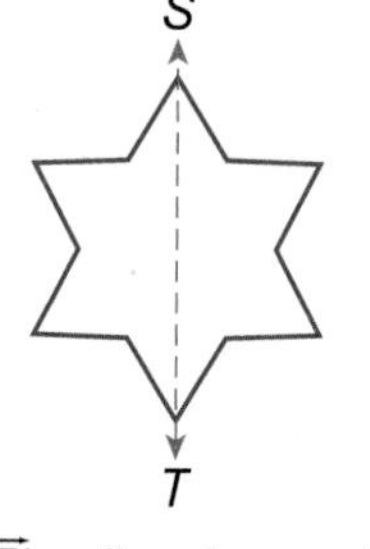

$\overleftrightarrow{ST}$ is a *line of symmetry.*

line plot a display of data that shows the frequency of data on a number line *see 4•2 Displaying Data*

line segment a section of a line between two points

Example: A•———•B

$\overline{AB}$ is a *line segment.*

liter a metric unit of capacity *see 8•3 Area, Volume, and Capacity*

logic the mathematical principles that use existing theorems to prove new ones *see Chapter 5 Logic*

lower quartile the median of the lower half of a set of data, represented by LQ *see 4•4 Statistics*

lowest common multiple the smallest number that is a multiple of all the numbers in a given set; same as least common multiple

Example: The *lowest common multiple* of 6, 9, and 18 is 18.

Lucas numbers *see Patterns page 67*

M

magic square *see Patterns page 68*

maximum value the greatest value of a function or a set of numbers

mean the quotient obtained when the sum of the numbers in a set is divided by the number of addends *see average, 4•4 Statistics*

Example: The *mean* of 3, 4, 7, and 10 is $(3 + 4 + 7 + 10) \div 4 = 6$.

measurement units standard measures, such as the meter, the liter, and the gram, or the foot, the quart, and the pound *see 8•1 Systems of Measurement*

measures of variation numbers used to describe the distribution of spread of a set of data *see 4•4 Statistics*

median the middle number in an ordered set of numbers *see 4•4 Statistics*

Example: 1, 3, 9, 16, 22, 25, 27
16 is the *median.*

meter the metric unit of length

metric system a decimal system of weights and measurements based on the meter as its unit of length, the kilogram as its unit of mass, and the liter as its unit of capacity
see 8•1 Systems of Measurement

midpoint the point on a line segment that divides it into two equal segments

Example:

A, M, B
$AM = MB$

M is the *midpoint* of $\overline{AB}$.

minimum value the least value of a function or a set of numbers

mixed number a number composed of a whole number and a fraction *see 2•1 Fractions*

Example: $5\frac{1}{4}$ is a *mixed number.*

mode the number or element that occurs most frequently in a set of data *see 4•4 Statistics*

Example: 1, 1, 2, 2, 3, 5, 5, 6, 6, 6, 8
6 is the *mode.*

monomial an algebraic expression consisting of a single term

Example: $5x^3y$, xy, and $2y$ are three *monomials.*

multiple the product of a given number and an integer
see 1•2 Factors and Multiples

Examples: 8 is a *multiple* of 4.
3.6 is a *multiple* of 1.2.

multiplication one of the four basic arithmetical operations, involving the repeated addition of numbers

multiplication growth number a number that when used to multiply a given number a given number of times results in a given goal number

Example: Grow 10 into 40 in two steps by multiplying
$(10 \cdot 2 \cdot 2 = 40)$
2 is the *multiplication growth number.*

Multiplication Property of Equality the mathematical rule that states that if each side of an equation is multiplied by the same number, the two sides remain equal *see 6•4 Solving Linear Equations*

Example: If $a = b$, then $a \cdot c = b \cdot c$.

multiplicative inverse two numbers are *multiplicative inverses* if their product is 1 *see 2•2 Operations with Fractions*

Example: $10 \cdot \frac{1}{10} = 1$
$\frac{1}{10}$ is the *multiplicative inverse* of 10.

natural variability the difference in results in a small number of experimental trials from the theoretical probabilities

negative integers the set of all integers that are less than zero $\{-1, -2, -3, -4, -5, \ldots\}$ *see 1•3 Integer Operations*

negative numbers the set of all real numbers that are less than zero $\{-1, -1.36, -\sqrt{2}, -\pi\}$

net a two-dimensional plan that can be folded to make a three-dimensional model of a solid *see 7•6 Surface Area*

Example:

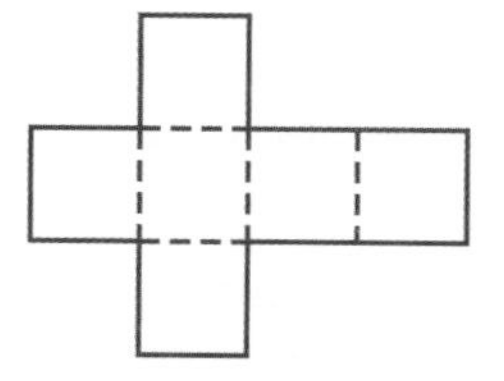

the *net* of a cube

nonagon a polygon with nine angles and nine sides

Example:

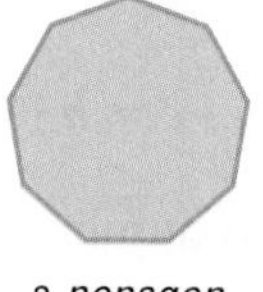

a *nonagon*

noncollinear points not lying on the same line

noncoplanar points or lines not lying on the same plane

normal distribution represented by a bell curve, the most common distribution of most qualities across a given population *see 4•3 Analyzing Data*

Example:

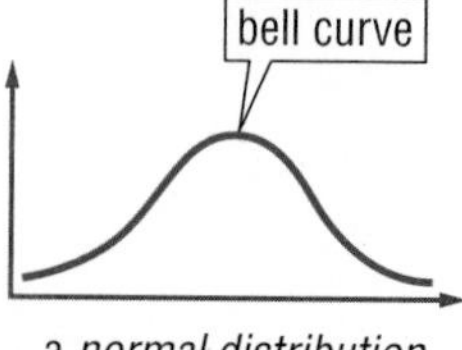

a *normal distribution*

number line a line showing numbers at regular intervals on which any real number can be indicated

Example:

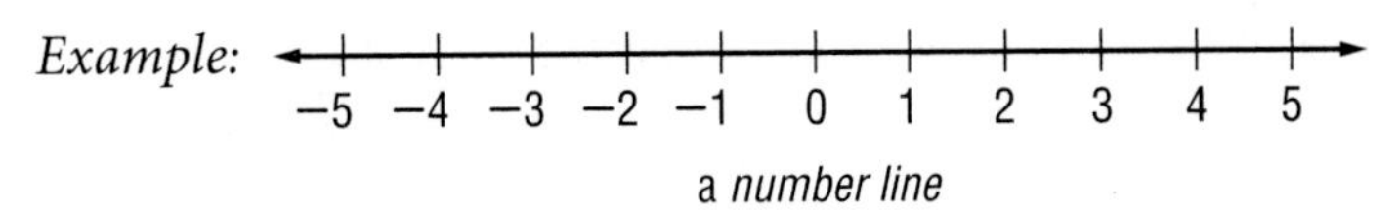

a *number line*

number symbols the symbols used in counting and measuring

Examples: 1, $-\frac{1}{4}$, 5, $\sqrt{2}$, $-\pi$

numerator the top number in a fraction representing the number of equal parts being considered *see 2·1 Fractions*

Example: In the fraction $\frac{a}{b}$, a is the *numerator.*

O

obtuse angle any angle that measures greater than 90° but less than 180°

Example:

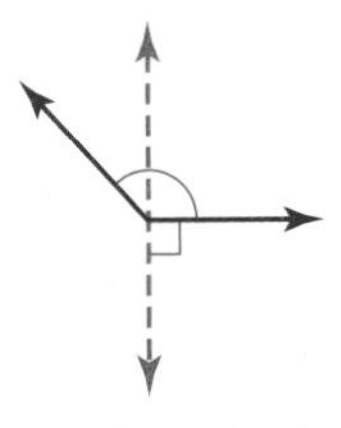

an *obtuse angle*

obtuse triangle a triangle that has one obtuse angle

Example:

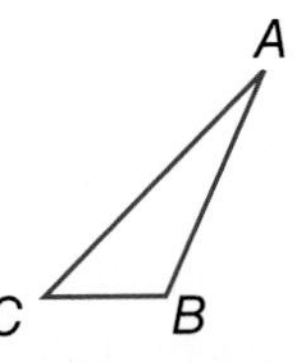

$\triangle ABC$ is an *obtuse triangle.*

octagon a polygon with eight angles and eight sides

Example:

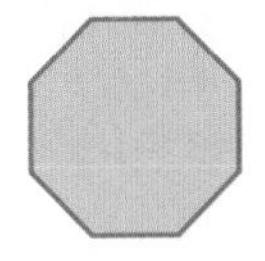

an *octagon*

octagonal prism a prism that has two octagonal bases and eight rectangular faces

Example:

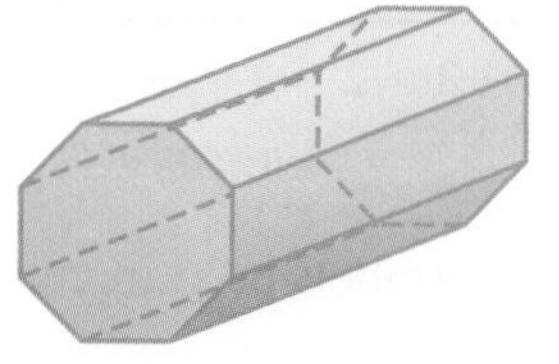

an *octagonal prism*

odd numbers the set of all integers that are not multiples of 2

odds against the ratio of the number of unfavorable outcomes to the number of favorable outcomes

odds for the ratio of the number of favorable outcomes to the number of unfavorable outcomes

one-dimensional having only one measurable quality

Example: A line and a curve are *one-dimensional.*

operations arithmetical actions performed on numbers, matrices, or vectors

opposite angle in a triangle, a side and an angle are said to be *opposite* if the side is not used to form the angle

Example:

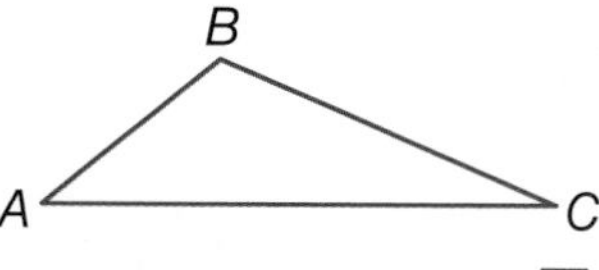

In $\triangle ABC$, $\angle A$ is opposite of $\overline{BC}$.

ordered pair two numbers that tell the x-coordinate and y-coordinate of a point *see 6•7 Graphing on the Coordinate Plane*

Example: The coordinates (3, 4) are an *ordered pair.* The x-coordinate is 3, and the y-coordinate is 4.

order of operations to simplify an expression, follow this four-step process: 1) do all operations within parentheses; 2) simplify all numbers with exponents; 3) multiply and divide in order from left to right; 4) add and subtract in order from left to right *see 1•1 Order of Operations, 3•4 Laws of Exponents*

origin the point (0, 0) on a coordinate graph where the x-axis and the y-axis intersect *see 6•7 Graphing on the Coordinate Plane*

orthogonal drawing always shows three views of an object—top, side, and front; the views are drawn straight-on

Example:

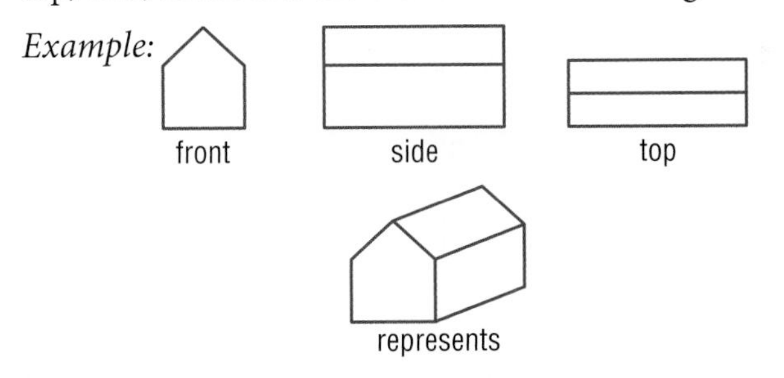

outcome a possible result in a probability experiment *see 4•5 Combinations and Permutations, 4•6 Probability*

outcome grid a visual model for analyzing and representing theoretical probabilities that shows all the possible outcomes of two independent events *see 4•6 Probability*

Example:

A grid used to find the sample space for rolling a pair of dice. The outcomes are written as ordered pairs.

	1	2	3	4	5	6
1	(1, 1)	(2, 1)	(3, 1)	(4, 1)	(5, 1)	(6, 1)
2	(1, 2)	(2, 2)	(3, 2)	(4, 2)	(5, 2)	(6, 2)
3	(1, 3)	(2, 3)	(3, 3)	(4, 3)	(5, 3)	(6, 3)
4	(1, 4)	(2, 4)	(3, 4)	(4, 4)	(5, 4)	(6, 4)
5	(1, 5)	(2, 5)	(3, 5)	(4, 5)	(5, 5)	(6, 5)
6	(1, 6)	(2, 6)	(3, 6)	(4, 6)	(5, 6)	(6, 6)

There are 36 possible outcomes.

outlier data that are more than 1.5 times the interquartile range from the upper or lower quartiles *see 4·4 Statistics*

P

parabola the curve formed by a quadratic equation such as $y = x^2$

Example:

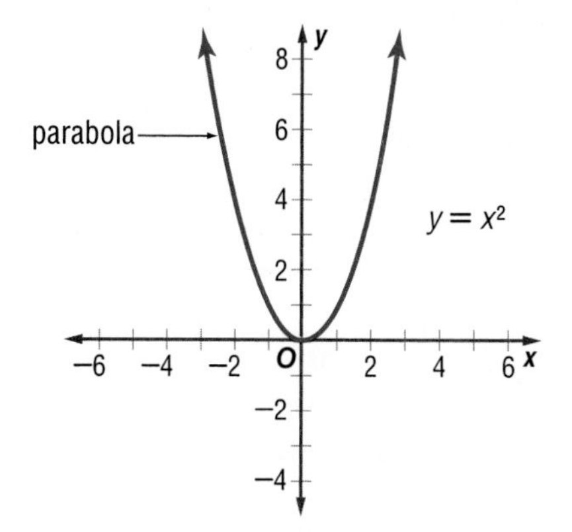

parallel straight lines or planes that remain a constant distance from each other and never intersect, represented by the symbol ||

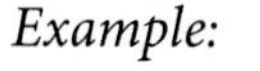

Example:

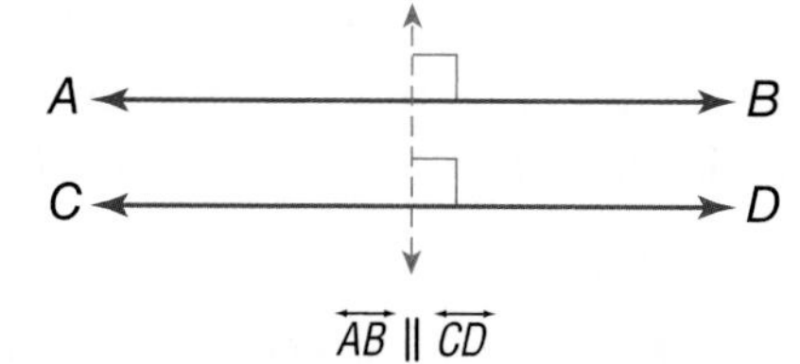

parallelogram a quadrilateral with two pairs of parallel sides *see 7·2 Naming and Classifying Polygons and Polyhedrons*

Example:

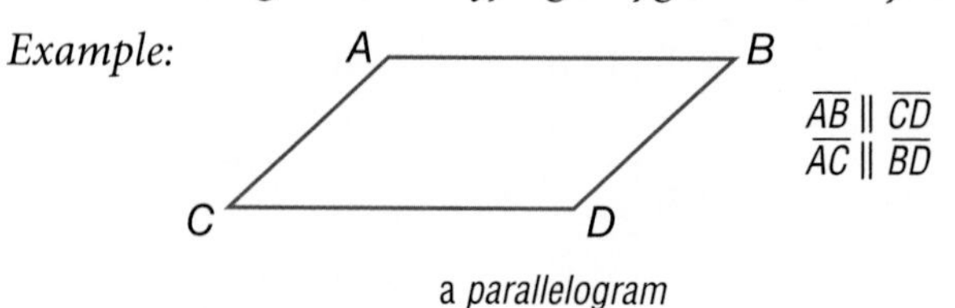

a parallelogram

parentheses the enclosing symbols ( ), which indicate that the terms within are a unit

Example: $(2 + 4) \div 2 = 3$

Pascal's Triangle *see Patterns page 68*

pattern a regular, repeating design or sequence of shapes or numbers *see Patterns pages 67–69*

PEMDAS an acronym for the order of operations: 1) do all operations within **p**arentheses; 2) simplify all numbers with **e**xponents; 3) **m**ultiply and **d**ivide in order from left to right; 4) **a**dd and **s**ubtract in order from left to right *see 1•1 Order of Operations*

pentagon a polygon with five angles and five sides

Example:

a *pentagon*

percent a number expressed in relation to 100, represented by the symbol % *see 2•6 Percents*

Example: 76 out of 100 students use computers.
76 *percent* or 76% of students use computers.

percent grade the ratio of the rise to the run of a hill, ramp, or incline expressed as a percent

Example:

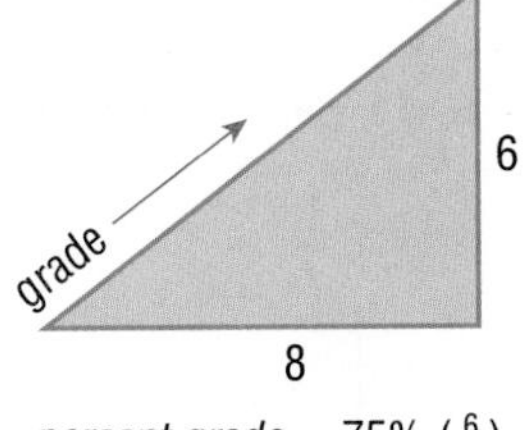

percent grade = 75% ($\frac{6}{8}$)

percent proportion compares part of a quantity to the whole quantity using a percent *see 2·7 Using and Finding Percents*

$$\frac{part}{whole} = \frac{percent}{100}$$

perfect cube a number that is the cube of an integer

Example: 27 is a *perfect cube* since $27 = 3^3$.

perfect number an integer that is equal to the sum of all its positive whole number divisors, excluding the number itself

Example: $1 \cdot 2 \cdot 3 = 6$ and $1 + 2 + 3 = 6$
6 is a *perfect number.*

perfect square a number that is the square of an integer *see 3·2 Square and Cube Roots*

Example: 25 is a *perfect square* since $25 = 5^2$.

perimeter the distance around the outside of a closed figure *see Formulas page 64*

Example:

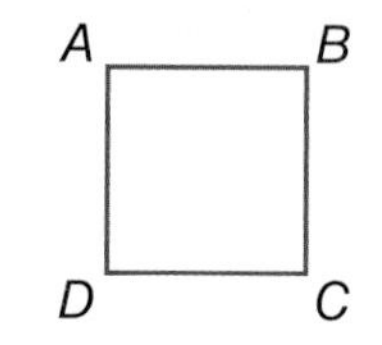

$AB + BC + CD + DA =$ *perimeter*

permutation a possible arrangement of a group of objects; the number of possible arrangements of *n* objects is expressed by the term $n!$ *see factorial, 4·5 Combinations and Permutations*

perpendicular two lines or planes that intersect to form a right angle

Example:

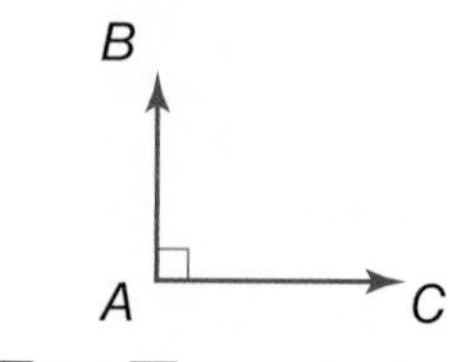

$\overline{AB}$ and $\overline{AC}$ are *perpendicular.*

pi the ratio of a circle's circumference to its diameter; *pi* is shown by the symbol π, and is approximately equal to 3.14 *see 7·7 Volume*

picture graph a display of data that uses pictures or symbols to represent numbers

place value the value given to a place a digit occupies in a numeral

place-value system a number system in which values are given to the places digits occupy in the numeral; in the decimal system, the value of each place is 10 times the value of the place to its right

point one of four undefined terms in geometry used to define all other terms; a *point* has no size *see 6·7 Graphing on the Coordinate Plane*

polygon a simple, closed plane figure, having three or more line segments as sides *see 7·1 Classifying Angles and Triangles*

Examples:

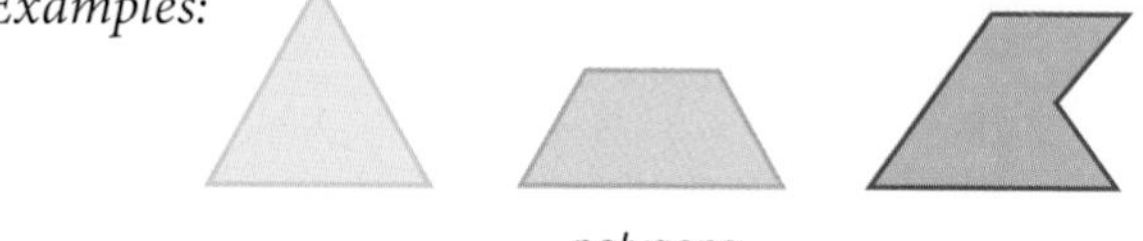

polygons

polyhedron a solid geometrical figure that has four or more plane faces *see 7·2 Naming and Classifying Polygons and Polyhedrons*

Examples:

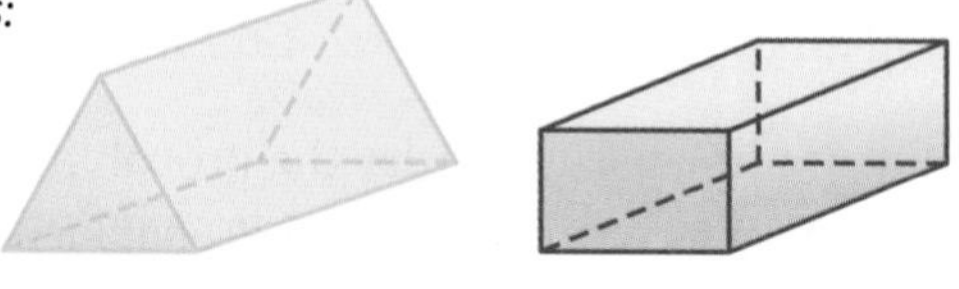

polyhedrons

population the universal set from which a sample of statistical data is selected *see 4·1 Collecting Data*

positive integers the set of all integers that are greater than zero {1, 2, 3, 4, 5, . . .} *see 1·3 Integer Operations*

positive numbers the set of all real numbers that are greater than zero $\{1, 1.36, \sqrt{2}, \pi\}$

power represented by the exponent *n*, to which a number is used as a factor *n* times *see 3•1 Powers and Exponents*

Example: 7 raised to the fourth *power.*
$7^4 = 7 \cdot 7 \cdot 7 \cdot 7 = 2{,}401$

predict to anticipate a trend by studying statistical data

prime factorization the expression of a composite number as a product of its prime factors *see 1•2 Factors and Multiples*

Examples: $504 = 2^3 \cdot 3^2 \cdot 7$
$30 = 2 \cdot 3 \cdot 5$

prime number a whole number greater than 1 whose only factors are 1 and itself *see 1•2 Factors and Multiples*

Examples: 2, 3, 5, 7, 11

prism a solid figure that has two parallel, congruent polygonal faces (called *bases*) *see 7•2 Naming and Classifying Polygons and Polyhedrons*

Examples:

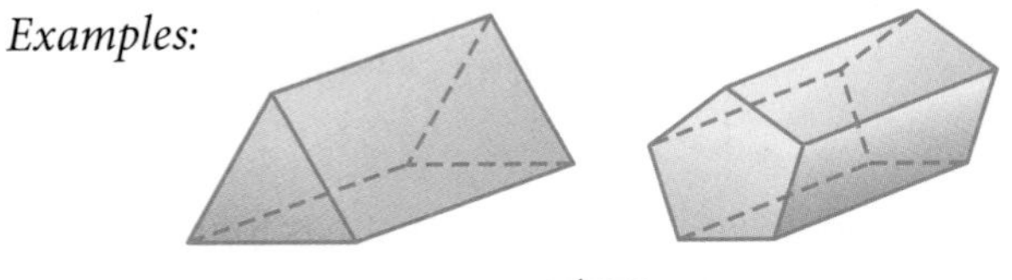

prisms

probability the study of likelihood or chance that describes the possibility of an event occurring *see 4•6 Probability*

probability line a line used to order the probability of events from least likely to most likely *see 4•6 Probability*

probability of events the likelihood or chance that events will occur

product the result obtained by multiplying two numbers or variables *see 6•1 Writing Expressions and Equations*

profit the gain from a business; what is left when the cost of goods and of carrying on the business is subtracted from the amount of money taken in

project to extend a numerical model, to either greater or lesser values, in order to predict likely quantities in an unknown situation

proportion a statement that two ratios are equal *see 6•5 Ratio and Proportion*

pyramid a solid geometrical figure that has a polygonal base and triangular faces that meet at a common vertex
see 7•2 Naming and Classifying Polygons and Polyhedrons

Examples:

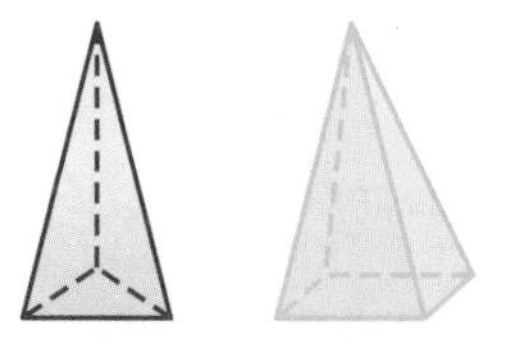

pyramids

Pythagorean Theorem a mathematical idea stating that the sum of the squared lengths of the two legs of a right triangle is equal to the squared length of the hypotenuse
see 7•9 Pythagorean Theorem

Example:

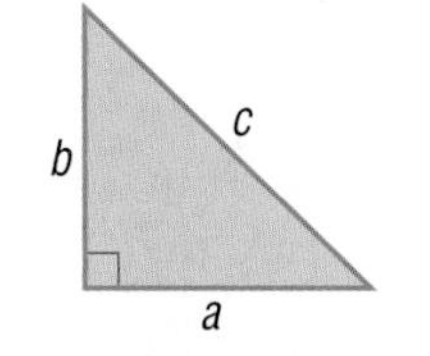

For a right triangle, $a^2 + b^2 = c^2$.

Pythagorean triple a set of three positive integers *a*, *b*, and *c*, such that $a^2 + b^2 = c^2$ *see 7•9 Pythagorean Theorem*

Example: The Pythagorean triple {3, 4, 5}

$3^2 + 4^2 = 5^2$

$9 + 16 = 25$

HOTWORDS

Q

quadrant [1] one quarter of the circumference of a circle; [2] on a coordinate graph, one of the four regions created by the intersection of the x-axis and the y-axis *see 6•7 Graphing on the Coordinate Plane*

quadratic equation a polynomial equation of the second degree, generally expressed as $ax^2 + bx + c = 0$, where a, b, and c are real numbers and a is not equal to zero

quadrilateral a polygon that has four sides *see 7•2 Naming and Classifying Polygons and Polyhedrons*

Examples:

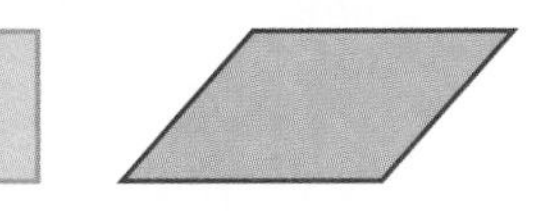

quadrilaterals

qualitative graphs a graph with words that describes such things as a general trend of profits, income, and expenses over time; it has no specific numbers

quantitative graphs a graph that, in contrast to a qualitative graph, has specific numbers

quartiles values that divide a set of data into four equal parts *see 4•2 Displaying Data*

quotient the result obtained from dividing one number or variable (the divisor) into another number or variable (the dividend) *see 6•1 Writing Expressions and Equations*

Example: $24 \div 4 = 6$

dividend (24), divisor (4), quotient (6)

R

radical the indicated root of a quantity

Examples: $\sqrt{3}$, $\sqrt[4]{14}$, $\sqrt[12]{23}$

radical sign the root symbol $\sqrt{\ }$

radius a line segment from the center of a circle to any point on its perimeter *see 7•8 Circles*

random sample a population sample chosen so that each member has the same probability of being selected *see 4•1 Collecting Data*

range in statistics, the difference between the largest and smallest values in a sample *see 4•4 Statistics*

rank to order the data from a statistical sample on the basis of some criterion—for example, in ascending or descending numerical order

ranking the position on a list of data from a statistical sample based on some criterion

rate [1] fixed ratio between two things; [2] a comparison of two different kinds of units, for example, miles per hour or dollars per hour *see 6•5 Ratio and Proportion*

ratio a comparison of two numbers *see 6•5 Ratio and Proportion*

Example: The *ratio* of consonants to vowels in the alphabet is 21:5.

rational numbers the set of numbers that can be written in the form $\frac{a}{b}$, where a and b are integers and b does not equal zero *see 2•1 Fractions*

Examples: $1 = \frac{1}{1}, \frac{2}{9}, 3\frac{2}{7} = \frac{23}{7}, -0.333 = -\frac{1}{3}$

ray the part of a straight line that extends infinitely in one direction from a fixed point

Example:

a *ray*

real numbers the set consisting of zero, all positive numbers, and all negative numbers; *real numbers* include all rational and irrational numbers

real-world data information processed by people in everyday situations

reciprocal one of a pair of numbers that have a product of 1 *see 2·2 Operations with Fractions*

Examples: The *reciprocal* of 2 is $\frac{1}{2}$; of $\frac{3}{4}$ is $\frac{4}{3}$; of x is $\frac{1}{x}$.

rectangle a parallelogram with four right angles *see 7·2 Naming and Classifying Polygons and Polyhedrons*

Example:

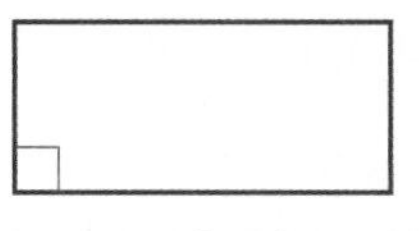

a *rectangle*

rectangular prism a prism that has rectangular bases and four rectangular faces *see 7·2 Naming and Classifying Polygons and Polyhedrons*

reflection a transformation that produces the mirror image of a figure *see 7·3 Symmetry and Transformations*

Example:

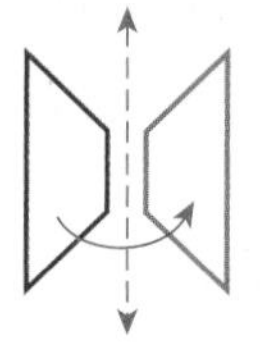

the *reflection* of a trapezoid

reflex angle any angle with a measure that is greater than 180° but less than 360°

Example:

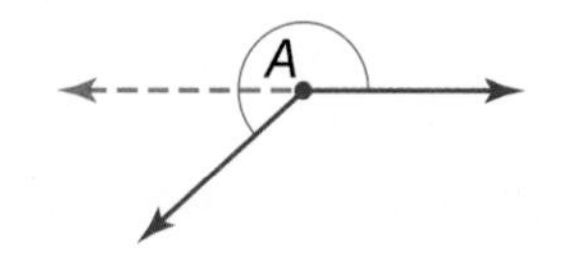

∠*A* is a *reflex angle.*

HOTWORDS

regular polygon a polygon in which all sides are equal and all angles are congruent *see 7·2 Naming and Classifying Polygons and Polyhedrons*

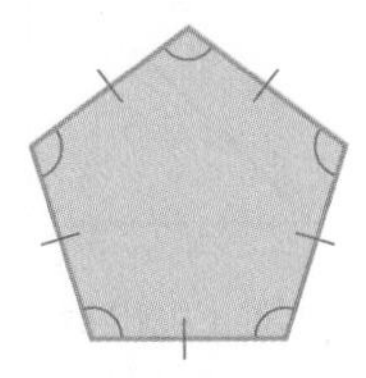

a *regular polygon*

relationship a connection between two or more objects, numbers, or sets; a mathematical *relationship* can be expressed in words or with numbers and letters

repeating decimal a decimal in which a digit or a set of digits repeat infinitely *see 2·4 Fractions and Decimals*

Example: 0.121212 . . . is a *repeating decimal.*

rhombus a parallelogram with all sides of equal length *see 7·2 Naming and Classifying Polygons and Polyhedrons*

Example:

A B

C D

$AB = CD = AC = BD$

a *rhombus*

right angle an angle that measures 90°

Example:

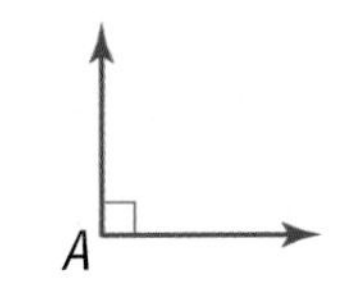

$\angle A$ is a *right angle.*

right triangle a triangle with one right angle *see 7•4 Perimeter*

Example:

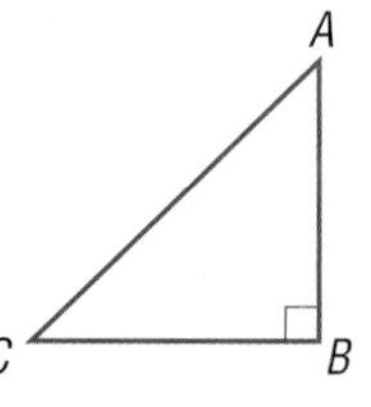

$\triangle ABC$ is a *right triangle.*

rise the vertical distance between two points *see 6•8 Slope and Intercept*

Roman numerals the numeral system consisting of the symbols I (1), V (5), X (10), L (50), C (100), D (500), and M (1,000); when a Roman symbol is preceded by a symbol of equal or greater value, the values of a symbol are added (XVI = 16); when a symbol is preceded by a symbol of lesser value, the values are subtracted (IV = 4)

root [1] the inverse of an exponent; [2] the radical sign $\sqrt{\ }$ indicates square root *see 3•2 Square and Cube Roots*

rotation a transformation in which a figure is turned a certain number of degrees around a fixed point or line *see 7•3 Symmetry and Transformations*

Example:

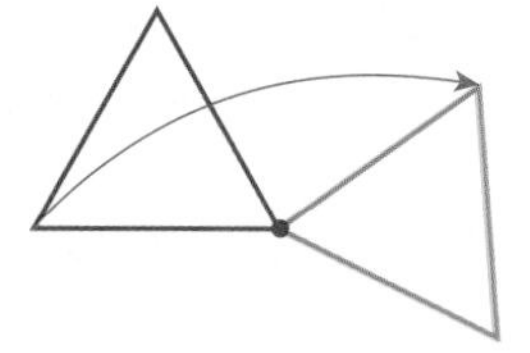

the *turning* of a triangle

round to approximate the value of a number to a given decimal place

Examples: 2.56 rounded to the nearest tenth is 2.6;
2.54 rounded to the nearest tenth is 2.5;
365 rounded to the nearest hundred is 400.

row a horizontal list of numbers or terms

rule a statement that describes a relationship between numbers or objects

run the horizontal distance between two points *see 6•8 Slope and Intercept*

S

sample a finite subset of a population, used for statistical analysis *see 4•1 Collecting Data*

sample space the set of all possible outcomes of a probability experiment *see 4•5 Combinations and Permutations*

sampling with replacement a sample chosen so that each element has the chance of being selected more than once *see 4•6 Probability*

Example: A card is drawn from a deck, placed back into the deck, and a second card is drawn. Since the first card is replaced, the number of cards remains constant.

scale the ratio between the actual size of an object and a proportional representation *see 8•5 Size and Scale*

scale drawing a proportionally correct drawing of an object or area at actual, enlarged, or reduced size

scale factor the factor by which all the components of an object are multiplied in order to create a proportional enlargement or reduction *see 8•5 Size and Scale*

scalene triangle a triangle with no sides of equal length

Example:

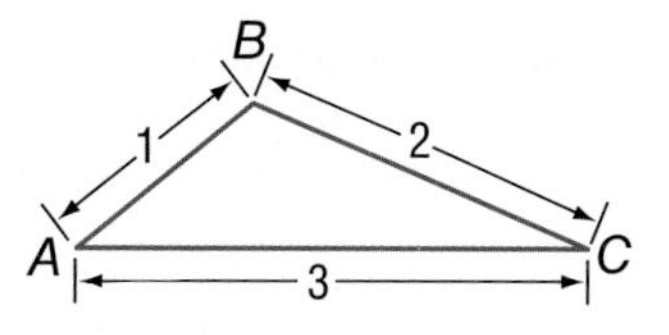

$\triangle ABC$ is a *scalene triangle.*

scale size the proportional size of an enlarged or reduced representation of an object or area *see 8•5 Size and Scale*

scatter plot (or scatter diagram) a display of data in which the points corresponding to two related factors are graphed and observed for correlation *see 4•3 Analyzing Data*

Example:

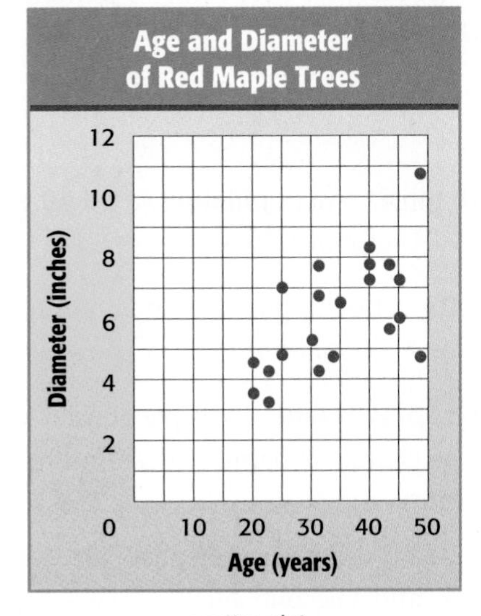

scatter plot

scientific notation a method of writing a number using exponents and powers of ten; a number in scientific notation is written as a number between 1 and 10 multiplied by a power of ten *see 3•3 Scientific Notation*

Examples: $9{,}572 = 9.572 \cdot 10^3$ and $0.00042 = 4.2 \cdot 10^{-4}$

segment two points and all the points on the line between them *see 7•2 Naming and Classifying Angles and Triangles*

sequence *see Patterns page 68*

series *see Patterns page 68*

set a collection of distinct elements or items *see 5•3 Sets*

side a line segment that forms an angle or joins the vertices of a polygon

sighting measuring a length or angle of an inaccessible object by lining up a measuring tool with one's line of vision

signed number a number preceded by a positive or negative sign

significant digit the number of digits in a value that indicate its precision and accuracy

Example: 297,624 rounded to three significant digits is 298,000; 2.97624 rounded to three significant digits is 2.98.

similar figures figures that have the same shape but are not necessarily the same size *see 8•5 Size and Scale*

Example:

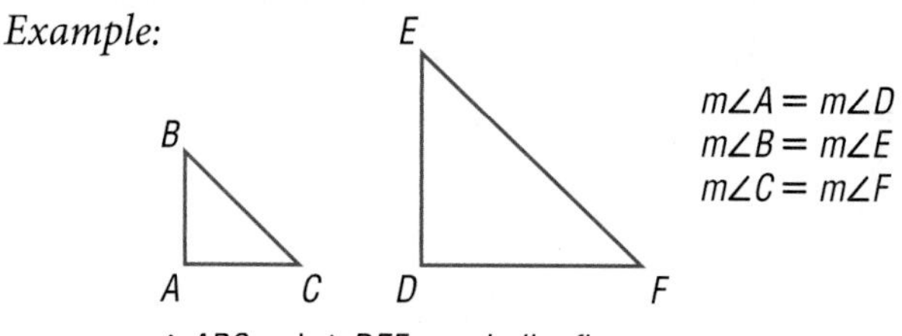

$\triangle ABC$ and $\triangle DEF$ are *similar figures.*

simple event an outcome or collection of outcomes

simulation a mathematical experiment that approximates real-world processes

skewed distribution an asymmetrical distribution curve representing statistical data that is not balanced around the mean *see 4•3 Analyzing Data*

Example:

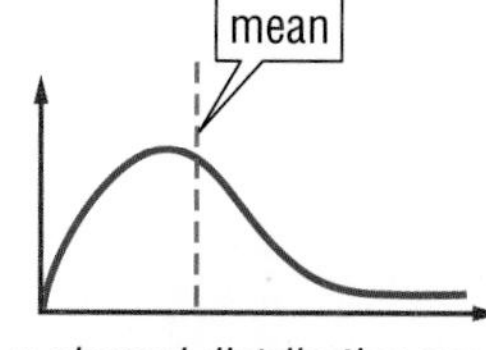

a *skewed distribution* curve

slide to move a shape to another position without rotating or reflecting it; also referred to as a translation *see 7•3 Symmetry and Transformations*

Example:

the *slide* of a trapezoid

HOTWORDS

slope [1] a way of describing the steepness of a line, ramp, hill, and so on; [2] the ratio of the rise to the run *see 6•8 Slope and Intercept*

slope angle the angle that a line forms with the x-axis or other horizontal

slope ratio the slope of a line as a ratio of the rise to the run

solid a three-dimensional figure

solution the answer to a mathematical problem; in algebra, a *solution* usually consists of a value or set of values for a variable

speed the rate at which an object moves

speed-time graph a graph used to chart how the speed of an object changes over time

sphere a perfectly round geometric solid, consisting of a set of points equidistant from a center point

Example:

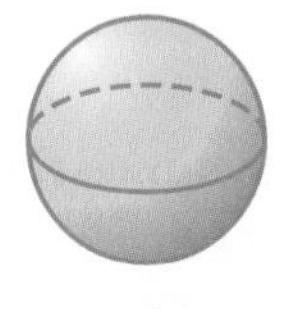

a *sphere*

spinner a device for determining outcomes in a probability experiment

Example:

a *spinner*

spiral *see Patterns page 69*

spreadsheet a computer tool where information is arranged into cells within a grid and calculations are performed within the cells; when one cell is changed, all other cells that depend on it automatically change *see 9•3 Spreadsheets*

square [1] a rectangle with congruent sides [2] the product of two equal terms *see 7•2 Naming and Classifying Polygons and Polyhedrons, 3•1 Powers and Exponents*

Example: [1]

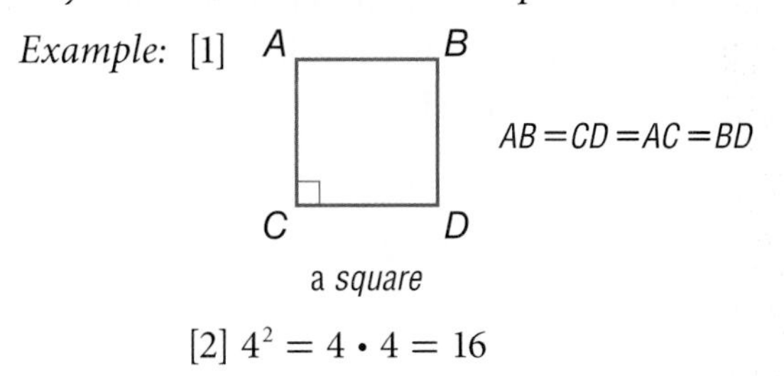

a square

[2] $4^2 = 4 \cdot 4 = 16$

square centimeter a unit used to measure the area of a surface; the area of a square measuring one centimeter on each side *see 8•3 Area, Volume, and Capacity*

square foot a unit used to measure the area of a surface; the area of a square measuring one foot on each side *see 8•3 Area, Volume, and Capacity*

square inch a unit used to measure the area of a surface; the area of a square measuring one inch on each side *see 8•3 Area, Volume, and Capacity*

square meter a unit used to measure the area of a surface; the area of a square measuring one meter on each side *see 8•3 Area, Volume, and Capacity*

square number *see Patterns page 69*

square pyramid a pyramid with a square base

square root a number that when multiplied by itself equals a given number *see 3•2 Square and Cube Roots*

Example: 3 is the *square root* of 9.
$\sqrt{9} = 3$

square root symbol the mathematical symbol $\sqrt{\ }$; indicates that the square root of a given number is to be calculated *see 3•2 Square and Cube Roots*

standard measurement commonly used measurements, such as the meter used to measure length, the kilogram used to measure mass, and the second used to measure time

statistics the branch of mathematics that investigates the collection and analysis of data *see 4•4 Statistics*

steepness a way of describing the amount of incline (or slope) of a ramp, hill, line, and so on

stem the tens digit of an item of numerical data between 1 and 99 *see stem-and-leaf plot, 4•2 Displaying Data*

stem-and-leaf plot a method of displaying numerical data between 1 and 99 by separating each number into its tens digit (stem) and its unit digit (leaf) and then arranging the data in ascending order of the tens digits *see 4•2 Displaying Data*

Example:

Average Points per Game

Stem	Leaf
0	6
1	1 8 2 2 5
2	6 1
3	7
4	3
5	8

2 | 6 = 26 points

a *stem-and-leaf plot* for the data set 11, 26, 18, 12, 12, 15, 43, 37, 58, 6, and 21

straight angle an angle that measures 180°; a straight line

strip graph a graph indicating the sequence of outcomes; a *strip graph* helps to highlight the differences among individual results and provides a strong visual representation of the concept of randomness

Example:

Outcomes of a coin toss
H = heads
T = tails

H	H	T	H	T	T	T	

a *strip graph*

subtraction one of the four basic arithmetical operations, taking one number or quantity away from another

Subtraction Property of Equality the mathematical rule that states that if the same number is subtracted from each side of the equation, then the two sides remain equal *see 6•4 Solving Linear Equations*

Example: If $a = b$, then $a - c = b - c$.

sum the result of adding two numbers or quantities *see 6•1 Writing Expressions and Equations*

Example: $6 + 4 = 10$

10 is the *sum* of the two addends, 6 and 4.

supplementary angles two angles that have measures whose sum is 180° *see 7•1 Classifying Angles and Triangles*

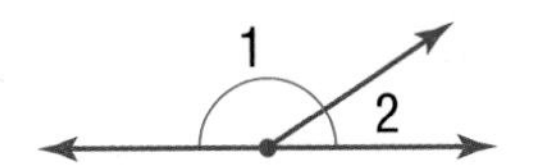

∠1 and ∠2 are *supplementary angles.*

surface area the sum of the areas of all the faces of a geometric solid, measured in square units *see 7•6 Surface Area*

Example:

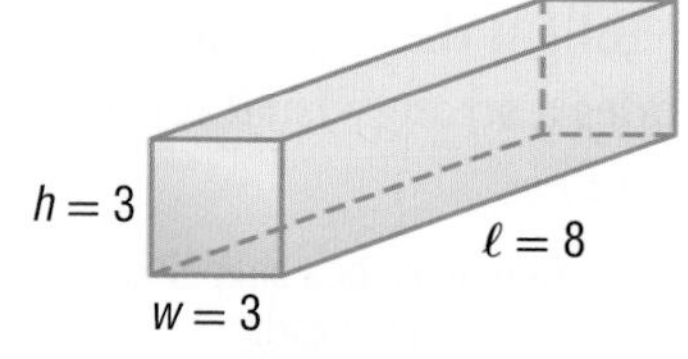

The *surface area* of this rectangular prism is $2(3 \cdot 3) + 4(3 \cdot 8) = 114$ square units.

survey a method of collecting statistical data in which people are asked to answer questions *see 4•1 Collecting Data*

symmetry *see line of symmetry, 7•3 Symmetry and Transformations*

Example:

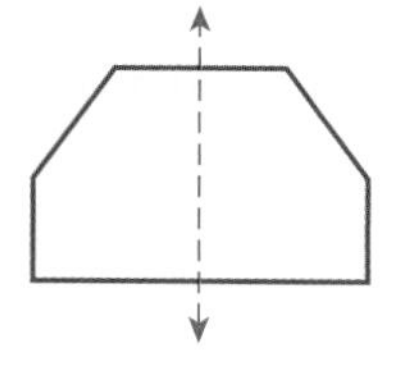

This hexagon has *symmetry* around the dotted line.

systems of equations a set of two or more equations with the same variables *see 6•10 Systems of Equations*

T

table a collection of data arranged so that information can be easily seen

tally marks marks made for certain numbers of objects in keeping account

Example: ~~||||~~ ||| = 8

tangent [1] a line that intersects a circle in exactly one point; [2] The *tangent* of an acute angle in a right triangle is the ratio of the length of the opposite side to the length of the adjacent side

Example:

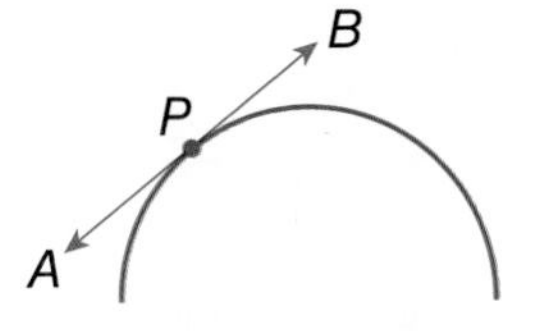

$\overleftrightarrow{AB}$ is *tangent* to the curve at point P.

tangent ratio the ratio of the length of the side opposite a right triangle's acute angle to the length of the side adjacent to it

Example:

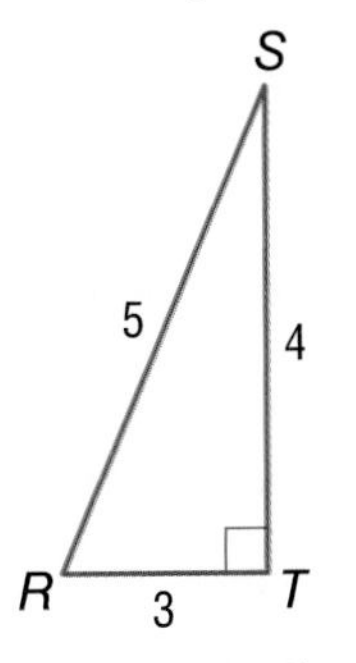

$\angle S$ has a *tangent ratio* of $\frac{3}{4}$.

term product of numbers and variables *see 6•1 Writing Expressions and Equations*

Example: x, ax^2, $2x^4y^2$, and $-4ab$ are all terms.

terminating decimal a decimal with a finite number of digits *see 2•4 Fractions and Decimals*

tessellation *see Patterns page 69*

tetrahedron a geometrical solid that has four triangular faces *see 7•2 Naming and Classifying Polygons and Polyhedrons*

Example:

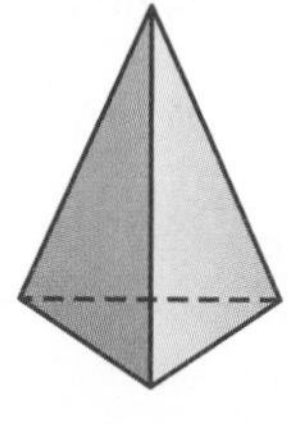

a *tetrahedron*

theoretical probability the ratio of the number of favorable outcomes to the total number of possible outcomes *see 4•6 Probability*

three-dimensional having three measurable qualities: length, height, and width

tiling completely covering a plane with geometric shapes *see tessellations*

time in mathematics, the element of duration, usually represented by the variable t

total distance the amount of space between a starting point and an endpoint, represented by d in the equation $d = s \text{ (speed)} \cdot t \text{ (time)}$

total distance graph a coordinate graph that shows cumulative distance traveled as a function of time

total time the duration of an event, represented by t in the equation $t = \frac{d \text{ (distance)}}{s \text{ (speed)}}$

transformation a mathematical process that changes the shape or position of a geometric figure *see 7•3 Symmetry and Transformations*

translation a transformation in which a geometric figure is slid to another position without rotation or reflection *see 7·3 Symmetry and Transformations*

transversal a line that intersects two or more other lines at different points *see 7·1 Classifying Angles and Triangles*

trapezoid a quadrilateral with only one pair of parallel sides *see 7·2 Naming and Classifying Polygons and Polyhedrons*

Example:

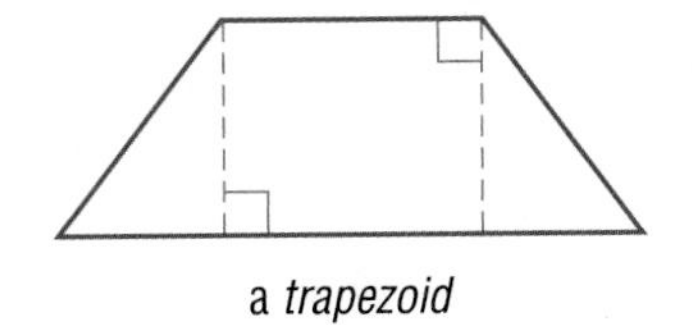

a *trapezoid*

tree diagram a connected, branching graph used to diagram probabilities or factors *see 1·2 Factors and Multiples, 4·5 Combinations and Permutations*

Example:

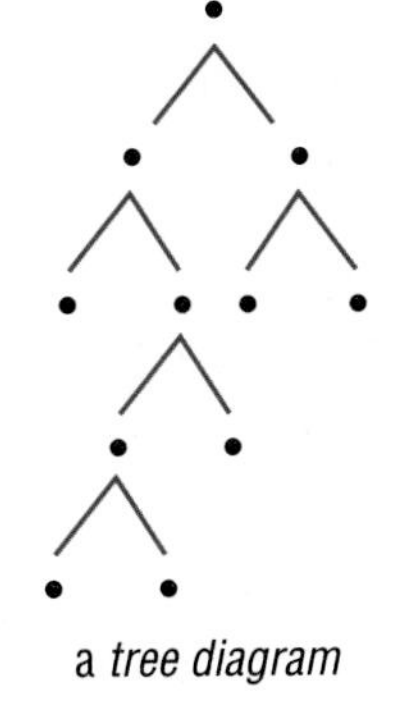

a *tree diagram*

trend a consistent change over time in the statistical data representing a particular population

triangle a polygon with three angles and three sides

triangular numbers *see Patterns page 69*

triangular prism a prism with two triangular bases and three rectangular sides *see 7·6 Surface Area*

turn to move a geometric figure by rotating it around a point *see 7•3 Symmetry and Transformations*

Example:

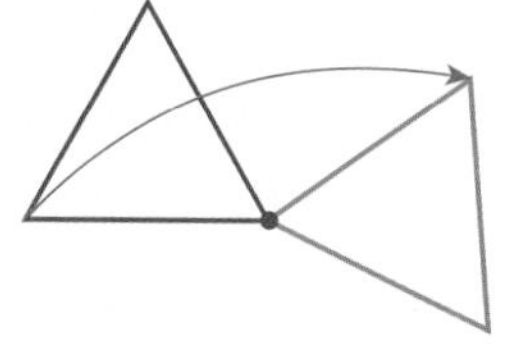

the *turning* of a triangle

two-dimensional having two measurable qualities: length and width

U

unbiased sample a sample representative of the entire population

unequal probabilities different likelihoods of occurrence; two events have *unequal probabilities* if one is more likely to occur than the other

unfair where the probability of each outcome is not equal

union a set that is formed by combining the members of two or more sets, as represented by the symbol ∪; the *union* contains all members previously contained in both sets *see Venn diagram, 5•3 Sets*

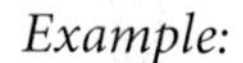

Example:

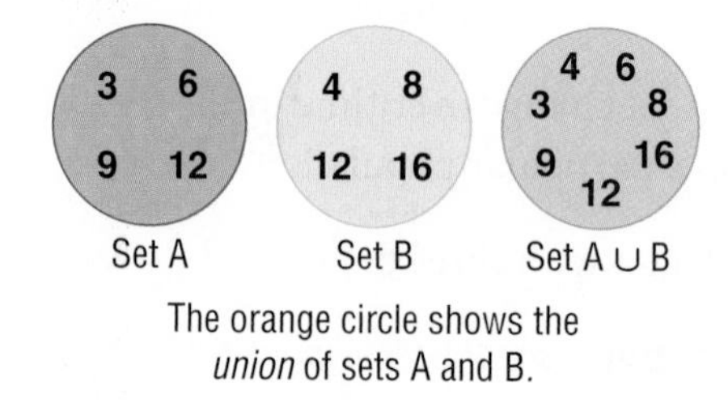

The orange circle shows the *union* of sets A and B.

unit price the price of a single item or amount

unit rate the rate in lowest terms *see 6•5 Ratio and Proportion*

Example: 120 miles in two hours is equivalent to a *unit rate* of 60 miles per hour.

upper quartile the median of the upper half of a set of data, represented by UQ *see 4•4 Statistics*

V

variable a letter or other symbol that represents a number or set of numbers in an expression or an equation *see 6•1 Writing Expressions and Equations*

Example: In the equation $x + 2 = 7$, the variable is x.

variation a relationship between two variables; direct variation, represented by the equation $y = kx$, exists when the increase in the value of one variable results in an increase in the value of the other; inverse variation, represented by the equation $y = \frac{k}{x}$, exists when an increase in the value of one variable results in a decrease in the value of the other

Venn diagram a pictorial means of representing the relationships between sets *see 5•3 Sets*

Example:

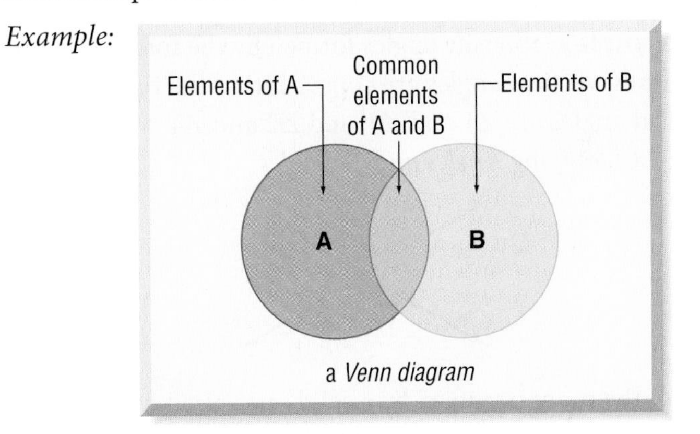

a *Venn diagram*

vertex (pl. *vertices*) the common point of two rays of an angle, two sides of a polygon, or three or more faces of a polyhedron

Examples:

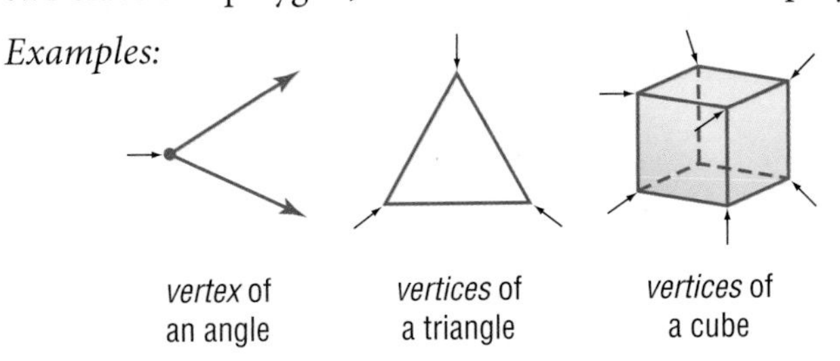

vertex of an angle

vertices of a triangle

vertices of a cube

vertex of tessellation the point where three or more tessellating figures come together

Example:

vertex of tessellation
(in the circle)

vertical a line that is perpendicular to a horizontal base line
see 6•7 Graphing on the Coordinate Plane

Example:

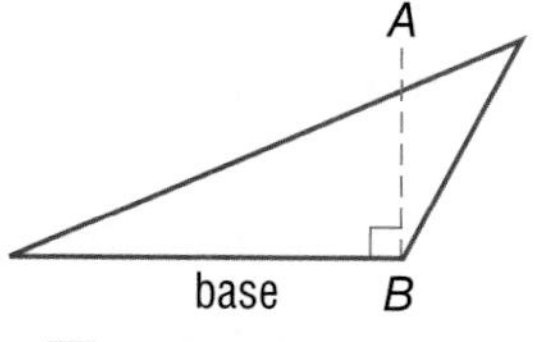

$\overline{AB}$ is *vertical* to the base of this triangle.

vertical angles opposite angles formed by the intersection of two lines; vertical angles are congruent; in the figure, the vertical angles are ∠1 and ∠3, and ∠2 and ∠4
see 7•1 Classifying Angles and Triangles

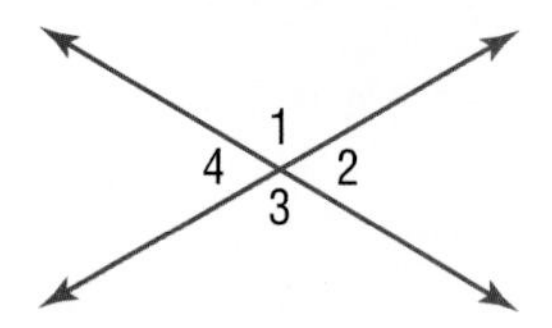

volume the space occupied by a solid, measured in cubic units
see Formulas page 64, 7•7 Volume, 8•3 Area, Volume, and Capacity

Example:

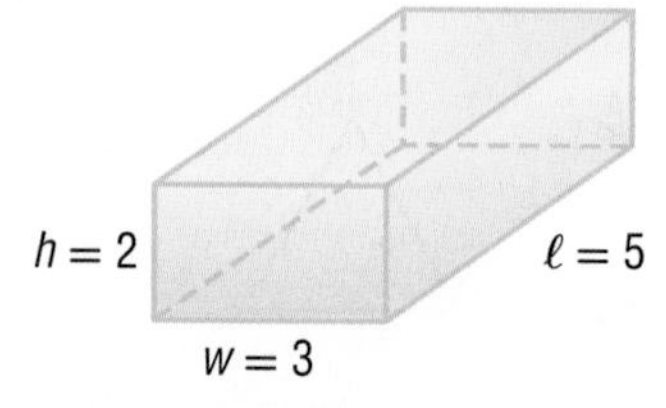

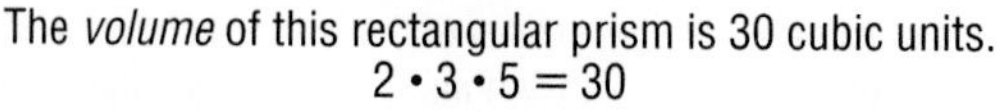
The *volume* of this rectangular prism is 30 cubic units.
$2 \cdot 3 \cdot 5 = 30$

W

weighted average a statistical average in which each element in the sample is given a certain relative importance, or weight *see 4•4 Statistics*

Example: To find the accurate average percentage of people who own cars in three towns with different-sized populations, the largest town's percentage would have to be *weighted.*

whole numbers the set of all counting numbers plus zero {0, 1, 2, 3, 4, 5 . . .}

width a measure of the distance of an object from side to side

X

***x*-axis** the horizontal reference line in the coordinate graph *see 6•7 Graphing on the Coordinate Plane*

***x*-intercept** the point at which a line or curve crosses the x-axis *see 6•8 Slope and Intercept*

Y

***y*-axis** the vertical reference line in the coordinate graph *see 6•7 Graphing on a Coordinate Plane*

***y*-intercept** the point at which a line or curve crosses the y-axis *see 6•8 Slope and Intercept*

Z

zero-pair one positive cube and one negative cube used to model signed-number arithmetic

Formulas

Area *(see 7•5)*

circle $A = \pi r^2$ (pi • square of the radius)

parallelogram $A = bh$ (base • height)

rectangle $A = \ell w$ (length • width)

square $A = s^2$ (side squared)

trapezoid $A = \frac{1}{2}h(b_1 + b_2)$

($\frac{1}{2}$ • height • sum of the bases)

triangle $A = \frac{1}{2}bh$

($\frac{1}{2}$ • base • height)

Volume *(see 7•7)*

cone $V = \frac{1}{3}\pi r^2 h$

($\frac{1}{3}$ • pi • square of the radius • height)

cylinder $V = \pi r^2 h$

(pi • square of the radius • height)

prism $V = Bh$ (area of the base • height)

pyramid $V = \frac{1}{3}Bh$

($\frac{1}{3}$ • area of the base • height)

rectangular prism $V = \ell wh$ (length • width • height)

sphere $V = \frac{4}{3}\pi r^3$

($\frac{4}{3}$ • pi • cube of the radius)

Perimeter *(see 7•4)*

parallelogram $P = 2a + 2b$

(2 • side a + 2 • side b)

rectangle $P = 2\ell + 2w$ (twice length + twice width)

square $P = 4s$

(4 • side)

triangle $P = a + b + c$ (side a + side b + side c)

Formulas

Circumference (see 7•8)

circle $C = \pi d$ (pi • diameter)

or

$C = 2\pi r$

(2 • pi • radius)

Probability (see 4•6)

The *Experimental Probability* of an event is equal to the total number of times a favorable outcome occurred, divided by the total number of times the experiment was done.

$$\textit{Experimental Probability} = \frac{\textit{favorable outcomes that occurred}}{\textit{total number of experiments}}$$

The *Theoretical Probability* of an event is equal to the number of favorable outcomes, divided by the total number of possible outcomes.

$$\textit{Theoretical Probability} = \frac{\textit{favorable outcomes}}{\textit{possible outcome}}$$

Other

Distance $d = rt$ (rate • time)

Interest $I = prt$ (principle • rate • time)

PIE Profit = Income − Expenses

Temperature $F = \frac{9}{5}C + 32$

($\frac{9}{5}$ • Temperature in °C + 32)

$C = \frac{5}{9}(F - 32)$

($\frac{5}{9}$ • (Temperature in °F − 32))

Symbols

Symbol	Meaning
{ }	set
∅	the empty set
⊆	is a subset of
∪	union
∩	intersection
$>$	is greater than
$<$	is less than
$\geq$	is greater than or equal to
$\leq$	is less than or equal to
$=$	is equal to
$\neq$	is not equal to
°	degree
%	percent
$f(n)$	function, f of n
$a{:}b$	ratio of a to b, $\frac{a}{b}$
$\lvert a \rvert$	absolute value of a
$P(E)$	probability of an event E
π	pi
$\perp$	is perpendicular to
$\parallel$	is parallel to
$\cong$	is congruent to
$\sim$	is similar to
$\approx$	is approximately equal to
$\angle$	angle
∟	right angle
$\triangle$	triangle
$\overline{AB}$	segment AB
$\overrightarrow{AB}$	ray AB
$\overleftrightarrow{AB}$	line AB
$\triangle ABC$	triangle ABC
$\angle ABC$	angle ABC
$m\angle ABC$	measure of angle ABC
$\overline{AB}$ or $m\overline{AB}$	length of segment AB
$\overset{\frown}{AB}$	arc AB
!	factorial
$_nP_r$	permutations of n things taken r at a time
$_nC_r$	combinations of n things taken r at a time
$\sqrt{\ }$	square root
$\sqrt[3]{\ }$	cube root
′	foot
″	inch
$\div$	divide
/	divide
*	multiply
$\times$	multiply
$\cdot$	multiply
$+$	add
$-$	subtract

Patterns

arithmetic sequence a sequence of numbers or terms that have a common difference between any one term and the next in the sequence; in the following sequence, the common difference is seven, so $8 - 1 = 7$; $15 - 8 = 7$; $22 - 15 = 7$, and so forth

Example: 1, 8, 15, 22, 29, 36, 43, . . .

Fibonacci numbers a sequence in which each number is the sum of its two predecessors; can be expressed as $x_n = x_{n-2} + x_{n-1}$; the sequence begins: 1, 1, 2, 3, 5, 8, 13, 21, 34, 55, . . .

Example:

1,	1,	2,	3,	5,	8,	13,	21,	34,	55,	. . .
1 + 1 = 2										
	1 + 2 = 3									
		2 + 3 = 5								
			3 + 5 = 8							

geometric sequence a sequence of terms in which each term is a constant multiple, called the *common ratio,* of the one preceding it; for instance, in nature, the reproduction of many single-celled organisms is represented by a progression of cells splitting in two in a growth progression of 1, 2, 4, 8, 16, 32, . . ., which is a geometric sequence in which the common ratio is 2

harmonic sequence a progression $a_1, a_2, a_3, \ldots$ for which the reciprocals of the terms, $\frac{1}{a_1}, \frac{1}{a_2}, \frac{1}{a_3}, \ldots$ form an arithmetic sequence

Lucas numbers a sequence in which each number is the sum of its two predecessors; can be expressed as $x_n = x_{n-2} + x_{n-1}$; the sequence begins: 2, 1, 3, 4, 7, 11, 18, 29, 47, . . .

magic square a square array of different integers in which the sum of the rows, columns, and diagonals are the same

Example:

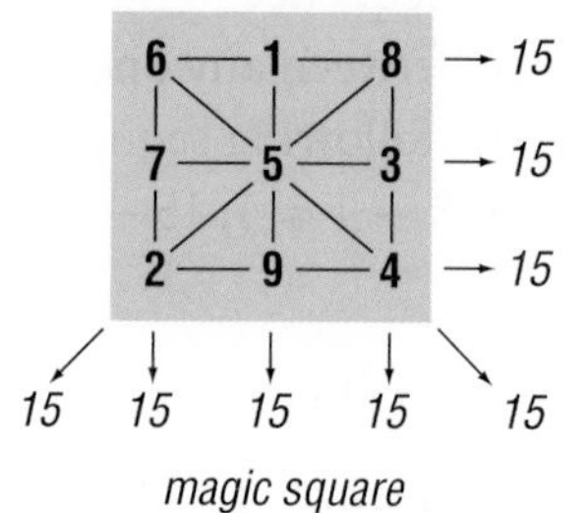

magic square

Pascal's triangle a triangular arrangement of numbers in which each number is the sum of the two numbers above it in the preceding row

Example:

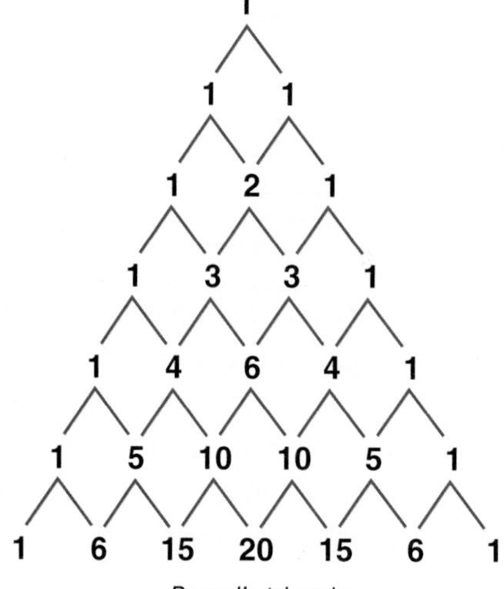

Pascal's triangle

sequence a set of elements, especially numbers, arranged in order according to some rule

series the sum of the terms of a sequence

spiral a plane curve traced by a point moving around a fixed point while continuously increasing or decreasing its distance from it

Example:

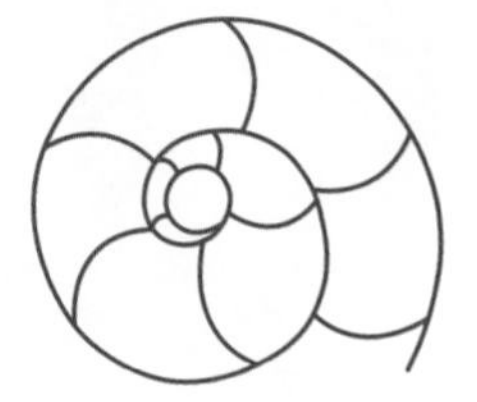

The shape of a chambered nautilus shell is a *spiral.*

square numbers a sequence of numbers that can be shown by dots arranged in the shape of a square; can be expressed as x^2; the sequence begins 1, 4, 9, 16, 25, 36, 49, . . .

Example:

1 4 9 16 25 36

square numbers

tessellation a tiling pattern made of repeating polygons that fills a plane completely, leaving no gaps

Example:

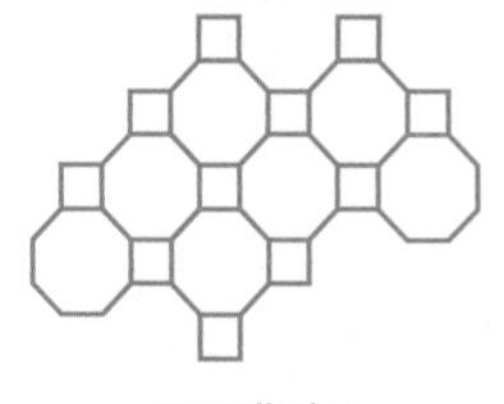

tessellation

triangular numbers a sequence of numbers that can be shown by dots arranged in the shape of a triangle; any number in the sequence can be expressed as $x_n = x_{n-1} + n$; the sequence begins 1, 3, 6, 10, 15, 21, . . .

Example:

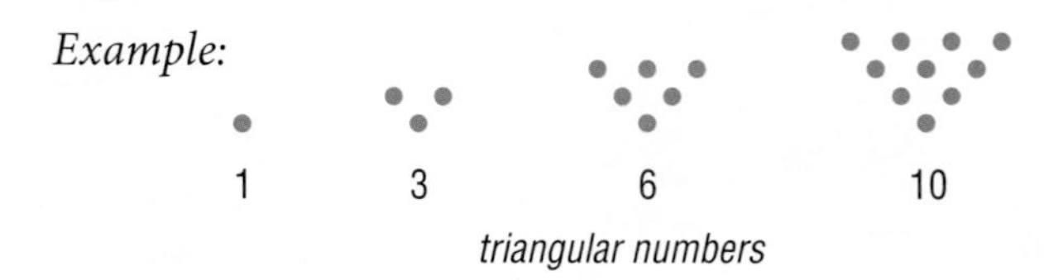

triangular numbers

hyperbola
slope
time

Part Two 2

Hot Topics

HotTopic 1

Numbers and Computation

You can use the problems and the list of words that follow to see what you already know about this chapter. The answers to the problems are in **HotSolutions** at the back of the book, and the definitions of the words are in **HotWords** at the front of the book. You can find out more about a particular problem or word by referring to the topic number (*for example,* Lesson 1·2).

Problem Set

Use parentheses to make each expression true. (Lesson 1·1)

1. $4 + 7 \cdot 3 = 33$

2. $30 + 15 \div 5 + 5 = 14$

Is it a prime number? Write *yes* or *no*. (Lesson 1·2)

3. 77 **4.** 111 **5.** 131 **6.** 301

Write the prime factorization for each number. (Lesson 1·2)

7. 40 **8.** 110 **9.** 230

Find the GCF for each pair of numbers. (Lesson 1·2)

10. 12 and 40 **11.** 15 and 50 **12.** 18 and 171

Find the LCM for each pair of numbers. (Lesson 1·2)

13. 5 and 12 **14.** 15 and 8 **15.** 18 and 30

16. A mystery number is a common multiple of 2, 4, and 15. It is also a factor of 120 but does not equal 120. What is the number? (Lesson 1·2)

Give the absolute value of the integer. Then write the opposite of the original integer. (Lesson 1·3)

17. −7

18. 15

19. −12

20. 10

Graph each integer on a number line. Write > or <. (Lesson 1·3)

21. 3 □ −1

22. −8 □ 4

23. −2 □ −4

24. −3 □ −7

Add or subtract. (Lesson 1·3)

25. $9 + (-7)$

26. $4 - 8$

27. $-5 + (-6)$

28. $8 - (-8)$

29. $-6 - (-6)$

30. $-3 + 9$

Compute. (Lesson 1·3)

31. $(-6) \cdot (-7)$

32. $48 \div (-12)$

33. $-56 \div (-8)$

34. $(-4 \cdot 3) \cdot (-2)$

35. $3 \cdot [-8 + (-4)]$

36. $-5\,[4 - (-6)]$

37. What can you say about the product of a negative integer and a positive integer? (Lesson 1·3)

38. What can you say about the sum of two positive integers? (Lesson 1·3)

HotWords

absolute value (Lesson 1·3)
common factor (Lesson 1·2)
composite number (Lesson 1·2)
coordinate (Lesson 1·3)
divisible (Lesson 1·2)
factor (Lesson 1·2)
greatest common factor (Lesson 1·2)
least common multiple (Lesson 1·2)
multiple (Lesson 1·2)
negative integer (Lesson 1·3)
positive integer (Lesson 1·3)
prime factorization (Lesson 1·2)
prime number (Lesson 1·2)

1•1 Order of Operations

Understanding the Order of Operations

Solving a problem may involve using more than one operation. Your answer can depend on the order in which you do the operations.

For instance, consider the expression $3^2 + 5 \cdot 7$.

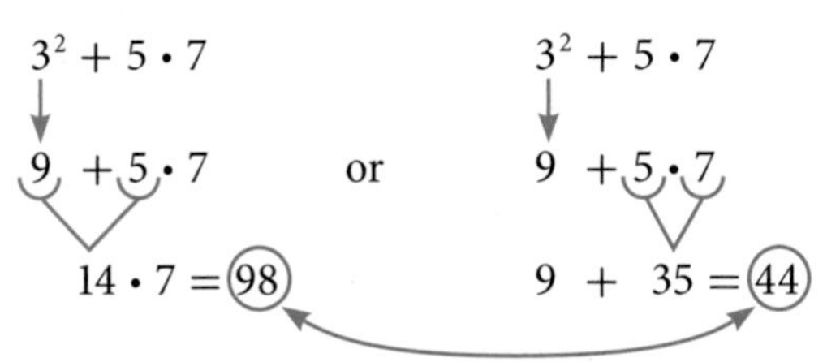

The order in which you perform operations makes a difference.

To make sure that there is just one answer to a series of computations, mathematicians have agreed upon an order in which to complete the operations.

EXAMPLE **Using the Order of Operations**

How can you simplify $(4 + 5) \cdot 3^2 - 5$?

$= (9) \cdot 3^2 - 5$ • Simplify within the parentheses.

$= 9 \cdot (9) - 5$ • Evaluate the power (p. 153).

$= (81) - 5$ • Multiply and divide from left to right.

$= 76$ • Add and subtract from left to right.

So, $(4 + 5) \cdot 3^2 - 5 = 76$.

Check It Out

Simplify.

1. $24 - 4 \cdot 3$
2. $3 \cdot (4 + 5^2)$

1•1 Exercises

Is each expression true? Write *yes* or *no*.

1. $7 \cdot 4 + 5 = 33$

2. $3 + 4 \cdot 8 = 56$

3. $6 \cdot (4 + 6 \div 2) = 30$

4. $4^2 - 1 = 9$

5. $(3 + 5)^2 = 64$

6. $(2^3 + 3 \cdot 4) + 5 = 49$

7. $25 - 4^2 = 9$

8. $(4^2 \div 2)^2 = 64$

Simplify.

9. $24 - (4 \cdot 5)$

10. $2 \cdot (6 + 5^2)$

11. $(2^4) \cdot (12 - 8)$

12. $5^2 + (5 - 3)^2$

13. $(16 - 10)^2 \cdot 5$

14. $12 + 4 \cdot 3^2$

15. $(4^2 + 4)^2$

16. $60 \div (12 + 3)$

17. $30 - (10 - 7)^2$

18. $44 + 5 \cdot (4^2 \div 8)$

Use parentheses to make each expression true.

19. $4 + 4 \cdot 7 = 56$

20. $5 \cdot 20 + 80 = 500$

21. $48 \div 8 - 2 = 8$

22. $10 + 10 \div 2 - 3 = 12$

23. $12 \cdot 3^2 + 7 = 192$

24. $6^2 - 15 \div 3 \cdot 2^2 = 124$

25. Use five 2s, a set of parentheses (as needed), and any of the operations to make the numbers 1 through 5.

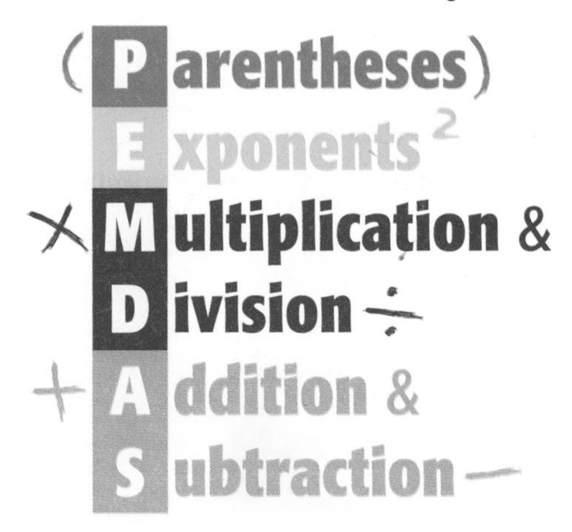

1•2 Factors and Multiples

Factors

Two numbers multiplied together to produce 12 are considered **factors** of 12. So, the factors of 12 are 1, 2, 3, 4, 6, and 12.

To decide whether one number is a factor of another, divide. If there is a remainder of 0, the number is a factor.

EXAMPLE Finding the Factors of a Number

What are the factors of 18?

$1 \cdot 18 = 18$
$2 \cdot 9 = 18$
$3 \cdot 6 = 18$

- Find all pairs of numbers that multiply to give the product.

1, 2, 3, 6, 9, 18

- List the factors in order, starting with 1.

So, the factors of 18 are 1, 2, 3, 6, 9, and 18.

Check It Out

Find the factors of each number.

1. 8
2. 48

Common Factors

Factors that are the same for two or more numbers are **common factors**.

EXAMPLE Finding Common Factors

What numbers are factors of both 12 and 40?

1, 2, 3, 4, 6, 12	• List the factors of the first number.
1, 2, 4, 5, 8, 10, 20, 40	• List the factors of the second number.
1, 2, 4	• List the common factors that are in both lists.

So, the common factors of 12 and 40 are 1, 2, and 4.

Check It Out

List the common factors of each set of numbers.

3. 8 and 18
4. 10, 30, and 45

Greatest Common Factor

The **greatest common factor** (GCF) of two whole numbers is the greatest number that is a factor of both the numbers.

One way to find the GCF is to list all the factors of each number, then list the common factors and choose the greatest common factor.

What is the GCF of 24 and 60?

- The factors of 24 are 1, 2, 3, 4, 6, 8, 12, 24.
- The factors of 60 are 1, 2, 3, 4, 5, 6, 10, 12, 15, 20, 30, 60.
- The common factors that are in both lists are 1, 2, 3, 4, 6, 12.

The greatest common factor of 24 and 60 is 12.

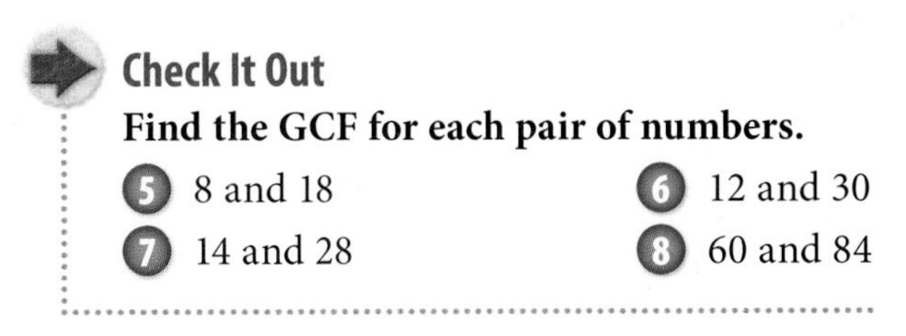

Check It Out

Find the GCF for each pair of numbers.

5. 8 and 18
6. 12 and 30
7. 14 and 28
8. 60 and 84

Divisibility Rules

Sometimes you may wish to know if a number is a factor of a much larger number. For example, if you want to form teams of 3 from a group of 246 basketball players, you will need to know whether 246 is divisible by 3. A number is **divisible** by another number if the remainder of their quotient is 0.

You can quickly figure out whether 246 is divisible by 3 if you know the divisibility rule for 3. A number is divisible by 3 if the sum of the digits is divisible by 3. For example, 246 is divisible by 3 because $2 + 4 + 6 = 12$, and 12 is divisible by 3.

It can be helpful to know other divisibility rules.

A number is divisible by:		
2	if	the last digit is an even number or 0.
3		the sum of the digits is divisible by 3.
4		the last two digits are divisible by 4.
5		the last digit is 0 or 5.
6		the number is divisible by both 2 and 3.
8		the last three digits are divisible by 8.
9		the sum of the digits is divisible by 9.
10		the number ends in 0.

Check It Out

Check by using divisibility rules.

9. Is 424 divisible by 4?
10. Is 199 divisible by 9?
11. Is 534 divisible by 6?
12. Is 1,790 divisible by 5?

Prime and Composite Numbers

A **prime number** is a whole number greater than 1 that has exactly two factors, 1 and itself. Here are the first 10 prime numbers:

2, 3, 5, 7, 11, 13, 17, 19, 23, 29

Twin primes are pairs of primes that have a difference of 2. The pairs of prime numbers (3, 5), (5, 7), and (11, 13) are examples of twin primes.

A number with more than two factors is called a **composite number**. When two composite numbers have no common factors greater than 1, they are said to be *relatively prime.*

The numbers 12 and 25 are relatively prime.

- The factors of 12 are 1, 2, 3, 4, 6, 12.
- The factors of 25 are 1, 5, 25.

Since 12 and 25 do not have a common factor greater than 1, they are relatively prime.

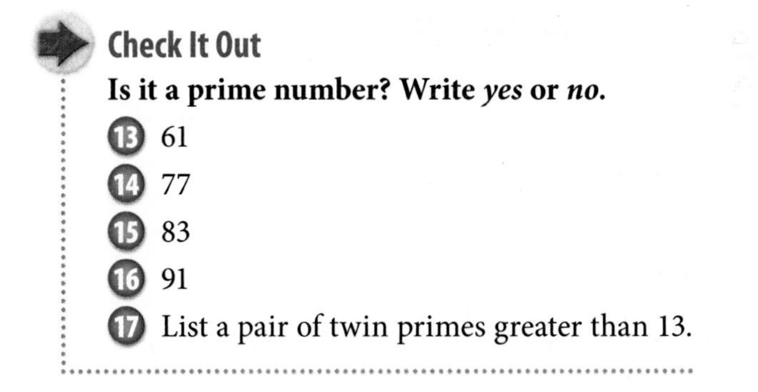

Check It Out

Is it a prime number? Write *yes* or *no*.

13. 61
14. 77
15. 83
16. 91
17. List a pair of twin primes greater than 13.

Prime Factorization

Every composite number can be expressed as a product of prime factors.

You can use a factor tree to find the prime factors. The one below shows the **prime factorization** of 60. Although the order of the factors may be different because you can start with different pairs of factors, every factor tree for 60 has the same prime factorization. You also can write the prime factorization using exponents (p. 152).

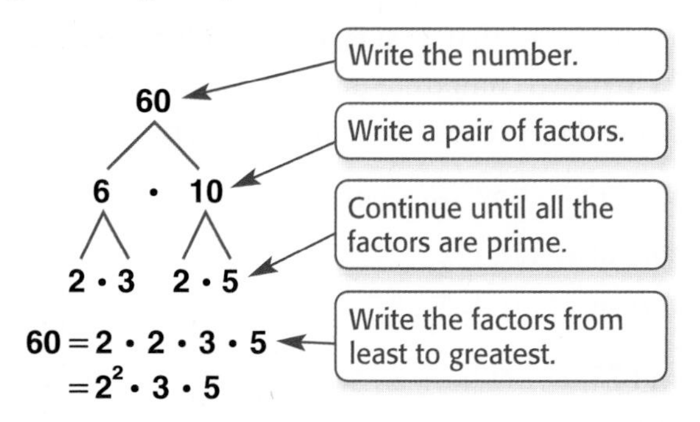

EXAMPLE Finding Prime Factorization

Find the prime factorization of 264.

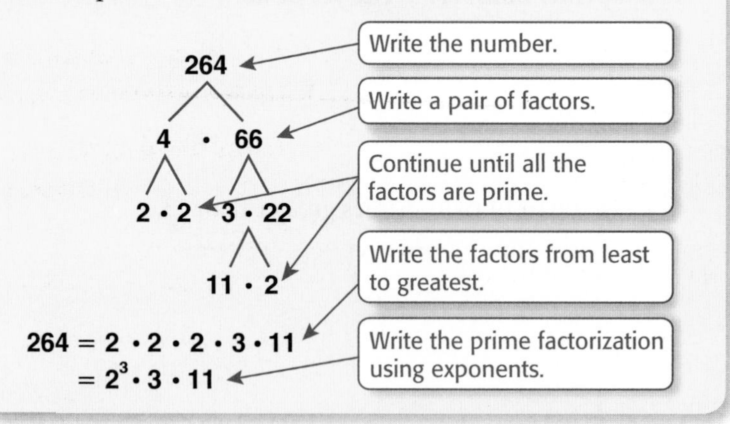

Check It Out

Write the prime factorization for each number.

18. 80

19. 120

Shortcut to Finding GCF

You can use prime factorization to find the greatest common factor.

EXAMPLE **Using Prime Factorization to Find the GCF**

Find the greatest common factor of 12 and 20.

$12 = (2) \cdot (2) \cdot 3$ $20 = (2) \cdot (2) \cdot 5$	• Find the prime factors of each number. Use a factor tree if necessary.
$2 \cdot 2 = 4$	• Identify the prime factors common to both numbers and find their product.

So, the GCF of 12 and 20 is 2^2, or 4.

Check It Out

Use prime factorization to find the GCF of each pair of numbers.

20. 6 and 24
21. 24 and 56
22. 14 and 28
23. 60 and 84

Multiples and Least Common Multiples

The **multiples** of a number are the whole-number products when that number is a factor. In other words, you can find a multiple of a number by multiplying it by 1, 2, 3, and so on.

The **least common multiple** (LCM) is the smallest positive number that is a multiple of two or more whole numbers. One way to find the LCM of a pair of numbers is to first list positive multiples of each number, and then identify the smallest multiple common to both. For instance, to find the LCM of 6 and 8:

- List multiples of 6: 6, 12, 18, 24, 30, . . .
- List multiples of 8: 8, 16, 24, 32, . . .
- The LCM of 6 and 8 is 24.

Another way to find the LCM is to use prime factorization.

EXAMPLE Using Prime Factorization to Find the LCM

Use prime factorization to find the least common multiple of 6 and 8.

$6 = 2 \cdot 3$
$8 = 2 \cdot 2 \cdot 2$

- Find the prime factors of each number. Both numbers have one 2 in their lists. The 6 has an extra 3 and the 8 has two extra 2s.

$2 \cdot 3 \cdot 2 \cdot 2 = 24$

- Multiply the common factors and the extra factors.

common factor — extra factors

So, the LCM of 6 and 8 is 24.

Check It Out

Find the LCM for each pair of numbers.

24. 6 and 9
25. 20 and 35
26. 9 and 4
27. 25 and 75

1•2 Exercises

Find the factors of each number.

1. 16

2. 21

3. 36

4. 54

Is it a prime number? Write *yes* or *no*.

5. 71

6. 87

7. 103

8. 291

Write the prime factorization for each number.

9. 50

10. 130

11. 180

12. 320

Find the GCF for each pair of numbers.

13. 75 and 125

14. 8 and 40

15. 18 and 60

16. 20 and 25

17. 16 and 50

18. 15 and 32

Find the LCM for each pair of numbers.

19. 9 and 15

20. 12 and 60

21. 18 and 24

22. 6 and 32

23. What is the divisibility rule for 9? Is 118 divisible by 9?

24. Describe how to use prime factorization to find the GCF of two numbers.

25. A mystery number is a factor of 100 and a common multiple of 2 and 5. The sum of its digits is 5. What is the number?

1•3 Integer Operations

Positive and Negative Integers

Some quantities can be expressed using negative numbers. For example, negative numbers show below-zero temperatures, drops in the value of stocks, or business losses.

Whole numbers less than zero are called **negative integers**. Whole numbers greater than zero are called **positive integers**.

Here is the set of all integers:

{. . ., −5, −4, −3, −2, −1, 0, 1, 2, 3, 4, 5, . . .}

Check It Out

Write an integer to describe the situation.

1. 6° below zero
2. a gain of $200

Opposites of Integers and Absolute Value

Integers can describe opposite ideas. Each integer has an opposite.

The opposite of a gain of 3 inches is a loss of 3 inches.
The opposite of spending $8 is earning $8.
The opposite of −6 is +6.

The **absolute value** of an integer is its distance from 0 on the number line. You write the absolute value of −7 as $|-7|$.

−7 is 7 units away from 0.

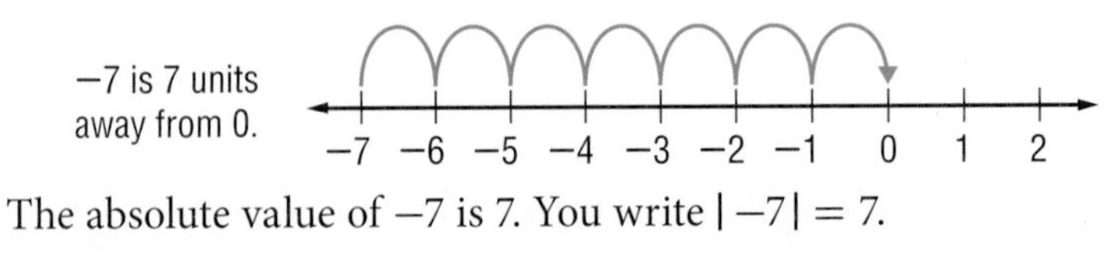

The absolute value of −7 is 7. You write $|-7| = 7$.

Check It Out

Give the absolute value of the integer. Then write the opposite of the original integer.

3. −12
4. 5
5. 0

Comparing and Ordering Integers

You can compare integers by graphing them on a number line. The number that is assigned to a point on a number line is called a **coordinate**. To graph an integer, locate the point corresponding to the integer on a number line.

−10 −9 −8 −7 −6 −5 −4 −3 −2 −1 0 1 2 3 4 5 6 7 8 9 10

The graph shows the points with coordinates −7 and 5 on the number line. When comparing integers on a number line, the integer farthest to the right is greater than the integers to the left on the number line. This means that 5 is greater than −7. When comparing a positive and a negative integer, the positive integer is always greater than the negative integer. When comparing negative integers, the negative integer closest to 0 is the greatest integer.

You can use inequality symbols ($<$, $>$) to compare integers. When comparing −4 and 2, 2 is to the right of −4. Therefore, $-4 < 2$ or $2 > -4$.

EXAMPLE Comparing and Ordering Integers

Compare and order −2, 6, −5, and 3.

- Graph the integers −2, 6, −5, and 3 on the number line.

−8 −7 −6 −5 −4 −3 −2 −1 0 1 2 3 4 5 6 7 8

- Place the integers in order from left to right and compare.

$-5, -2, 3,$ and 6; $-2 < 6$, $-2 > 5$, $6 > -5$, $6 > 3$, $-5 < 3$

Check It Out

Graph each integer on a number line. Write $>$ or $<$.

6. 2 □ −5
7. −4 □ −2

Place the integers in order from least to greatest.

8. 6, −1, 3, −4
9. −5, −2, 7, 4

Adding and Subtracting Integers

Use a number line to model addition and subtraction of integers.

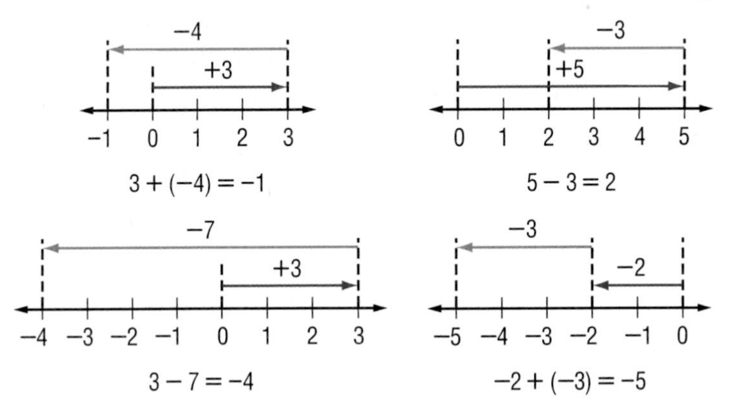

When you add integers of the same sign, add the absolute values of the integers and use the sign of the addends. When adding integers of opposite signs, find the absolute values, subtract the smaller integer from the larger integer, and give the sum the sign of the integer that has the larger absolute value.

EXAMPLE **Adding Integers**

Add $-4 + (-8)$.

$\lvert -4 \rvert + \lvert -8 \rvert =$	• Add the absolute values.
$4 + 8 = 12$	
$-4 + (-8) = -12$	• Give the sum the same sign as the addends.

So, $-4 + (-8) = -12$.

Add $8 + (-3)$.

$\lvert 8 \rvert + \lvert -3 \rvert =$	• Find the absolute values.
$8 - 3 = 5$	• Subtract the smaller absolute value from the larger.
$8 + (-3) = 5$	• Give the sum the sign of the integer that has the larger absolute value.

So, $8 + (-3) = 5$.

Subtracting an integer is the same as adding its opposite. When subtracting integers, change the subtracted integer to its opposite and add the two integers following the rules for addition.

EXAMPLE Subtracting Integers

Subtract $5 - 8$.

$5 + (-8) =$
- Change to an addition problem and convert the subtracted integer to its opposite.

$5 + (-8) = -3$
- Add the two integers by following the rules for addition.

So, $5 - 8 = -3$.

Subtract $9 - (-4)$.

$9 + 4 =$
- Change to an addition problem and convert the subtracted integer to its opposite.

$9 + 4 = 13$
- Add the two integers by following the rules for addition.

So, $9 - (-4) = 13$.

Subtract $-7 - 2$.

$-7 + (-2) =$
- Change to an addition problem and convert the subtracted integer to its opposite.

$-7 + (-2) = -9$
- Add the two integers by following the rules for addition.

So, $-7 - 2 = -9$.

Check It Out

Solve.

10. $5 - 7$
11. $4 + (-4)$
12. $-9 - (-4)$
13. $0 + (-3)$

Multiplying and Dividing Integers

Multiply and divide integers as you would whole numbers, then use these rules for writing the sign of the answer. The product and quotient of two integers with like signs are positive.

EXAMPLE Multiplying and Dividing Integers with Like Signs

Multiply $(-4) \cdot (-3)$.

$(-4) \cdot (-3) = 12$

- When the signs of the two integers are the same, the product is positive.

So, $(-4) \cdot (-3) = 12$.

Divide $-24 \div (-4)$.

$-24 \div (-4) = 6$

- When the signs of the two integers are the same, the quotient is positive.

So, $-24 \div (-4) = 6$.

When the signs of two integers are different, the product and quotient are negative.

EXAMPLE Multiplying and Dividing Integers with Unlike Signs

Multiply $(-3) \cdot (6)$.

$(-3) \cdot (6) = -18$

- When the signs of the two integers are different, the product is negative.

So, $(-3) \cdot (6) = -18$.

Divide $-8 \div 2$.

$-8 \div 2 = -4$

- When the signs of the two integers are different, the quotient is negative.

So, $-8 \div 2 = -4$.

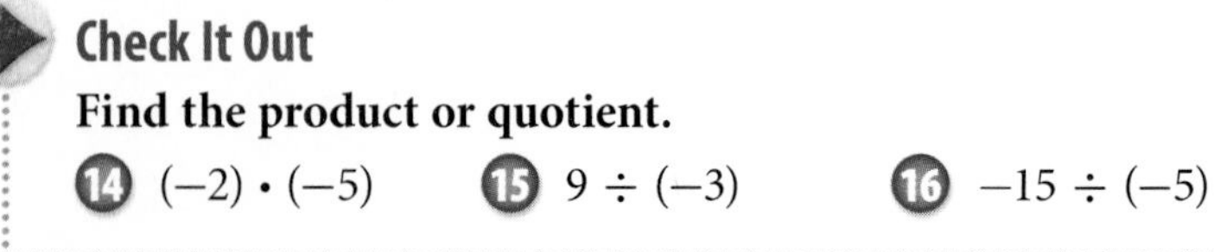

Check It Out

Find the product or quotient.

14 $(-2) \cdot (-5)$ 15 $9 \div (-3)$ 16 $-15 \div (-5)$

1•3 Exercises

Give the absolute value of the integer. Then write its opposite.

1. −14 **2.** 6 **3.** −8 **4.** 1

Graph each integer on a number line. Write > or <.

5. −7 □ 3 **6.** −5 □ −2 **7.** 8 □ −5 **8.** −1 □ 6

Add or subtract.

9. $5 - 3$

10. $4 + (-6)$

11. $-7 - (-4)$

12. $0 + (-5)$

13. $-2 + 6$

14. $0 - 8$

15. $0 - (-8)$

16. $-2 - 8$

17. $4 + (-4)$

18. $-9 - (-5)$

19. $-5 - (-5)$

20. $-7 + (-8)$

Find the product or quotient.

21. $(-2) \cdot (-6)$

22. $8 \div (-4)$

23. $-15 \div 5$

24. $(-6)(7)$

25. $(4) \cdot (-9)$

26. $-24 \div 8$

27. $-18 \div (-3)$

28. $(3) \cdot (-7)$

Compute.

29. $[(-3)(-2)] \cdot 4$

30. $6 \cdot [(3)(-4)]$

31. $[(-2)(-5)] \cdot -3$

32. $-4 \cdot [3 + (-5)]$

33. $(-8 - 2) \cdot 3$

34. $-4 \cdot [6 - (-3)]$

35. Is the absolute value of a negative integer positive or negative?

36. If you know that the absolute value of an integer is 4, what are the possible values for that integer?

37. What can you say about the sum of two negative integers?

38. What can you say about the product of two negative integers?

39. The temperature at noon was 18°F. For the next 4 hours it dropped at a rate of 3 degrees an hour. First express this change as an integer. Then give the temperature at 4 P.M.

Numbers and Computation

What have you learned?

You can use the problems and the list of words that follow to see what you learned in this chapter. You can find out more about a particular problem or word by referring to the topic number (*for example,* Lesson 1·2).

Problem Set

Use parentheses to make each expression true. (Lesson 1·1)

1. $4 + 9 \cdot 2 = 26$

2. $25 + 10 \div 2 + 7 = 37$

3. $2 \cdot 3 + 4^2 = 38$

4. $6 + 7 \cdot 5^2 - 7 = 318$

5. $14 + 9 \cdot 6 \div 3^2 = 20$

Is it a prime number? Write *yes* or *no*. (Lesson 1·2)

6. 87 **7.** 102 **8.** 143 **9.** 401

Write the prime factorization for each number. (Lesson 1·2)

10. 35 **11.** 150 **12.** 320

Find the GCF for each pair of numbers. (Lesson 1·2)

13. 16 and 30

14. 12 and 50

15. 10 and 160

Find the LCM for each pair of numbers. (Lesson 1·2)

16. 5 and 12

17. 15 and 8

18. 18 and 30

19. What is the divisibility rule for 6? Is 246 a multiple of 6? (Lesson 1·2)

Give the absolute value of the integer. Then write the opposite of the original integer. (Lesson 1·3)

20. −9 **21.** 13 **22.** −10 **23.** 20

Graph each integer on a number line. Write > or <. (Lesson 1·3)

24. 5 □ −3 **25.** −7 □ 2

26. −1 □ −8 **27.** −4 □ −6

Add or subtract. (Lesson 1·3)

28. $9 + (-8)$ **29.** $6 - 7$ **30.** $-8 + (-9)$

31. $5 - (-5)$ **32.** $-7 - (-7)$ **33.** $-4 + 12$

Compute. (Lesson 1·3)

34. $(-8) \cdot (-9)$ **35.** $64 \div (-32)$

36. $-36 \div (-9)$ **37.** $(-4 \cdot 5) \cdot (-3)$

38. $4 \cdot [-3 + (-8)]$ **39.** $-6 \cdot [5 - (-8)]$

40. What is true of the product of two positive integers? (Lesson 1·3)

41. What is true of the difference of two negative integers? (Lesson 1·3)

HotWords

Write definitions for the following words.

absolute value (Lesson 1·3)
common factor (Lesson 1·2)
composite number (Lesson 1·2)
coordinate (Lesson 1·3)
divisible (Lesson 1·2)
factor (Lesson 1·2)
greatest common factor (Lesson 1·2)
least common multiple (Lesson 1·2)
multiple (Lesson 1·2)
negative integer (Lesson 1·3)
positive integer (Lesson 1·3)
prime factorization (Lesson 1·2)
prime number (Lesson 1·2)

HotTopic 2

Rational Numbers

What do you know?

You can use the problems and the list of words that follow to see what you already know about this chapter. The answers to the problems are in **HotSolutions** at the back of the book, and the definitions of the words are in **HotWords** at the front of the book. You can find out more about a particular problem or word by referring to the topic number (*for example,* Lesson 2·2).

Problem Set

1. It takes Mr. Chen about $1\frac{1}{2}$ work days to install a tile floor in an average-size kitchen. How many days would it take him to install floors for 6 kitchens? **(Lesson 2·2)**
2. Leslie has $7\frac{1}{2}$ cups of cooked pasta. She wants each serving to be $\frac{3}{4}$ cup. How many servings does she have? **(Lesson 2·2)**
3. In one basketball game, Julian scored $\frac{3}{7}$ of his free throws. In a second basketball game, he scored $\frac{1}{2}$ of his free throws. In which game did he perform better? **(Lesson 2·4)**
4. Nalani got 17 out of 20 questions correct on her science test. What percent did she get correct? **(Lesson 2·6)**
5. Which fraction is not equivalent to $\frac{9}{12}$? **(Lesson 2·1)**

   A. $\frac{3}{4}$ **B.** $\frac{6}{8}$ **C.** $\frac{8}{11}$ **D.** $\frac{75}{100}$
6. Find the improper fraction and write it as a mixed number. **(Lesson 2·1)**

   A. $\frac{6}{12}$ **B.** $\frac{4}{3}$ **C.** $3\frac{5}{6}$

Add or subtract as indicated. Write your answers in simplest form. (Lesson 2·2)

7. $\frac{2}{3} + \frac{1}{2}$ **8.** $3\frac{3}{8} - 1\frac{5}{8}$ **9.** $6 - 2\frac{3}{4}$ **10.** $3\frac{1}{2} + 4\frac{4}{5}$

Multiply or divide. (Lesson 2·2)

11. $\frac{4}{5} \cdot \frac{1}{2}$ **12.** $\frac{3}{4} \div 1\frac{1}{2}$ **13.** $3\frac{3}{8} \cdot \frac{2}{9}$ **14.** $7\frac{1}{2} \div 2\frac{1}{2}$

Solve. (Lesson 2·3)

15. $3.604 + 12.55$ **16.** $11.4 - 10.08$

17. $6.05 \cdot 5.1$ **18.** $67.392 \div 9.6$

19. Write the following numbers in order from least to greatest: $1\frac{13}{20}$, 1.605, 1.065, $\frac{33}{200}$. (Lesson 2·4)

Name all sets of numbers to which each real number belongs. (Lesson 2·5)

20. $\sqrt{19}$ **21.** -3.56

Solve the following. Round answers to the nearest tenth. (Lesson 2·7)

22. What percent of 80 is 24? **23.** Find 23% of 121.

24. 44 is 80% of what number?

HotWords

benchmark (Lesson 2·6)
discount (Lesson 2·7)
improper fraction (Lesson 2·1)
irrational number (Lesson 2·5)
mixed number (Lesson 2·1)
multiplicative inverse (Lesson 2·2)
percent (Lesson 2·6)
percent proportion (Lesson 2·7)
rational numbers (Lesson 2·1)
reciprocal (Lesson 2·2)
repeating decimal (Lesson 2·4)
terminating decimal (Lesson 2·4)

2•1 Fractions

Recall that **rational numbers** are numbers that can be written as fractions such as $\frac{a}{b}$, where a and b are integers and $b \neq 0$. Because -5 can be written as $-\frac{5}{1}$, and $3\frac{3}{4}$ can be written as $\frac{15}{4}$, -5 and $3\frac{3}{4}$ are rational numbers.

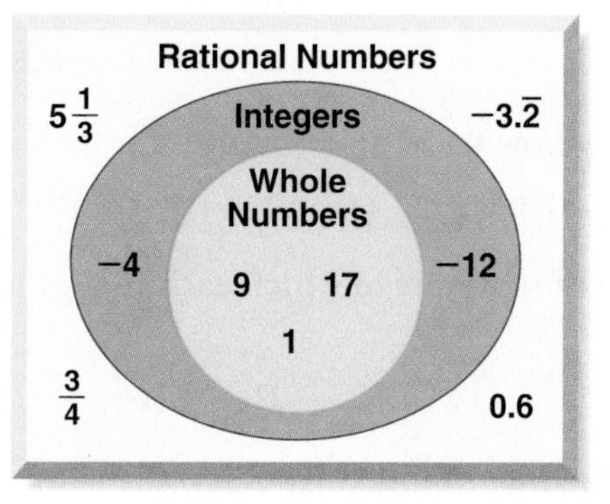

All whole numbers and integers can also be classified as rational numbers. There are some numbers that are not rational, such as $\sqrt{2}$ and π because they cannot be expressed as a fraction.

Equivalent Fractions

Finding Equivalent Fractions

To find a fraction that is equivalent to another fraction, you can multiply or divide the original fraction by a form of 1.

EXAMPLE Methods for Finding Equivalent Fractions

Find a fraction equal to $\frac{6}{12}$.

- Multiply or divide by a form of 1.

Multiply	or	Divide
$\frac{6}{12} \cdot \frac{5}{5} = \frac{30}{60}$		$\frac{6}{12} \div \frac{2}{2} = \frac{3}{6}$
$\frac{6}{12} = \frac{30}{60}$		$\frac{6}{12} = \frac{3}{6}$

So, $\frac{6}{12} = \frac{3}{6}$.

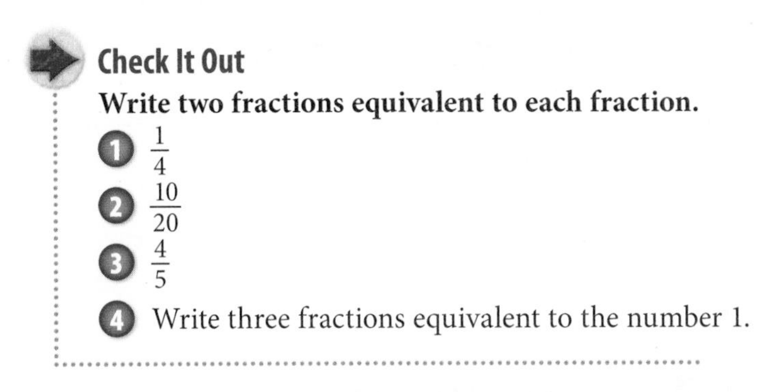

Check It Out

Write two fractions equivalent to each fraction.

1. $\frac{1}{4}$
2. $\frac{10}{20}$
3. $\frac{4}{5}$
4. Write three fractions equivalent to the number 1.

Deciding Whether Two Fractions Are Equivalent

Two fractions are equivalent if each fraction names the same amount. There are many fractional names for the same amount.

Such as: $\frac{1}{2} = \frac{3}{6} = \frac{4}{8} = \frac{6}{12}$

You can identify equivalent fractions by comparing the *cross products* (p. 282) of the fractions. If the cross products are equal, the fractions are equivalent.

EXAMPLE Deciding Whether Two Fractions Are Equivalent

Determine whether $\frac{2}{4}$ is equivalent to $\frac{10}{20}$.

$\frac{2}{4} \overset{?}{\times} \frac{10}{20}$ • Find the cross products of the fractions.

$40 = 40$ • Compare the cross products.

$\frac{2}{4} = \frac{10}{20}$ • The cross products are the same; so the fractions are equivalent.

So, $\frac{2}{4}$ is equivalent to $\frac{10}{20}$.

Check It Out

Use the cross products method to determine whether each pair of fractions is equivalent.

5. $\frac{15}{20}, \frac{30}{20}$
6. $\frac{4}{5}, \frac{24}{30}$
7. $\frac{3}{4}, \frac{15}{24}$

Writing Fractions in Simplest Form

When the numerator and the denominator of a fraction have no common factor other than 1, the fraction is in simplest form.

To express a fraction in simplest form, you can divide the numerator and denominator by their greatest common factor (GCF).

EXAMPLE Finding Simplest Form of Fractions

Express $\frac{18}{24}$ in simplest form.

The factors of 18 are: 1, 2, 3, 6, 9, 18
- List the factors of the numerator.

The factors of 24 are: 1, 2, 3, 4, 6, 8, 12, 24
- List the factors of the denominator.

The GCF is 6.
- Find the greatest common factor (GCF).

$\frac{18 \div 6}{24 \div 6} = \frac{3}{4}$
- Divide the numerator and the denominator by the GCF.

So, $\frac{18}{24}$ in simplest form is $\frac{3}{4}$.

Check It Out

Express each fraction in simplest form.

8. $\frac{8}{10}$

9. $\frac{12}{16}$

10. $\frac{24}{60}$

Writing Improper Fractions and Mixed Numbers

An **improper fraction**, such as $\frac{7}{2}$, is a fraction in which the numerator is greater than the denominator. Improper fractions represent quantities that are greater than 1.

A number composed of a whole number and a fraction, such as $3\frac{1}{2}$, is a **mixed number**.

You can write any mixed number as an improper fraction and any improper fraction as a mixed number.

$$\frac{7}{2} = 3\frac{1}{2}$$

You can use division to change an improper fraction to a mixed number.

EXAMPLE Changing an Improper Fraction to a Mixed Number

Change $\frac{17}{5}$ to a mixed number.

$$\begin{array}{r} 3 \leftarrow \text{quotient} \\ \text{divisor} \rightarrow 5\overline{)17} \\ -15 \\ \hline 2 \leftarrow \text{remainder} \end{array}$$

- Divide the numerator by the denominator. $\frac{17}{5}$

quotient $\rightarrow 3\frac{2}{5}$ $\leftarrow$ remainder / $\leftarrow$ divisor

- Write the mixed number.

So, $\frac{17}{5}$ written as a mixed number is $3\frac{2}{5}$.

Check It Out

Write a mixed number for each improper fraction.

11. $\frac{43}{6}$
12. $\frac{34}{3}$
13. $\frac{32}{5}$
14. $\frac{37}{4}$

You can use multiplication to change a mixed number to an improper fraction. Rename the whole-number part as an improper fraction with the same denominator as the fraction part. Then add the two parts.

EXAMPLE **Changing a Mixed Number to an Improper Fraction**

Change $3\frac{1}{4}$ to an improper fraction.

$3 \cdot \frac{4}{4} = \frac{12}{4}$

- Multiply the whole-number part by a representation of 1 that has the same denominator as the fraction part.

$3\frac{1}{4} = \frac{12}{4} + \frac{1}{4} = \frac{13}{4}$

- Add the two parts.

So, $3\frac{1}{4}$ written as an improper fraction is $\frac{13}{4}$.

Check It Out

Write an improper fraction for each mixed number.

15. $4\frac{5}{8}$
16. $12\frac{5}{6}$
17. $24\frac{1}{2}$
18. $32\frac{2}{3}$

2-1 Exercises

Write one fraction equivalent to the given fraction.

1. $\frac{1}{2}$

2. $\frac{7}{8}$

3. $\frac{40}{60}$

4. $\frac{18}{48}$

Express each fraction in simplest form.

5. $\frac{45}{90}$

6. $\frac{24}{32}$

7. $\frac{12}{34}$

8. $3\frac{12}{60}$

9. $\frac{38}{14}$

10. $\frac{82}{10}$

Find the GCF of each pair of numbers.

11. 16, 21

12. 81, 27

13. 18, 15

Write a mixed number for each improper fraction.

14. $\frac{25}{4}$

15. $\frac{12}{10}$

16. $\frac{11}{4}$

Write an improper fraction for each mixed number.

17. $5\frac{1}{6}$

18. $8\frac{3}{5}$

19. $13\frac{4}{9}$

2•2 Operations with Fractions

Adding and Subtracting Fractions with Like Denominators

When you add or subtract fractions that have the same denominator, you add or subtract the numerators. The denominator stays the same.

EXAMPLE **Adding and Subtracting Fractions with Like Denominators**

Add $\frac{3}{4} + \frac{2}{4}$.

$3 + 2 = 5$ • Add or subtract the numerators.

$\frac{3}{4} + \frac{2}{4} = \frac{5}{4}$ • Write the result over the like denominator.

$\frac{5}{4} = 1\frac{1}{4}$ • Simplify, if possible.

So, $\frac{3}{4} + \frac{2}{4} = 1\frac{1}{4}$.

Check It Out

Add or subtract. Simplify, if possible.

1. $\frac{12}{15} + \frac{6}{15}$
2. $\frac{24}{34} + \frac{13}{34}$
3. $\frac{11}{12} - \frac{5}{12}$
4. $\frac{7}{10} - \frac{2}{10}$

Adding and Subtracting Fractions with Unlike Denominators

To add or subtract fractions with unlike denominators, you need to change the fractions to equivalent fractions with common denominators.

EXAMPLE **Adding and Subtracting Fractions with Unlike Denominators**

Add $\frac{4}{5} + \frac{3}{4}$.

20 is the LCD of 4 and 5.
- Find the least common denominator.

$\frac{4}{5} = \frac{4}{5} \cdot \frac{4}{4} = \frac{16}{20}$ and $\frac{3}{4} = \frac{3}{4} \cdot \frac{5}{5} = \frac{15}{20}$
- Write equivalent fractions with the LCD.

$\frac{16}{20} + \frac{15}{20} = \frac{31}{20}$
- Add or subtract the numerators. Write the result over the common denominator.

$\frac{31}{20} = 1\frac{11}{20}$
- Simplify, if possible.

So, $\frac{4}{5} + \frac{3}{4} = 1\frac{11}{20}$.

Check It Out

Add or subtract. Simplify, if possible.

5. $\frac{9}{10} + \frac{1}{2}$
6. $\frac{1}{2} + \frac{5}{7}$
7. $\frac{4}{5} - \frac{3}{4}$
8. $\frac{5}{8} - \frac{1}{6}$

Adding and Subtracting Mixed Numbers

Adding and subtracting mixed numbers is similar to adding and subtracting fractions.

Adding Mixed Numbers with Common Denominators

You add mixed numbers with like fractions by adding the fraction part and then the whole numbers.

EXAMPLE **Adding Mixed Numbers with Common Denominators**

Add $5\frac{3}{8} + 2\frac{3}{8}$.

$$\begin{array}{r} 5\frac{3}{8} \\ +\ 2\frac{3}{8} \\ \hline 7\frac{6}{8} \end{array}$$

- Add the fractions, and then add the whole numbers.

$7\frac{6}{8} = 7\frac{3}{4}$

- Simplify, if possible.

So, $5\frac{3}{8} + 2\frac{3}{8} = 7\frac{3}{4}$.

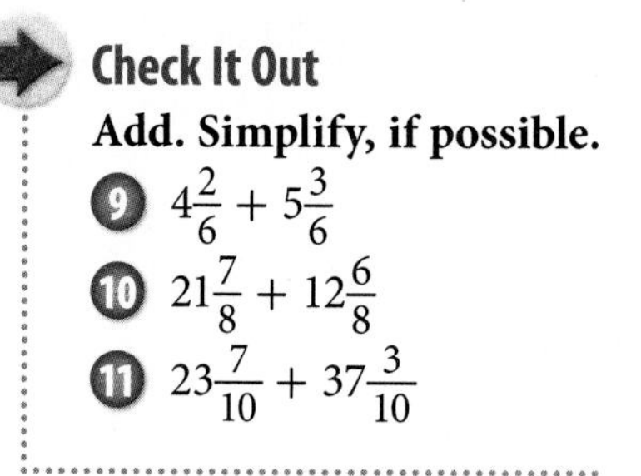

Check It Out

Add. Simplify, if possible.

9. $4\frac{2}{6} + 5\frac{3}{6}$
10. $21\frac{7}{8} + 12\frac{6}{8}$
11. $23\frac{7}{10} + 37\frac{3}{10}$

Adding Mixed Numbers with Unlike Denominators

You can add mixed numbers with unlike fractions by writing equivalent fractions with a common denominator. Sometimes you will have to simplify an improper fraction in the answer.

EXAMPLE Adding Mixed Numbers with Unlike Denominators

Add $4\frac{2}{3} + 1\frac{3}{5}$.

$\frac{2}{3} = \frac{10}{15}$ and $\frac{3}{5} = \frac{9}{15}$

- Write equivalent fractions with the LCD.

$$\begin{array}{r} 4\frac{10}{15} \\ +\ 1\frac{9}{15} \\ \hline 5\frac{19}{15} \end{array}$$

- Add the fractions and then add the whole numbers.

$5\frac{19}{15} = 6\frac{4}{15}$

- Simplify, if possible.

So, $4\frac{2}{3} + 1\frac{3}{5} = 6\frac{4}{15}$.

Add $3 + 8\frac{5}{7}$.

$$\begin{array}{r} 3 \\ +\ 8\frac{5}{7} \\ \hline 11\frac{5}{7} \end{array}$$

- Add the fractions and then add the whole numbers.

So, $3 + 8\frac{5}{7} = 11\frac{5}{7}$.

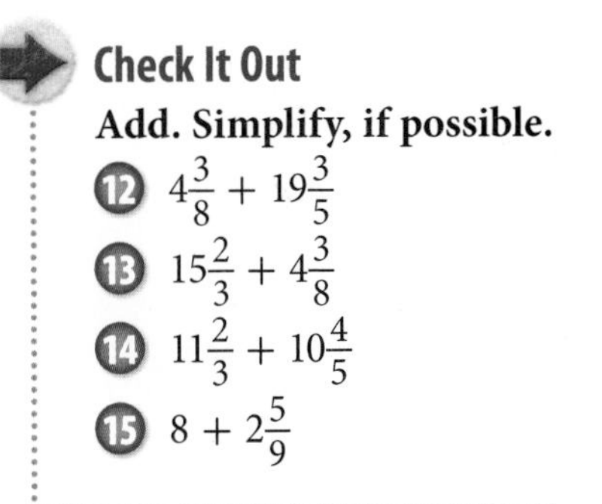

Check It Out

Add. Simplify, if possible.

12. $4\frac{3}{8} + 19\frac{3}{5}$
13. $15\frac{2}{3} + 4\frac{3}{8}$
14. $11\frac{2}{3} + 10\frac{4}{5}$
15. $8 + 2\frac{5}{9}$

Subtracting Mixed Numbers with Common Denominators

To subtract mixed numbers with common denominators, you write the difference of the numerators over the common denominator.

EXAMPLE Subtracting Mixed Numbers with Common Denominators

Subtract $18\frac{3}{9} - 4\frac{2}{9}$.

$$\begin{array}{r} 18\frac{3}{9} \\ -\ 4\frac{2}{9} \\ \hline 14\frac{1}{9} \end{array}$$

- Subtract the fractions, and then subtract the whole numbers.

So, $18\frac{3}{9} - 4\frac{2}{9} = 14\frac{1}{9}$.

Subtract $13\frac{1}{4} - 9\frac{3}{4}$.

$$\begin{array}{rcr} 13\frac{1}{4} & \rightarrow & 12\frac{5}{4} \\ 9\frac{3}{4} & \rightarrow & -\ 9\frac{3}{4} \\ \hline & & 3\frac{2}{4} = 3\frac{1}{2} \end{array}$$

- Regroup before subtracting. $13\frac{1}{4} = 12 + \frac{4}{4} + \frac{1}{4} = 12\frac{5}{4}$
- Subtract the fractions, and then subtract the whole numbers.
- Simplify, if possible.

So, $13\frac{1}{4} - 9\frac{3}{4} = 3\frac{1}{2}$.

Check It Out

Subtract. Write in simplest form.

16. $7\frac{6}{11} - 4\frac{2}{11}$
17. $10\frac{1}{12} - 4\frac{7}{12}$
18. $-1\frac{7}{8} - 5\frac{7}{8}$

Subtracting Mixed Numbers with Unlike Denominators

To subtract mixed numbers, you need to have like fractions.

EXAMPLE **Subtracting Mixed Numbers with Unlike Denominators**

Subtract $6\frac{1}{2} - 1\frac{5}{6}$.

$6\frac{1}{2} \rightarrow 6\frac{3}{6}$
$1\frac{5}{6} \rightarrow -1\frac{5}{6}$

- Write equivalent fractions with a common denominator.

- Rename, if necessary.
$6\frac{3}{6} = 5 + \frac{6}{6} + \frac{3}{6} = 5\frac{9}{6}$

$$\begin{array}{r} 5\frac{9}{6} \\ -\ 1\frac{5}{6} \\ \hline 4\frac{4}{6} = 4\frac{2}{3} \end{array}$$

- Subtract the fractions, and then subtract the whole numbers.
- Simplify, if possible.

So, $6\frac{1}{2} - 1\frac{5}{6} = 4\frac{2}{3}$.

Check It Out

Subtract.

19. $12 - 4\frac{1}{2}$
20. $9\frac{1}{10} - 5\frac{4}{7}$
21. $14\frac{7}{8} - 3\frac{3}{4}$

Multiplying Fractions

You know that $5 \cdot 4$ means "5 groups of 4." Multiplying fractions involves the same concept: $3 \cdot \frac{1}{2}$ means "3 groups of $\frac{1}{2}$." You may find it helpful to know that in math, the word *of* frequently means *multiply.*

The same is true when you are multiplying a fraction by a fraction. For example, $\frac{1}{2} \cdot \frac{1}{3}$ means that you actually find $\frac{1}{2}$ of $\frac{1}{3}$.

To multiply fractions, you multiply the numerators and then the denominators. There is no need to find a common denominator.

$$\frac{1}{2} \cdot \frac{1}{4} = \frac{1}{8}$$

EXAMPLE Multiplying Fractions

Multiply $\frac{3}{4}$ and $2\frac{2}{5}$.

$\frac{3}{4} \cdot 2\frac{2}{5} = \frac{3}{4} \cdot \frac{12}{5}$
- Convert the mixed number to an improper fraction.

$\frac{3}{4} \cdot \frac{12}{5} = \frac{3 \cdot 12}{4 \cdot 5} = \frac{36}{20}$
- Multiply the numerators and the denominators.

$\frac{36}{20} = 1\frac{16}{20} = 1\frac{4}{5}$
- Write the product in simplest form, if necessary.

So, $\frac{3}{4} \cdot 2\frac{2}{5} = 1\frac{4}{5}$.

Check It Out

Multiply.

22. $\frac{2}{5} \cdot \frac{5}{6}$
23. $\frac{3}{8} \cdot \frac{2}{9}$
24. $5\frac{1}{3} \cdot \frac{3}{8}$
25. $3\frac{2}{3} \cdot 4\frac{1}{5}$

Shortcut for Multiplying Fractions

You can use a shortcut when you multiply fractions. Instead of multiplying across and then writing the product in simplest form, you can simplify *factors* first.

EXAMPLE Simplifying Factors

Simplify the factors and then multiply $\frac{5}{8}$ and $1\frac{1}{5}$.

$\frac{5}{8} \cdot \frac{6}{5}$
- Write the mixed number as an improper fraction. $1\frac{1}{5} = \frac{6}{5}$

$\frac{^1\cancel{5}}{_1\cancel{2} \cdot 4} \cdot \frac{^1\cancel{2} \cdot 3}{\cancel{5}_1} = \frac{1}{4} \cdot \frac{3}{1}$
- Simplify factors if you can.

$= \frac{3}{4}$
- Write the product in simplest form, if necessary.

So, $\frac{5}{8} \cdot 1\frac{1}{5} = \frac{3}{4}$.

Check It Out

Simplify the factors and then multiply.

26. $\frac{3}{5} \cdot \frac{1}{6}$

27. $\frac{4}{7} \cdot \frac{14}{15}$

28. $1\frac{1}{2} \cdot 1\frac{1}{3}$

Finding the Reciprocal of a Number

Two numbers with a product of 1 are **multiplicative inverses**, or **reciprocals**, of each other. For example, $\frac{5}{1}$ and $\frac{1}{5}$ are multiplicative inverses because $\frac{5}{1} \cdot \frac{1}{5} = 1$.

EXAMPLE Finding the Multiplicative Inverse

Write the multiplicative inverse of $-7\frac{1}{3}$.

$-7\frac{1}{3} = -\frac{22}{3}$
- Write as an improper fraction.

Since $-\frac{22}{3}\left(-\frac{3}{22}\right) = 1$, the multiplicative inverse of $-7\frac{1}{3}$ is $-\frac{3}{22}$.

The number 0 does not have a reciprocal.

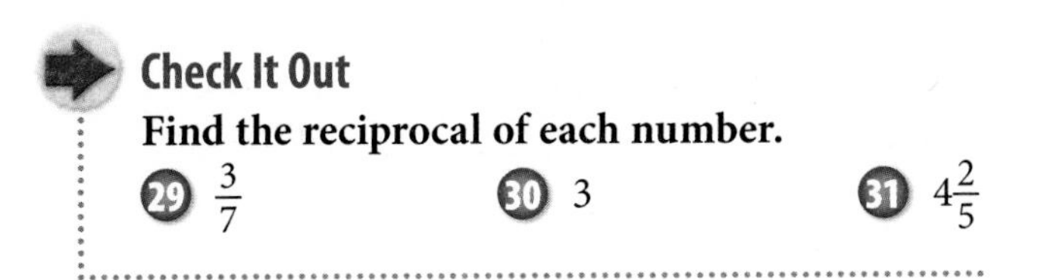

Check It Out

Find the reciprocal of each number.

29. $\frac{3}{7}$

30. 3

31. $4\frac{2}{5}$

Dividing Fractions

When you divide a fraction by a fraction, such as $\frac{1}{2} \div \frac{1}{4}$, you are really finding out how many $\frac{1}{4}$s are in $\frac{1}{2}$. That's why the answer is 2. To divide fractions, you replace the divisor with its reciprocal and then multiply to get your answer.

$$\frac{1}{2} \div \frac{1}{4} = \frac{1}{2} \cdot \frac{4}{1} = 2$$

EXAMPLE **Dividing Fractions**

Divide $\frac{5}{8} \div 3\frac{3}{4}$.

$\frac{5}{8} \div \frac{15}{4}$

- Write the mixed number as an improper fraction.

$\frac{5}{8} \cdot \frac{4}{15} = \frac{{}^{1}\cancel{5}}{{}_{2}\cancel{8}} \cdot \frac{\cancel{4}^{1}}{\cancel{15}_{3}} = \frac{1}{2} \cdot \frac{1}{3}$

- Replace the divisor with its reciprocal and simplify factors.

$\frac{1}{2} \cdot \frac{1}{3} = \frac{1}{6}$

- Multiply.

So, $\frac{5}{8} \div 3\frac{3}{4} = \frac{1}{6}$.

Check It Out

Divide.

32. $\frac{3}{4} \div \frac{1}{2}$

33. $\frac{5}{7} \div 10$

34. $1\frac{1}{8} \div 4\frac{1}{2}$

2•2 Exercises

Add or subtract.

1. $\frac{7}{9} - \frac{4}{9}$ **2.** $\frac{3}{8} - \frac{1}{4}$ **3.** $\frac{5}{6} + \frac{3}{4}$ **4.** $\frac{7}{12} + \frac{9}{16}$

5. $1\frac{1}{2} + \frac{1}{6}$ **6.** $8\frac{3}{8} + 2\frac{1}{3}$ **7.** $12 - 11\frac{5}{9}$ **8.** $4\frac{1}{2} + 2\frac{1}{2}$

9. $4\frac{3}{4} - 2\frac{1}{4}$ **10.** $13\frac{7}{12} - 2\frac{5}{8}$ **11.** $7\frac{3}{8} - 2\frac{2}{3}$

Multiply.

12. $\frac{1}{2} \cdot \frac{8}{9}$ **13.** $\frac{2}{5} \cdot 3$ **14.** $4\frac{1}{5} \cdot \frac{5}{6}$ **15.** $3\frac{2}{3} \cdot 6$

16. $3\frac{2}{5} \cdot 2\frac{1}{2}$ **17.** $6 \cdot 2\frac{1}{2}$ **18.** $2\frac{2}{3} \cdot 3\frac{1}{2}$ **19.** $4\frac{3}{8} \cdot 1\frac{3}{5}$

Find the reciprocal of each number.

20. $\frac{5}{8}$ **21.** 2 **22.** $3\frac{1}{5}$ **23.** $2\frac{2}{5}$ **24.** $\frac{7}{9}$

Divide.

25. $\frac{3}{4} \div \frac{3}{2}$ **26.** $\frac{1}{3} \div 2$ **27.** $\frac{1}{2} \div 1\frac{1}{2}$ **28.** $2 \div 2\frac{1}{3}$

29. $0 \div \frac{1}{4}$ **30.** $3\frac{1}{5} \div \frac{1}{10}$ **31.** $1\frac{2}{3} \div 3\frac{1}{5}$ **32.** $\frac{2}{9} \div 2\frac{2}{3}$

33. Last week Gabriel worked $9\frac{1}{4}$ hours babysitting and $6\frac{1}{2}$ hours giving gymnastic lessons. How many hours did he work in all?

34. Bill's Burger Palace had its grand opening on Tuesday. They had $164\frac{1}{2}$ pounds of ground beef in stock. They had $18\frac{1}{4}$ pounds left at the end of the day. Each burger requires $\frac{1}{4}$ pound of ground beef. How many hamburgers did they sell?

35. Girls make up $\frac{5}{8}$ of the eighth-grade enrollment at Marshall Middle School. If $\frac{1}{5}$ of the girls try out for the basketball team, what fractional part of the entire class is this?

36. Of the cafeteria dessert selections, $\frac{1}{3}$ are baked goods. Each day before lunch, the cafeteria workers divide the desserts so that an equal number of baked goods is available on both serving lines. On each line, what fraction of the total desserts are baked goods?

2•3 Operations with Decimals

Adding and Subtracting Decimals

Adding and subtracting decimals is similar to adding and subtracting whole numbers.

EXAMPLE **Adding and Subtracting Decimals**

Add 6.75 + 29.49 + 16.9.

$$\begin{array}{r} 6.75 \\ 29.49 \\ +\ 16.9 \\ \hline \end{array}$$

- Line up the decimal points.

$$\begin{array}{r} {}^{1} \\ 6.7\mathbf{5} \\ 29.4\mathbf{9} \\ +\ 16.9 \\ \hline \mathbf{4} \end{array}$$

- Add or subtract the place farthest right. Rename, if necessary.

$$\begin{array}{r} {}^{2} \\ 6.\mathbf{7}5 \\ 29.\mathbf{4}9 \\ +\ 16.\mathbf{9} \\ \hline \mathbf{14} \end{array}$$

- Add or subtract the next place left. Rename, if necessary.

$$\begin{array}{r} 6.75 \\ 29.49 \\ +\ 16.9 \\ \hline 53.14 \end{array}$$

- Continue with the whole numbers. Place the decimal point in the result.

So, 6.75 + 29.49 + 16.9 = 53.14.

Check It Out

Solve.

1. 1.387 + 2.3444 + 3.45
2. 0.7 + 87.8 + 8.174
3. 56.13 − 17.59
4. 826.7 − 24.6444

Multiplying Decimals

Multiplying decimals is much the same as multiplying whole numbers.

EXAMPLE **Multiplying Decimals**

Multiply 42.8 • 0.06.

$$\begin{array}{r} 42.8 \\ \times\ 0.06 \\ \hline 2568 \end{array} \qquad \begin{array}{r} 428 \\ \times\quad 6 \\ \hline 2568 \end{array}$$

- Multiply as with whole numbers.

$$\begin{array}{rl} 42.8 & \longleftarrow \text{1 decimal place} \\ \times\ 0.06 & \longleftarrow \text{2 decimal places} \\ \hline 2.568 & \longleftarrow 1 + 2 = 3 \text{ decimal places} \end{array}$$

- Add the number of decimal places for the factors, and place the decimal point in the product.

So, 42.8 • 0.06 = 2.568.

Check It Out

Multiply.

5. 22.03 • 2.7
6. 9.655 • 8.33
7. 11.467 • 5.49

Estimating Decimal Products

To estimate decimal products, you can replace given numbers with compatible numbers. Compatible numbers are estimates you choose because they are easier to work with mentally.

Estimate 26.2 • 52.3.

- Replace the factors with compatible numbers.
  26.2 → 30   52.3 → 50
- Multiply mentally.
  30 • 50 = 1,500

Check It Out

Estimate each product by using compatible numbers.

8. 12.75 • 91.3
9. 3.76 • 0.61
10. 25.25 • 1.95

Dividing Decimals

Dividing decimals is similar to dividing whole numbers.

EXAMPLE Dividing Decimals

Divide 38.35 ÷ 6.5.

$6.5 \cdot 10 = 65$	• Multiply the divisor by a power of ten to make it a whole number.
$38.35 \cdot 10 = 383.5$	• Multiply the dividend by the same power of ten.
$$\begin{array}{r} \mathbf{5.9} \\ 65.\overline{)383.5} \\ -\ 325 \\ \hline 58\ 5 \\ -\ 58\ 5 \\ \hline 0 \end{array}$$	• Divide. Place the decimal point in the quotient.

So, 38.35 ÷ 6.5 = 5.9.

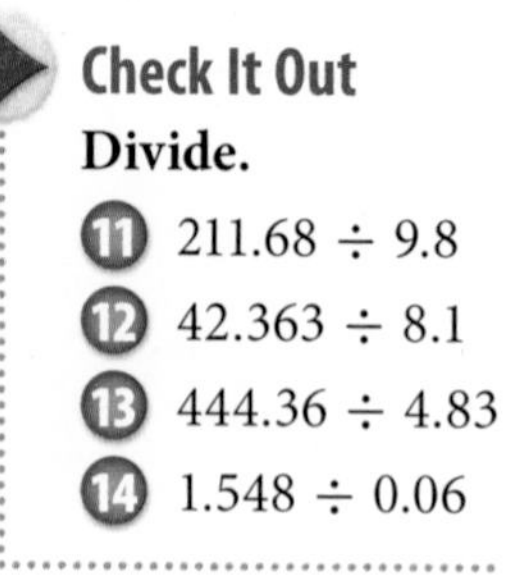

Check It Out

Divide.

11. 211.68 ÷ 9.8
12. 42.363 ÷ 8.1
13. 444.36 ÷ 4.83
14. 1.548 ÷ 0.06

Zeros in Division

You can use zeros as placeholders in the dividend when you are dividing decimals.

EXAMPLE **Zeros in Division**

Divide 375.1 ÷ 6.2.

$6.2 \cdot 10 = 62$

$375.1 \cdot 10 = 3{,}751$

- Multiply the divisor and the dividend by a power of ten. Place the decimal point.

$$\begin{array}{r} 60. \\ 62.\overline{)3751.} \\ \underline{-\ 372} \\ 31 \\ \underline{-\ 0} \\ 31 \end{array}$$

- Divide. Place the decimal point.

$$\begin{array}{r} 60.5 \\ 62.\overline{)3751.0} \\ \underline{-\ 372} \\ 31 \\ \underline{-\ 0} \\ 310 \\ \underline{-\ 310} \\ 0 \end{array}$$

- Use zeros as placeholders in the dividend. Continue to divide.

So, 375.1 ÷ 6.2 = 60.5.

Check It Out

Divide until the remainder is zero.

15. 0.7042 ÷ 0.07
16. 37.2 ÷ 1.5
17. 246.1 ÷ 0.8

Rounding Decimal Quotients

You can use a calculator to divide decimals and round quotients.

EXAMPLE Rounding Decimals on a Calculator

Divide 6.3 by 2.6. Round to the nearest hundredth.

6.3 [÷] 2.6 [=] [2.4230769]	• Use your calculator to divide.
2.4230769 (arrow under the 3)	• Look at the digit one place to the right of the hundredths place.
2.4230769 rounds to 2.42.	• Round.

Some calculators have a "fix" function. Press [FIX] and the number of decimal places you want. The calculator will then display all numbers rounded to that number of places. Consider the example above. Again, enter 6.3 [÷] 2.6 [=] in a calculator. Since you want to round to the nearest hundredth press [FIX] [2]. The answer [2.42] is shown in the display.

Check It Out

Solve with a calculator. Round to the nearest hundredth.

18. $0.0258 \div 0.345$
19. $0.817 \div 1.25$
20. $0.4369 \div 0.267$
21. $0.3112 \div 0.4$

2•3 Exercises

Add.

1. $256.3 + 0.624$

2. $78.239 + 38.6$

3. $7.02396 + 4.88$

4. $\$250.50 + \$385.16$

5. $2.9432 + 1.9 + 3 + 1.975$

Subtract.

6. $43 - 28.638$

7. $58.543 - 0.768$

8. $435.2 - 78.376$

9. $38.3 - 16.254$

10. $11.01 - 2.0063$

Multiply.

11. $0.66 \cdot 17.3$

12. $0.29 \cdot 6.25$

13. $7.526 \cdot 0.33$

14. $37.82 \cdot 9.6$

15. $22.4 \cdot 9.4$

Divide until the remainder is zero.

16. $29.38 \div 0.65$

17. $62.55 \div 4.5$

18. $84.6 \div 4.7$

19. $0.657 \div 0.6$

Divide. Round to the nearest hundredth.

20. $142.7 \div 7$

21. $2.55 \div 1.6$

22. $22.9 \div 6.2$

23. $15.25 \div 2.3$

24. The Moon orbits Earth in 27.3 days. How many orbits does the Moon make in 365.25 days? Round your answer to the nearest hundredth.

25. *Apollo 15* astronauts drove the lunar rover about 27.8 kilometers on the Moon. Their average speed was 3.3 kilometers/hour. How long did they drive the lunar rover?

2•4 Fractions and Decimals

Writing Fractions as Decimals

A fraction can be written as either a **terminating** or a **repeating decimal**.

Fraction	Decimal	Terminating or Repeating
$\frac{1}{2}$	0.5	terminating
$\frac{1}{3}$	$0.333333\overline{3}$	repeating
$\frac{1}{6}$	$0.16666\overline{6}$	repeating
$\frac{3}{4}$	0.75	terminating
$\frac{2}{5}$	0.4	terminating
$\frac{3}{22}$	$0.13636\overline{36}$	repeating

EXAMPLE Changing Fractions to Decimals

Write $-\frac{3}{25}$ as a decimal.

$-3 \div 25 = -0.12$
- Divide the numerator of the fraction by the denominator.

So, $-\frac{3}{25} = -0.12$. The remainder is zero. The decimal is a terminating decimal.

Write $\frac{1}{6}$ and $\frac{5}{22}$ as decimals.

$1 \div 6 = 0.1666\ldots$

$5 \div 22 = 0.22727\ldots$
- Divide the numerator of each fraction by the denominator.

$0.1\overline{6}$

$0.2\overline{27}$
- Place a bar over any digit or digits that repeat.

So, $\frac{1}{6} = 0.1\overline{6}$ and $\frac{5}{22} = 0.2\overline{27}$. Both decimals are repeating decimals.

A mixed number can be expressed as a decimal by changing the mixed number to an improper fraction and dividing the numerator by the denominator.

EXAMPLE Changing Mixed Numbers to Decimals

Write $2\frac{3}{4}$ as a decimal.

$2\frac{3}{4} = \frac{11}{4}$ • Change the mixed number to an improper fraction.

$11 \div 4 = 2.75$ • Then divide the numerator of the fraction by the denominator.

So, $2\frac{3}{4} = 2.75$.

Check It Out

Use a calculator to find a decimal for each fraction or mixed number.

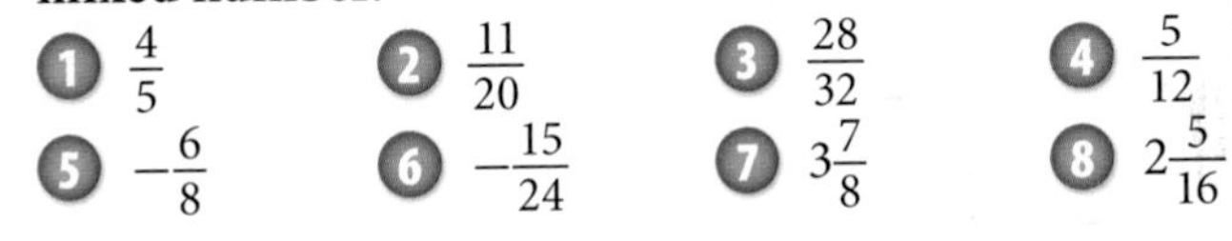

1. $\frac{4}{5}$
2. $\frac{11}{20}$
3. $\frac{28}{32}$
4. $\frac{5}{12}$
5. $-\frac{6}{8}$
6. $-\frac{15}{24}$
7. $3\frac{7}{8}$
8. $2\frac{5}{16}$

Writing Decimals as Fractions

Terminating decimals are rational numbers because you can write them as fractions.

EXAMPLE Changing Decimals to Fractions

Write 0.55 as a fraction.

$0.55 = \frac{55}{100}$ • Write the decimal as a fraction.

$\frac{55}{100} = \frac{55 \div 5}{100 \div 5} = \frac{11}{20}$ • Write the fraction in simplest form.

So, $0.55 = \frac{11}{20}$.

Repeating decimals are also rational numbers because you can write them as fractions.

EXAMPLE **Changing Repeating Decimals to Fractions**

Write $0.\overline{2}$ as a fraction in simplest form.

Let $n = 0.222$.

$10(n) = 10(0.222)$ • Multiply each side of the equation by 10 because one digit repeats.

$10n = 2.222$ • Multiplying by 10 moves the decimal point one place to the right.

$$\begin{aligned} 10n &= 2.222 \\ -n &= 0.222 \\ \hline 9n &= 2 \end{aligned}$$

• Subtract $n = 0.222$ to eliminate the repeating part.

$\frac{9n}{9} = \frac{2}{9}$ • Simplify.

$n = \frac{2}{9}$ • Divide each side by 9.

So, the decimal $0.\overline{2}$ can be written as $\frac{2}{9}$.

A decimal greater than 1 can be expressed as a mixed number by writing the decimal part as a fraction.

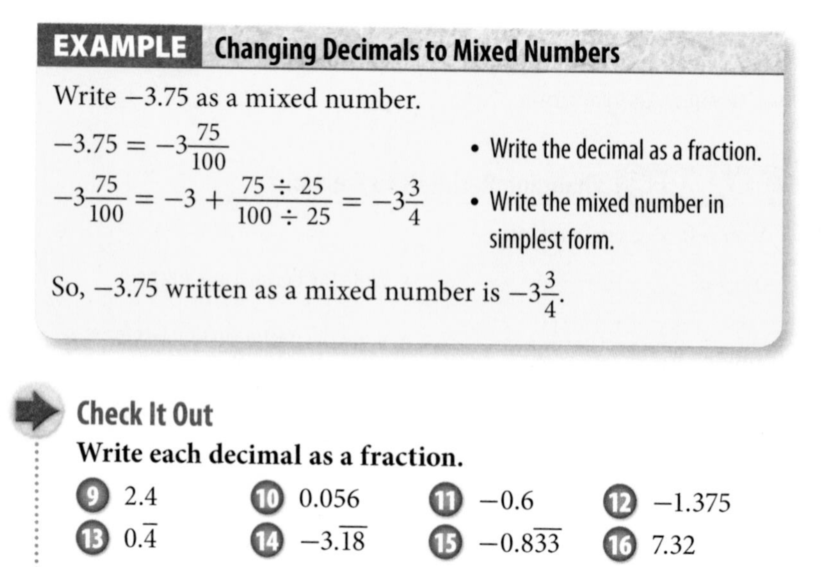

EXAMPLE **Changing Decimals to Mixed Numbers**

Write −3.75 as a mixed number.

$-3.75 = -3\frac{75}{100}$ • Write the decimal as a fraction.

$-3\frac{75}{100} = -3 + \frac{75 \div 25}{100 \div 25} = -3\frac{3}{4}$ • Write the mixed number in simplest form.

So, −3.75 written as a mixed number is $-3\frac{3}{4}$.

Check It Out

Write each decimal as a fraction.

9. 2.4
10. 0.056
11. −0.6
12. −1.375
13. $0.\overline{4}$
14. $-3.\overline{18}$
15. $-0.8\overline{33}$
16. 7.32

Comparing and Ordering Rational Numbers

You can compare rational numbers by renaming each fraction by using the least common denominator and then compare the numerators.

EXAMPLE Comparing Rational Numbers

Replace □ with <, >, or = to make $\frac{3}{8}$ □ $\frac{2}{3}$ a true sentence.

$\frac{3}{8} = \frac{3 \cdot 3}{8 \cdot 3}$ or $\frac{9}{24}$ • Rename the fractions using the LCD.

$\frac{2}{3} = \frac{2 \cdot 8}{3 \cdot 8}$ or $\frac{16}{24}$ • The LCD is 24.

Since $\frac{9}{24} < \frac{16}{24}$, then $\frac{3}{8} < \frac{2}{3}$.

Replace □ with <, >, or = to make $\frac{5}{9}$ □ 0.7 a true sentence.

$\frac{5}{9} = 0.55\overline{5}$ • Express $\frac{5}{9}$ as a decimal.

$.55\overline{5}$ □ 0.7 • In the tenths place, $5 < 7$.

So, $\frac{5}{9} < 0.7$.

You can use a number line to help compare and order rational numbers.

EXAMPLE Comparing and Ordering Rational Numbers

Replace □ with a <, >, or = to make −2.35 □ −3.4 a true sentence.

• Graph the decimals on a number line.

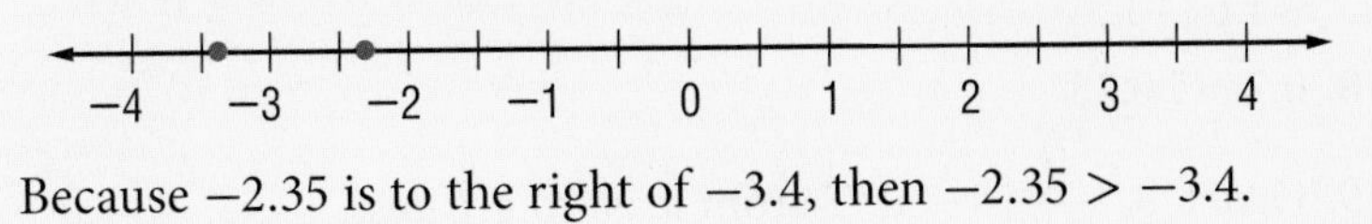

Because −2.35 is to the right of −3.4, then $-2.35 > -3.4$.

2•4 Exercises

Change each fraction to a decimal. Use bar notations to show repeating decimals.

1. $\frac{3}{18}$

2. $\frac{30}{111}$

3. $\frac{4}{18}$

4. $\frac{7}{15}$

5. $-\frac{5}{9}$

6. $4\frac{5}{6}$

Write each decimal as a fraction or a mixed number.

7. 0.4

8. 2.004

9. 3.42

10. 0.27

11. −0.3

12. $2.\overline{15}$

Replace each □ with <, >, or = to make a true sentence.

13. $\frac{2}{3}$ □ $\frac{8}{9}$

14. $\frac{5}{6}$ □ $\frac{7}{8}$

15. $-\frac{1}{4}$ □ $-\frac{5}{8}$

16. 0.7 □ 0.07

17. −1.6 □ 1.57

18. $0.\overline{24}$ □ 0.28

19. Order $\frac{2}{8}$, $-\frac{14}{8}$, $1\frac{1}{3}$, 0.75 from least to greatest.

2•5 The Real Number System

Irrational Numbers

An **irrational number** is a number that cannot be expressed as a quotient $\frac{a}{b}$, where a and b are integers and $b \neq 0$. The decimal representation of the $\sqrt{17} \approx 4.123105626\ldots$ does not terminate or repeat. Therefore, the $\sqrt{17}$ cannot be written as a fraction and is an irrational number.

The set of Real Numbers includes the set of rational numbers and the set of irrational numbers. Study the Venn diagram below.

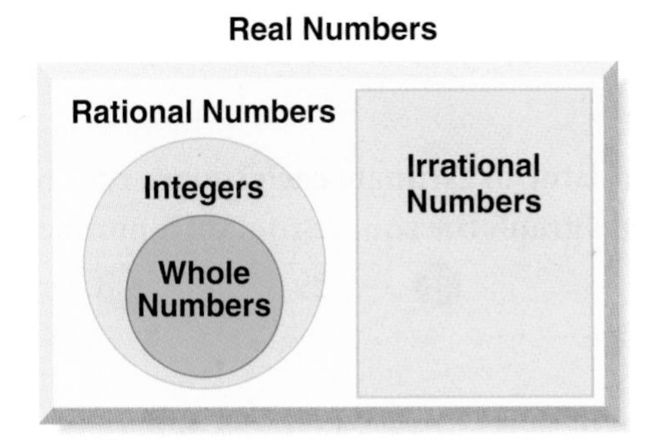

Check It Out

Name all sets of numbers to which each real number belongs.

1. $-\sqrt{49}$
2. $1.\overline{72}$
3. $\sqrt{15}$
4. π
5. $8.\overline{15}$
6. $\sqrt{23}$

Graphing Real Numbers

You can use a calculator to estimate irrational numbers and graph them on a number line.

EXAMPLE **Graphing Real Numbers**

Estimate $\sqrt{7}$, $-\sqrt{2}$, and 2.25 to the nearest tenth. Then graph $\sqrt{7}$, $-\sqrt{2}$, and 2.25 on a number line.

$\sqrt{7} \approx 2.645751311\ldots$ • Use a calculator.
or about 2.6

$-\sqrt{2} \approx -1.414213562\ldots$ • Use a calculator.
or about −1.4

$2.25 \approx 2.3$

−4 −3 −2 −1 0 1 2 3 4

Check It Out

Use a calculator to estimate each square root to the nearest tenth. Then graph the square root on a number line.

7. $\sqrt{11}$ 8. $-\sqrt{29}$ 9. $\sqrt{18}$

2•5 Exercises

Name all sets of numbers to which each real number belongs.

1. 0.909090 . . . **2.** $-\sqrt{81}$ **3.** $\sqrt{13}$
4. $\sqrt{23}$ **5.** $-2\frac{1}{4}$ **6.** $0.\overline{63}$
7. $-\frac{8}{9}$ **8.** $-\sqrt{41}$ **9.** $-9.7\overline{33}$

Estimate each square root to the nearest tenth. Then graph the square root on a number line.

10. $\sqrt{2}$ **11.** $\sqrt{5}$ **12.** $-\sqrt{3}$ **13.** $\sqrt{15}$
14. $-\sqrt{8}$ **15.** $\sqrt{53}$ **16.** $-\sqrt{24}$ **17.** $-\sqrt{10}$
18. $\sqrt{27}$ **19.** $\sqrt{17}$ **20.** $\sqrt{46}$ **21.** $\sqrt{67}$

2•6 Percents

The Meaning of Percents

A *ratio* of a number to 100 is called a **percent**. Percent means *per hundred* and is represented by the symbol %.

Any ratio can be expressed as a fraction, a decimal, and a percent. A quarter is 25% of \$1.00. You can express a quarter as 25¢, \$0.25, $\frac{1}{4}$ of a dollar, $\frac{25}{100}$ of a dollar, and 25% of a dollar.

You can use the following **benchmarks** to help you estimate percents.

None			Half		All
0					100%
$\frac{1}{100}$	$\frac{1}{10}$	$\frac{1}{4}$	$\frac{1}{2}$	$\frac{3}{4}$	
0.01	0.10	0.25	0.50	0.75	
1%	10%	25%	50%	75%	

EXAMPLE Estimating Percents

Estimate 26% of 200.

26% is close to 25%.
$\frac{1}{4}$ is equal to 25%.

- Choose a benchmark, or a combination of benchmarks, close to the target percent and find a fraction or decimal equivalent.

$\frac{1}{4}$ of 200 is 50.

- Use the benchmark equivalent to estimate the percent.

So, 26% of 200 is about 50.

Check It Out

Use fractional benchmarks to estimate the percents.

1. 47% of 300
2. 22% of 400
3. 72% of 200
4. 99% of 250

PERCENTS 2•6

Use Mental Math to Estimate Percent

You can use fraction or decimal benchmarks to help you quickly estimate the percent of something, such as a tip in a restaurant.

EXAMPLE **Mental Math to Estimate Percent**

Estimate a 15% tip for a bill of \$5.45.

\$5.45 rounds to \$5.50. • Round to a convenient number.

15% = 10% + 5% • Think of the percent as a combination of benchmarks.

10% of \$5.50 = 0.55 • Multiply mentally.

5% of \$5.50 = about 0.25

0.55 + 0.25 = about 0.80

So, the tip is about \$0.80.

Check It Out

Estimate the amount of each tip.

5. 20% of \$4.75
6. 15% of \$40
7. 10% of \$94.89
8. 18% of \$50

Percents and Fractions

Recall that percents describe a ratio out of 100. A percent can be written as a fraction with a denominator of 100. The table shows how some percents are written as fractions.

Percent	Fraction
10 out of 100 = 10%	$\frac{10}{100} = \frac{1}{10}$
15 out of 100 = 15%	$\frac{15}{100} = \frac{3}{20}$
25 out of 100 = 25%	$\frac{25}{100} = \frac{1}{4}$
30 out of 100 = 30%	$\frac{30}{100} = \frac{3}{10}$
60 out of 100 = 60%	$\frac{60}{100} = \frac{3}{5}$
75 out of 100 = 75%	$\frac{75}{100} = \frac{3}{4}$

You can write fractions as percents and percents as fractions.

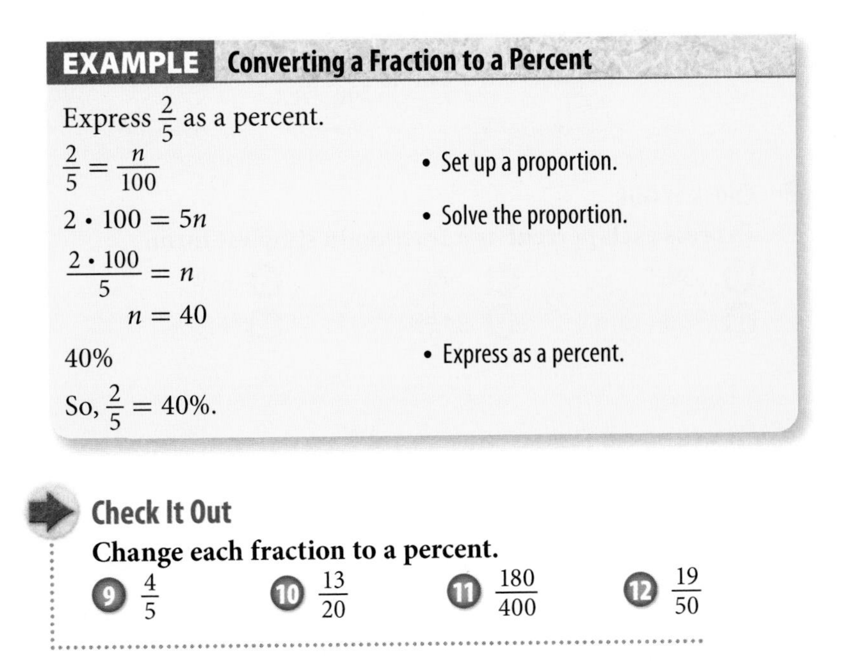

EXAMPLE **Converting a Fraction to a Percent**

Express $\frac{2}{5}$ as a percent.

$\frac{2}{5} = \frac{n}{100}$ • Set up a proportion.

$2 \cdot 100 = 5n$ • Solve the proportion.

$\frac{2 \cdot 100}{5} = n$

$n = 40$

40% • Express as a percent.

So, $\frac{2}{5} = 40\%$.

Check It Out

Change each fraction to a percent.

9. $\frac{4}{5}$
10. $\frac{13}{20}$
11. $\frac{180}{400}$
12. $\frac{19}{50}$

Changing Percents to Fractions

To change a percent to a fraction, write the percent as the numerator of a fraction with a denominator of 100, and express in simplest form. Similarly, you can change a mixed number percent to a fraction.

EXAMPLE **Changing Percents to Fractions**

Express 45% as a fraction.

$45\% = \frac{45}{100}$
- Change the percent to a fraction with a denominator of 100. The percent becomes the numerator of the fraction.

$\frac{45}{100} = \frac{9}{20}$
- Simplify, if possible.

So, 45% expressed as a fraction is $\frac{9}{20}$.

Express $54\frac{1}{2}\%$ as a fraction.

$54\frac{1}{2} = \frac{109}{2}$
- Change the mixed number to an improper fraction.

$\frac{109}{2} \cdot \frac{1}{100} = \frac{109}{200}$
- Multiply the percent by $\frac{1}{100}$.

$54\frac{1}{2} = \frac{109}{200}$
- Simplify, if possible.

So, $54\frac{1}{2}\%$ expressed as a fraction is $\frac{109}{200}$.

Check It Out

Express each percent as a fraction in simplest form.

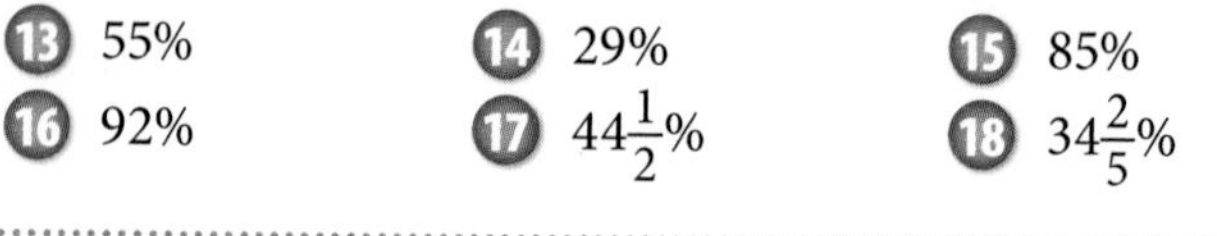

13. 55%
14. 29%
15. 85%
16. 92%
17. $44\frac{1}{2}\%$
18. $34\frac{2}{5}\%$

Percents and Decimals

Percents can be expressed as decimals, and decimals can be expressed as percents.

EXAMPLE Changing Decimals to Percents

Change 0.8 to a percent.

$0.8 \cdot 100 = 80$ • Multiply the decimal by 100.

$0.8 = 80\%$ • Add the percent sign.

So, $0.8 = 80\%$.

A Shortcut for Changing Decimals to Percents

Change 0.5 to a percent.

- Move the decimal point two places to the right. Add zeros, if necessary.

  0.5 → 50.

- Add the percent sign.

  0.5 → 50%

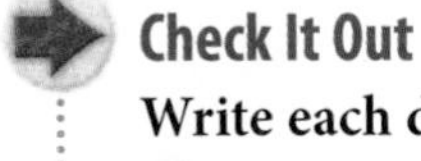

Write each decimal as a percent.

19. 0.08
20. 0.66
21. 0.398
22. 0.74

Because *percent* means part of a hundred, percents can be converted directly to decimals.

EXAMPLE **Changing Percents to Decimals**

Change 3% to a decimal.

$3\% = \frac{3}{100}$

- Express the percent as a fraction with 100 as the denominator.

$3 \div 100 = 0.03$

- Divide the numerator by 100.

So, $3\% = 0.03$.

A Shortcut for Changing Percents to Decimals

Change 8% to a decimal.

- Move the decimal point two places to the left.

  8% → . 8.

- Add zeros, if necessary.

  $8\% = 0.08$

Check It Out

Express each percent as a decimal.

23. 14.5%
24. 0.01%
25. 23%
26. 35%

2•6 Exercises

Use fractional benchmarks to estimate the percent of each number.

1. 15% of 200

2. 49% of 800

3. 2% of 50

4. 76% of 200

5. Estimate a 15% tip for a bill of $65.

6. Estimate a 20% tip for a bill of $49.

7. Estimate a 10% tip for a bill of $83.

8. Estimate an 18% tip for a bill of $79.

Change each fraction to a percent.

9. $\frac{17}{100}$ **10.** $\frac{19}{20}$ **11.** $\frac{13}{100}$ **12.** $\frac{19}{50}$ **13.** $\frac{24}{25}$

Change each percent to a fraction in simplest form.

14. 42% **15.** 60% **16.** 44% **17.** 12% **18.** 80%

Write each decimal as a percent.

19. 0.4 **20.** 0.41 **21.** 0.105

22. 0.83 **23.** 3.6

Write each percent as a decimal.

24. 35% **25.** 13.6% **26.** 18%

27. 4% **28.** 25.4%

29. One middle school survey said 40% of eighth-grade students preferred pizza for lunch. Another survey said $\frac{2}{5}$ of eighth-grade students preferred pizza for lunch. Could both surveys be correct? Explain.

30. Blades on Second is advertising $109 skateboards at 33% off. Skates on Seventh is advertising the same skateboard at $\frac{1}{3}$ off. Which is the better buy?

2•7 Using and Finding Percents

Finding a Percent of a Number

You have changed a percent to a decimal and a percent to a fraction. Now you will find the percent of a number. To find the percent of a number, you must first change the percent to a decimal or a fraction. Sometimes it is easier to change the percent to a decimal representation and other times to a fractional one.

To find 30% of 80, you can use either the fraction method or the decimal method.

EXAMPLE Finding the Percent of a Number: Two Methods

Find 30% of 80.

Decimal Method

- Change the percent to a decimal.

  $30\% = 0.3$

- Multiply.

  $80 \cdot 0.3 = 24$

Fraction Method

- Change the percent to a fraction in simplest form.

  $30\% = \frac{30}{100} = \frac{3}{10}$

- Multiply.

  $80 \cdot \frac{3}{10} = 24$

So, 30% of 80 = 24.

Check It Out

Find the percent of each number.

1. 80% of 75
2. 95% of 700
3. 21% of 54
4. 75% of 36

APPLICATION Go For It!

Calling all couch potatoes! To get in shape, you need to do an aerobic activity (such as walking, jogging, biking, or swimming) at least three times a week.

The goal is to get your heart beating at $\frac{1}{2}$ to $\frac{3}{4}$ of its maximum rate and to keep it there long enough to give it a good workout.

For example, if you want to exercise in the aerobic fitness zone, which is at 70%–80% of your maximum heart rate, first find your maximum heart rate (MHR) by subtracting your age from 220. Second, multiply your MHR by the higher percentage and lower percentage of your zone.

$$220 - 25 = 195$$
$$195 \cdot 0.80 = 156$$
$$195 \cdot 0.70 = 136.5$$

In order for 25-year-old people to maximize their workout and stay in the aerobic zone, they need to keep their heart rate between 137–156 beats per minute for 15 to 25 minutes.

Heart Rate Zone	% of Maximum Heart Rate	Number of Minutes You Need to Exercise
Health Zone	50%–60%	35–45
Conditioning Zone	60%–70%	25–35
Aerobic Fitness Zone	70%–80%	15–25

2•7 USING AND FINDING PERCENTS

The Percent Proportion

You can use proportions to help you find the percent of a number.

EXAMPLE Finding the Percent of a Number: Proportion Method

Pei works in a sporting goods store. He receives a commission of 12% on his sales. Last month he sold $9,500 worth of sporting goods. What was his commission?

$\frac{p}{w} = \frac{n}{100}$ } percent

- Use a proportion to find the percent of a number.
  p = part
  w = whole

p is the unknown, called x.
n is 12.
w is $9,500.

- Identify the given items before trying to find the unknown.

$\frac{p}{w} = \frac{n}{100} \qquad \frac{x}{9{,}500} = \frac{12}{100}$

- Set up the proportion.

$100x = 114{,}000$

- Cross multiply.

$\frac{114{,}000}{100} = \frac{100x}{100}$

- Divide both sides of the equation by 100.

$1{,}140 = x$

So, Pei received $1,140 in commission.

Check It Out

Use a proportion to find the percent of each number.

5. 95% of 700
6. 150% of 48
7. 65% of 200
8. 85% of 400

Finding Percent and Whole

You can use a *percent proportion* to solve percent problems. One ratio of a **percent proportion** compares a part to the whole. The other ratio is the percent written as a fraction.

$$\text{part} \rightarrow \frac{p}{w} = \frac{n}{100} \Big\} \text{percent}$$
$$\text{whole} \rightarrow$$

EXAMPLE Finding the Percent

What percent of 70 is 14?

- Set up a percent proportion. Let *n* represent the percent.
  $\frac{\text{part}}{\text{whole}} = \frac{n}{100}$
  (The number after the word *of* is the whole.)

$\frac{14}{70} = \frac{n}{100}$

$14 \cdot 100 = 70 \cdot n$ • Show the cross products of the proportion.

$1{,}400 = 70n$ • Find the products.

$\frac{1{,}400}{70} = \frac{70n}{70}$ • Divide both sides of the equation by the coefficient of *n*.

$n = 20$

So, 14 is 20% of 70.

Check It Out

Solve.

9. What percent of 240 is 80?
10. What percent of 64 is 288?
11. What percent of 2 is 8?
12. What percent of 55 is 33?

You can use a *percent proportion* to find the percent of the whole.

EXAMPLE Finding the Whole

12 is 48% of what number?

- Set up a percent proportion using this form: $\frac{\text{part}}{\text{whole}} = \frac{\text{percent}}{100}$

$\frac{12}{w} = \frac{48}{100}$ (The phrase *what number* after the word *of* is the whole.)

$12 \cdot 100 = 48 \cdot w$
- Show the cross products of the proportion.

$1200 = 48w$
- Find the products.

$\frac{1200}{48} = \frac{48w}{48}$
- Divide both sides of the equation by the coefficient of *w*.

$w = 25$

So, 12 is 48% of 25.

Check It Out

Solve.

13. 52 is 50% of what number?
14. 15 is 75% of what number?
15. 40 is 160% of what number?
16. 84 is 7% of what number?

Percent of Increase or Decrease

Sometimes it is helpful to keep a record of your monthly expenses. Keeping a record allows you to see the actual percent of increase or decrease in your expenses. You can make a chart to record expenses.

Expenses	September	October	Increase or Decrease (amount of)	(% of)
Food	225	189	36	16
Travel	75	95	20	
Rent	360	375	15	4
Clothing	155	62	93	60
Miscellaneous	135	108	27	20
Entertainment	80	44		
Total	1,030	871	159	15

You can use a calculator to find the percent of increase or decrease.

EXAMPLE Finding the Percent of Increase

What was the percent of increase in the amount spent on travel from September to October?

$95 - 75 = 20$ • Find the amount of change.

$\frac{20}{75} = 0.2\overline{66}$ • Find the percent of change by using the following formula:

$$\text{Percent of change} = \frac{\textit{amount of change}}{\textit{original amount}}$$

$0.27 = 27\%$ • Round to the nearest hundredth and convert to a percent.

So, the percent of increase from \$75 to \$95 is 27%.

Check It Out

Use a calculator to find the percent of increase.

17. 56 to 70
18. 20 to 39
19. 45 to 99
20. 105 to 126

EXAMPLE Finding the Percent of Decrease

During September, \$80 was spent on entertainment. In October, \$44 was spent on entertainment. What was the percent of decrease from September to October?

$80 - 44 = 36$ • Find the amount of change.

$\frac{36}{80} = 0.45$ • Percent of change $= \frac{\textit{amount of change}}{\textit{original amount}}$

$0.45 = 45\%$ • Round to the nearest hundredth and convert from a decimal to a percent.

So, the percent of decrease from \$80 to \$44 is 45%.

Check It Out

Use your calculator to find the percent of decrease.

21. 72 to 64
22. 46 to 23
23. 225 to 189
24. 120 to 84

Discounts and Sale Prices

A **discount** is the amount that is reduced from the regular price of an item. The sale price is the regular price minus the discount. Discount stores have regular prices marked below the suggested retail price. You can use percent to find discounts and resulting sale prices.

A CD player has a regular price of \$109.99. It is on sale for 25% off the regular price. How much money will you save by buying the item on sale?

You can use a calculator to help you find the discount and resulting sale price of an item.

EXAMPLE Finding Discounts and Sale Prices

The regular price of the item is \$109.99. It is marked 25% off. Find the discount and the sale price.

$d = 0.25 \cdot 109.99$ • Find the amount of discount.

$d = 27.50$ Amount of discount (d) = percent • whole

So, the discount is \$27.50.

$109.99 - 27.50 = s$ • Find the sale price.

$82.49 = s$ regular price − discount = sale price (s)

So, the sale price is \$82.49.

Check It Out

Use a calculator to find the discount and sale price.

25. regular price: \$813.25, discount percent: 20%
26. regular price: \$18.90, discount percent: 30%
27. regular price: \$79.99, discount percent: 15%

Estimating a Percent of a Number

You can use what you know about compatible numbers and simple fractions to estimate a percent of a number. You can use the table to help you estimate the percent of a number.

Percent	1%	5%	10%	20%	25%	$33\frac{1}{3}\%$	50%	$66\frac{2}{3}\%$	75%	100%
Fraction	$\frac{1}{100}$	$\frac{1}{20}$	$\frac{1}{10}$	$\frac{1}{5}$	$\frac{1}{4}$	$\frac{1}{3}$	$\frac{1}{2}$	$\frac{2}{3}$	$\frac{3}{4}$	1

EXAMPLE Estimating a Percent of a Number

Estimate 17% of 46.

17% is about 20%.
- Find the percent that is closest to the percent you are asked to find.

20% is equivalent to $\frac{1}{5}$.
- Find the fractional equivalent for the percent.

46 is about 50.
- Find a compatible number for the number you are asked to find the percent of.

$\frac{1}{5}$ of 50 is 10.
- Use the fraction to find the percent.

So, 17% of 46 is about 10.

Check It Out

Use compatible numbers to estimate.

28. 67% of 150
29. 35% of 6
30. 27% of 54
31. 32% of 89

Finding Simple Interest

When you have a savings account, the bank pays you for the use of your money. With a loan, you pay the bank for the use of their money. In both situations, the payment is called *interest.* The amount of money you borrow or save is called the *principal.* To find the total amount you pay or earn, you add the principal and interest.

You want to borrow $5,000 at 7% interest for 3 years. To find out how much interest you will pay, you can use the formula $I = p \cdot r \cdot t$. The table can help you understand the formula.

p	Principal—the amount of money you borrow or save
r	Interest Rate—a percent of the principal you pay or earn
t	Time—the length of time you borrow or save (in years)
I	Total Interest—interest you pay or earn for the entire time

EXAMPLE **Finding Simple Interest**

Find the interest and the total amount that you will pay if you borrow $5,000 at 7% interest for 3 years.

$\$5{,}000 \cdot 0.07 \cdot 3 = \$1{,}050$
$p \cdot r \cdot t = I$

- Multiply the principal (p) by the interest rate (r) by the time (t) to find the interest (I) you will pay.

So, the interest is $1,050.

$p + I = \textit{total amount}$
$\$5{,}000 + \$1{,}050 = \$6{,}050$

- To find the total amount you will pay back, add the principal and the interest.

So, the total amount of money to be paid back is $6,050.

Check It Out

Find the interest (I) and the total amount.

32. principal: $4,800
rate: 12.5%
time: 3 years

33. principal: $2,500
rate: 3.5%
time: $1\frac{1}{2}$ years

APPLICATION Oseola McCarty

Miss Oseola McCarty had to leave school after sixth grade. At first she charged $1.50 to do a bundle of laundry, later $10.00. But she always managed to save. By age 86, she had accumulated $250,000. In 1995, she decided to donate $150,000 to endow a scholarship. Miss McCarty said, "The secret to building a fortune is compounding interest. You've got to leave your investment alone long enough for it to increase."

2•7 Exercises

Find the percent of each number.

1. 2% of 50

2. 42% of 700

3. 125% of 34

4. 4% of 16.3

Solve.

5. What percent of 60 is 48?

6. 14 is what percent of 70?

7. 3 is what percent of 20?

8. What percent of 8 is 6?

Solve.

9. 82% of what number is 492?

10. 24% of what number is 18?

11. 3% of what number is 4.68?

12. 80% of what number is 24?

Find the percent of increase or decrease to the nearest percent.

13. 20 to 39

14. 175 to 91

15. 112 to 42

Estimate the percent of each number.

16. 48% of 70

17. 34% of 69

18. Mariko needed a helmet to snowboard on the half-pipe at Holiday Mountain. She bought a helmet for 45% off the regular price of $39.50. How much did she save? How much did she pay?

19. A snowboard is on sale for 20% off the regular price of $389.50. Find the discount and the sale price of the snowboard.

Find the discount and sale price.

20. regular price: $80
discount percent: 20%

21. regular price: $17.89
discount percent: 10%

22. regular price: $1,200
discount percent: 12%

23. regular price: $250
discount percent: 18%

Find the interest and total amount. Use a calculator.

24. $p = \$9,000$
$r = 7.5\%$ per year
$t = 2\frac{1}{2}$ years

25. $p = \$1,500$
$r = 9\%$ per year
$t = 2$ years

Rational Numbers

2

WHAT HAVE YOU LEARNED?

What have you learned?

You can use the problems and the list of words that follow to see what you learned in this chapter. You can find out more about a particular problem or word by referring to the topic number (*for example,* Lesson 2·2).

Problem Set

1. Of the 16 girls on the softball team, 12 play regularly. What percent of girls play regularly? (Lesson 2·6)
2. Itay missed 6 questions on a 25-question test. What percent did he get correct? (Lesson 2·6)
3. Fenway Park in Boston has a seating capacity of approximately 35,900 seats. 27% of the seats are held by season-ticket holders. About how many seats are taken by season-ticket holders? (Lesson 2·6)
4. Which fraction is equivalent to $\frac{14}{21}$? (Lesson 2·1)

   A. $\frac{2}{7}$ **B.** $\frac{7}{7}$ **C.** $\frac{2}{3}$ **D.** $\frac{3}{2}$
5. Which fraction is greater, $\frac{1}{12}$ or $\frac{3}{35}$? (Lesson 2·4)

Add or subtract. Write your answers in simplest form. (Lesson 2·2)

6. $\frac{5}{8} + \frac{3}{4}$
7. $2\frac{1}{5} - 1\frac{1}{2}$
8. $3 - 1\frac{1}{8}$
9. $7\frac{3}{4} + 2\frac{7}{8}$
10. Write the improper fraction $\frac{11}{4}$ as a mixed number. (Lesson 2·1)

In Exercises 11–14, multiply or divide as indicated. (Lesson 2·2)

11. $\frac{4}{5} \cdot \frac{5}{6}$
12. $\frac{3}{10} \div 4\frac{1}{2}$
13. $2\frac{5}{8} \cdot \frac{4}{7}$
14. $5\frac{1}{3} \div 2\frac{1}{6}$

15. Write the following numbers in order from least to greatest: $0.90, -0.\overline{33}, 1\frac{2}{3}, \frac{7}{8}$. (Lesson 2•4)

Solve. (Lesson 2•3)

16. $10.55 + 3.884$

17. $13.4 - 2.08$

18. $8.05 \cdot 6.4$

19. $69.69 \div 11.5$

Name all sets of numbers to which each real number belongs. (Lesson 2•5)

20. $\sqrt{22}$

21. -1.78

Solve the following. Round answers to the nearest tenth. (Lesson 2•7)

22. What percent of 125 is 30?

23. Find 18% of 85.

24. 36 is 40% of what number?

HotWords

Write definitions for the following words.

benchmark (Lesson 2•6)
discount (Lesson 2•7)
improper fraction (Lesson 2•1)
irrational number (Lesson 2•5)
mixed number (Lesson 2•1)
multiplicative inverse (Lesson 2•2)
percent (Lesson 2•6)
percent proportion (Lesson 2•7)
rational numbers (Lesson 2•1)
reciprocal (Lesson 2•2)
repeating decimal (Lesson 2•4)
terminating decimal (Lesson 2•4)

HotTopic 3

Powers and Roots

What do you know?

You can use the problems and the list of words that follow to see what you already know about this chapter. The answers to the problems are in **HotSolutions** at the back of the book, and the definitions of the words are in **HotWords** at the front of the book. You can find out more about a particular problem or word by referring to the topic number (*for example,* Lesson 3·2).

Problem Set

Write each product, using an exponent. (Lesson 3·1)

1. $5 \cdot 5 \cdot 5 \cdot 5 \cdot 5 \cdot 5 \cdot 5$ **2.** $a \cdot a \cdot a \cdot a \cdot a$

Evaluate each square. (Lesson 3·1)

3. 2^2 **4.** 9^2 **5.** 6^2

Evaluate each cube. (Lesson 3·1)

6. 2^3 **7.** 5^3 **8.** 7^3

Evaluate each power. (Lesson 3·1)

9. 6^4 **10.** 3^7 **11.** 2^9

Evaluate each power of 10. (Lesson 3·1)

12. 10^3 **13.** 10^7 **14.** 10^{11}

Evaluate each square root. (Lesson 3·2)

15. $\sqrt{16}$ **16.** $\sqrt{49}$ **17.** $\sqrt{121}$

Estimate each square root between two consecutive whole numbers. (Lesson 3·2)

18. $\sqrt{33}$ **19.** $\sqrt{12}$ **20.** $\sqrt{77}$

Estimate each square root to the nearest thousandth. (Lesson 3·2)

21. $\sqrt{15}$ **22.** $\sqrt{38}$

Evaluate each cube root. (Lesson 3·2)

23. $\sqrt[3]{8}$ **24.** $\sqrt[3]{64}$ **25.** $\sqrt[3]{343}$

Identify each number as very large or very small. (Lesson 3·3)

26. 0.00014

27. 205,000,000

Write each number in scientific notation. (Lesson 3·3)

28. 78,000,000

29. 200,000

30. 0.0028

31. 0.0000302

Write each number in standard form. (Lesson 3·3)

32. $8.1 \cdot 10^6$

33. $2.007 \cdot 10^8$

34. $4 \cdot 10^3$

35. $8.5 \cdot 10^{-4}$

36. $9.06 \cdot 10^{-6}$

37. $7 \cdot 10^{-7}$

Evaluate each expression. (Lesson 3·4)

38. $8 + (9 - 5)^2 - 3 \cdot 4$

39. $3^2 + 6^2 \div 9$

40. $(10 - 8)^3 + 4 \cdot 3 - 2$

base (Lesson 3·1)
cube (Lesson 3·1)
cube root (Lesson 3·2)
exponent (Lesson 3·1)
order of operations (Lesson 3·4)
perfect squares (Lesson 3·2)
power (Lesson 3·1)
scientific notation (Lesson 3·3)
square (Lesson 3·1)
square root (Lesson 3·2)

3•1 Powers and Exponents

Exponents

Multiplication is the shortcut for showing a repeated addition: $5 \cdot 3 = 3 + 3 + 3 + 3 + 3$. A shortcut for the repeated multiplication $3 \cdot 3 \cdot 3 \cdot 3 \cdot 3$ is the **power** 3^5. The 3, the factor to be multiplied, is called the **base**. The 5 is the **exponent**, which tells how many times the base is to be multiplied. The expression can be read as "three to the fifth power." When you write an exponent, it is written slightly higher than the base, and the size is usually a little smaller.

EXAMPLE Writing Products Using Exponents

Write $2 \cdot 2 \cdot 2 \cdot 2 \cdot 2 \cdot 2 \cdot 2$ using an exponent.

All the factors are 2.	• Check that the same factor is being used in the expression.
There are 7 factors of 2.	• Count the number of times 2 is being multiplied.
The factor 2 is being multiplied 7 times, write 2^7.	• Write the product using an exponent.

So, $2 \cdot 2 \cdot 2 \cdot 2 \cdot 2 \cdot 2 \cdot 2 = 2^7$.

Check It Out

Write each product, using an exponent.

1. $4 \cdot 4 \cdot 4$
2. $6 \cdot 6 \cdot 6 \cdot 6 \cdot 6 \cdot 6 \cdot 6 \cdot 6 \cdot 6$
3. $x \cdot x \cdot x \cdot x$
4. $y \cdot y \cdot y \cdot y \cdot y \cdot y$

Evaluating the Square of a Number

The **square** of a number applies the exponent 2 to the base. The square of 4 is written 4^2. To evaluate 4^2, identify 4 as the base and 2 as the exponent. Remember that the exponent tells you how many times to use the base as a factor. So 4^2 means to use 4 as a factor 2 times.

$4^2 = 4 \cdot 4 = 16$

The expression 4^2 can be read as "four to the second power." It can also be read as "four squared."

EXAMPLE Evaluating the Square of a Number

Evaluate $(-9)^2$.

The base is (-9), and the exponent is 2.	• Identify the base and the exponent.
$(-9)^2 = (-9)(-9)$	• Write as an expression using multiplication.
$(-9)(-9) = 81$	• Evaluate.

So, $(-9)^2 = 81$.

Check It Out

Evaluate each square.

5. 5^2
6. $(-10)^2$
7. 3 squared
8. $\left(\frac{1}{4}\right)^2$

APPLICATION Squaring Triangles

As you can see, some numbers can be pictured by using arrays of dots that form geometric figures. You might have already noticed that this sequence shows the first five square numbers: 1^2, 2^2, 3^2, 4^2, and 5^2.

Can you think of places where you have seen numbers that form a triangular array? Think of cans stacked in a pyramid supermarket display, bowling pins, and 15 pool balls before the break. What are the next two triangular numbers?

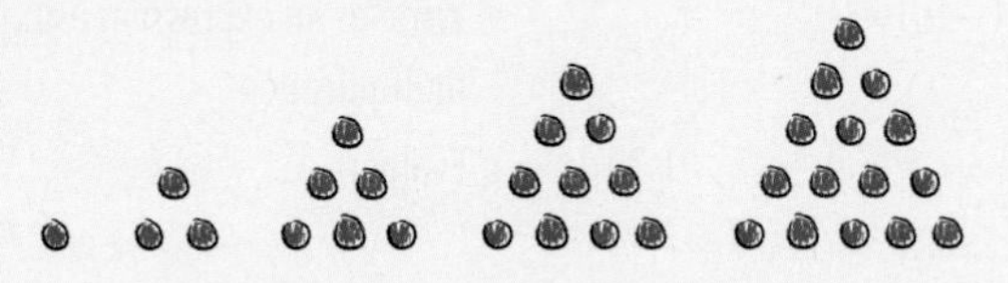

Add each pair of consecutive triangular numbers to form a new sequence as shown here. What do you notice about this sequence?

1   3   6   10   15   . . .

\ / \ / \ / \ / \ /

4   9   . . .

Think about how you could use the dot arrays for the square numbers to show the same result. *Hint:* What line could you draw in each array? See **HotSolutions** for the answers.

Evaluating the Cube of a Number

The **cube** of a number applies the exponent 3 to a base. The cube of 2 is written 2^3.

Evaluating cubes is very similar to evaluating squares. For example, if you evaluate 2^3, 2 is the base and 3 is the exponent. Remember that the exponent tells you how many times to use the base as a factor. So, 2^3 means to use 2 as a factor 3 times:

$$2^3 = 2 \cdot 2 \cdot 2 = 8$$

The expression 2^3 can be read as "two to the third power." It can also be read as "two cubed."

EXAMPLE Evaluating the Cube of a Number

Evaluate $(-5)^3$.

The base is (-5), and the exponent is 3. • Identify the base and the exponent.

$(-5)^3 = (-5)(-5)(-5)$ • Write as an expression using multiplication.

$(-5)(-5)(-5) = -125$ • Evaluate.

So, $(-5)^3 = -125$.

Check It Out

Evaluate each cube.

9. 4^3
10. $(-6)^3$
11. 3 cubed
12. (-8) cubed

Evaluating Higher Powers

You have evaluated the second power of numbers (squares) and the third power of numbers (cubes). You can evaluate higher powers of numbers as well.

To evaluate 5^4, identify 5 as the base and 4 as the exponent. The exponent tells you how many times to use the base as a factor. So, 5^4 means to use 5 as a factor 4 times:

$$5^4 = 5 \cdot 5 \cdot 5 \cdot 5 = 625$$

Powers with exponents of four and higher do not have special names. Therefore, 5^4 is read as "five to the fourth power."

EXAMPLE Evaluating Higher Powers

Evaluate 4^6.

The base is 4, and the exponent is 6.	• Identify the base and the exponent.
$4^6 = 4 \cdot 4 \cdot 4 \cdot 4 \cdot 4 \cdot 4$	• Write as an expression using multiplication.
$4 \cdot 4 \cdot 4 \cdot 4 \cdot 4 \cdot 4 = 4{,}096$	• Evaluate.

So, $4^6 = 4{,}096$.

Check It Out

Evaluate each power.

13. $(-2)^7$
14. 9^5
15. (−3) to the fourth power
16. 5 to the eighth power

Zero and Negative Exponents

Powers may contain negative exponents or exponents equal to zero.

Exponential Form	Standard Form
10^3	1,000
10^2	100
10^1	10
10^0	1
10^{-1}	$\frac{1}{10}$
10^{-2}	$\frac{1}{100}$

Any nonzero number to the zero power is 1.

$$x^0 = 1, x \neq 0$$

To evaluate 7^0, identify 7 as the base and 0 as the exponent.

$$7^0 = 1$$

Any nonzero number to the negative n power is equal to the multiplicative inverse of its nth power. For example, $10^{-2} = \frac{1}{10^2}$.

To evaluate 5^{-3}, identify 5 as the base and -3 as the exponent.

$$5^{-3} = \frac{1}{5^3} = \frac{1}{5} \cdot \frac{1}{5} \cdot \frac{1}{5} = \frac{1}{125}$$

EXAMPLE Evaluating Zero and Negative Exponents

Evaluate 4^0.

The base is 4, and the exponent is 0. • Identify the base and the exponent.

$4^0 = 1$ • Evaluate.

Evaluate 3^{-2}.

The base is 3, and the exponent is -2. • Identify the base and the exponent.

$3^{-2} = \frac{1}{3^2}$ • Write the power using a positive exponent.

$\frac{1}{3^2} = \frac{1}{3 \cdot 3} = \frac{1}{9}$ • Evaluate.

So, $3^{-2} = \frac{1}{9}$.

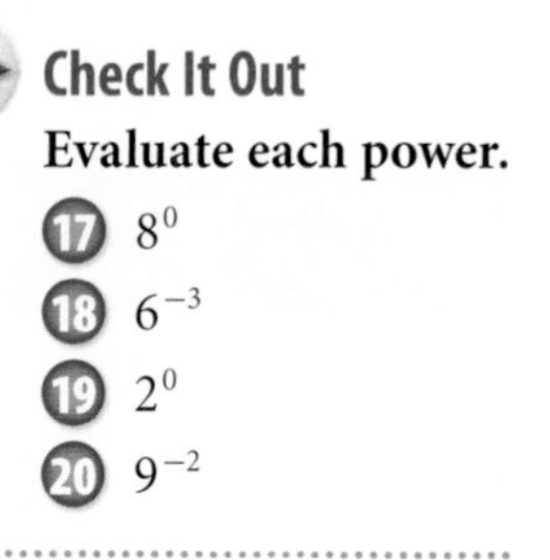

Check It Out

Evaluate each power.

17. 8^0
18. 6^{-3}
19. 2^0
20. 9^{-2}

Powers of Ten

Our decimal system is based on 10. For each factor of 10, the decimal point moves one place to the right.

$3.15 \rightarrow 31.5$ ($\times 10$) $\quad 3. \rightarrow 30$ ($\times 10$) $\quad 14.25 \rightarrow 1{,}425$ ($\times 100$)

When multiplying by a negative power of 10, the decimal point moves to the left. The number of places the decimal moves is the absolute value of the exponent.

$3.17 \rightarrow 0.0317$ ($\times 10^{-2}$) $\quad 5.8 \rightarrow 0.00058$ ($\times 10^{-4}$)

Negative powers are the result of repeated division.

$$10^{-1} = \frac{1}{10} = 1 \div 10$$

When the decimal point is at the end of a number and the number is multiplied by 10, a zero is added to the number.

Try to discover a pattern for the powers of 10.

Power	As a Multiplication	Results	Number of Zeros
10^3	$10 \cdot 10 \cdot 10$	1,000	3
10^7	$10 \cdot 10 \cdot 10 \cdot 10 \cdot 10 \cdot 10 \cdot 10$	10,000,000	7
10^{-3}	$\frac{1}{10} \cdot \frac{1}{10} \cdot \frac{1}{10}$	$\frac{1}{1000}$ or 0.001	3
10^{-5}	$\frac{1}{10} \cdot \frac{1}{10} \cdot \frac{1}{10} \cdot \frac{1}{10} \cdot \frac{1}{10}$	$\frac{1}{100000}$ or 0.00001	5
10^{-6}	$\frac{1}{10} \cdot \frac{1}{10} \cdot \frac{1}{10} \cdot \frac{1}{10} \cdot \frac{1}{10} \cdot \frac{1}{10}$	$\frac{1}{1000000}$ or 0.000001	6

Notice that the number of zeros before or after the 1 is the same as the power of 10. This means that if you want to evaluate 10^8, you simply write a 1 followed by 8 zeros: 100,000,000.

Check It Out

Evaluate each power of 10.

21. 10^{-4}
22. 10^6
23. 10^9
24. 10^{-8}

Using a Calculator to Evaluate Powers

You can use a calculator to evaluate powers. On a calculator, you can multiply a number any number of times just by using the [×] key.

Many calculators have the key [x^2]. This key is used to calculate the square of a number. Enter the number that is the base, and then press [x^2]. The display will show the square of the number. On some calculators, you will have to press [=] or [ENTER] to see the answer.

Some calculators have the key [y^x] or [x^y]. This key is used to calculate any power of a number. Enter the number that is the base, and press [y^x] or [x^y]. Then enter the number that is the exponent, and press [=].

Other calculators use the key [^] for calculating powers. Enter the number that is the base, and press [^]. Then enter the number that is the exponent, and press [ENTER].

EXAMPLE **Evaluating Powers**

Find 12^2.

- Enter 12 in the calculator.
- Press [x^2].
- Press [=] or [ENTER].

144 should be displayed on the calculator.

Find 9^3.

- Enter 9 in the calculator.
- Press [y^x] or [x^y] or [^].
- Enter 3.
- Press [ENTER].

729 should be displayed on the calculator.

For more information on calculators, see Topic 9·1.

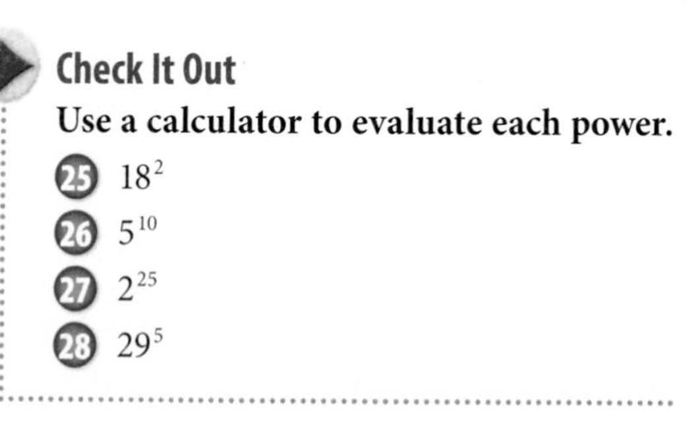

Check It Out

Use a calculator to evaluate each power.

25. 18^2
26. 5^{10}
27. 2^{25}
28. 29^5

3-1 Exercises

Write each product, using an exponent.

1. $7 \cdot 7 \cdot 7$
2. $9 \cdot 9 \cdot 9 \cdot 9 \cdot 9 \cdot 9 \cdot 9$
3. $a \cdot a \cdot a \cdot a \cdot a \cdot a$
4. $w \cdot w \cdot w \cdot w \cdot w \cdot w \cdot w \cdot w \cdot w \cdot w$
5. $16 \cdot 16$

Evaluate each square.

6. 8^2
7. 15^2
8. $(-7)^2$
9. 1 squared
10. $\left(\frac{2}{3}\right)^2$

Evaluate each cube.

11. 7^3
12. 11^3
13. $(-6)^3$
14. 3 cubed
15. $\left(\frac{2}{3}\right)^3$

Evaluate each power.

16. 6^4
17. $(-2)^5$
18. 5^5
19. 4 to the seventh power
20. 1 to the fifteenth power
21. 9^0
22. 3^{-2}
23. 2^{-4}

Evaluate each power of 10.

24. 10^2
25. 10^8
26. 10^{13}
27. 10^{-7}
28. 10^{-9}

Use a calculator to evaluate each power.

29. 8^6
30. 6^{10}

3•2 Square and Cube Roots

Square Roots

In mathematics, certain operations are opposites of each other; that is, one operation "undoes" the other. Addition undoes subtraction: $9 - 5 = 4$, so $4 + 5 = 9$. Multiplication undoes division: $12 \div 4 = 3$, so $3 \cdot 4 = 12$.

The opposite of squaring a number is finding the **square root**. You know that $5^2 = 25$. The square root of 25 is the number that can be multiplied by itself to get 25, which is 5. The symbol for square root is $\sqrt{}$. Therefore, $\sqrt{25} = 5$.

EXAMPLE **Finding the Square Root**

Find $\sqrt{81}$.

$9 \cdot 9 = 81$ • *Think:* What number times itself makes 81?

Since $9 \cdot 9 = 81$, the square root of 81 is 9. • Find the square root.

So, $\sqrt{81} = 9$.

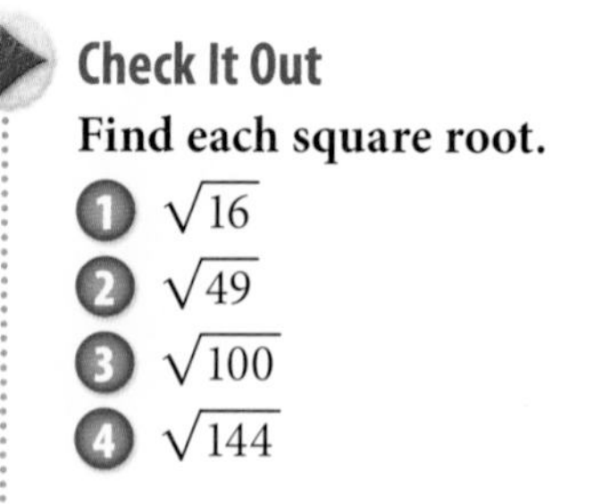

Check It Out

Find each square root.

1. $\sqrt{16}$
2. $\sqrt{49}$
3. $\sqrt{100}$
4. $\sqrt{144}$

Estimating Square Roots

The table shows the first ten **perfect squares** and their square roots.

Perfect square	1	4	9	16	25	36	49	64	81	100
Square root	1	2	3	4	5	6	7	8	9	10

You can estimate the value of a square root by finding the two consecutive numbers that the square root is between. Notice that in the table above, $\sqrt{40}$ lies between 36 and 49; therefore, the value of $\sqrt{40}$ is a number between 6 and 7.

EXAMPLE **Estimating a Square Root**

Estimate $\sqrt{70}$.

70 is between 64 and 81. • Identify the perfect squares that 70 is between.

$\sqrt{64} = 8$ and $\sqrt{81} = 9$. • Find the square roots of the perfect squares.

• Estimate the square root.

So, $\sqrt{70}$ is between 8 and 9.

Check It Out

Estimate each square root.

5. $\sqrt{55}$
6. $\sqrt{18}$
7. $\sqrt{7}$
8. $\sqrt{95}$

Better Estimates of Square Roots

You can use a calculator to find a better estimate for the value of a square root. Most calculators have a [√] key for finding square roots.

On some calculators, the $\sqrt{}$ function is shown on the same key as the [x^2] key on the calculator's keypad. If this is true for your calculator, you should then see a key that has either [INV] or [2nd] on it. To use the $\sqrt{}$ function, you would press [INV] or [2nd] and then the key with $\sqrt{}$ on it.

When finding the square root of a number that is not a perfect square, the answer will be a decimal, and the entire calculator display will be used. Generally, you round square roots to the nearest thousandth. Remember that the thousandths place is the third place after the decimal point.

EXAMPLE Estimating the Square Root of a Number

Estimate $\sqrt{42}$.

Press 42 [√], or 42 [INV] [x^2], or press [2nd] [x^2] 42 [ENTER].	• Use a calculator.
6.4807407 if your calculator shows 8 digits or 6.480740698 if your calculator shows 10 digits	• Read the display.
Locate the digit in the third place after the decimal, which is 0. Then look at the digit to its right, which is 7.	• Round to the nearest thousandth.
Because this digit is 5 or more, round up.	• Estimate the square root.

So, $\sqrt{42} = 6.481$.

See Topic 9•1 for more about calculators.

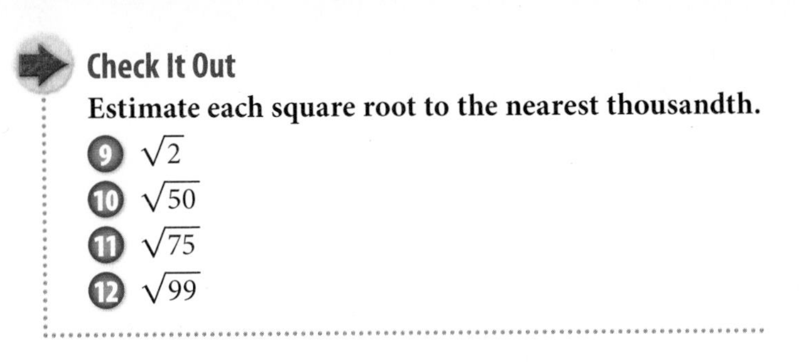

Check It Out

Estimate each square root to the nearest thousandth.

9. $\sqrt{2}$
10. $\sqrt{50}$
11. $\sqrt{75}$
12. $\sqrt{99}$

Cube Roots

In the same way that finding a square root is the opposite of squaring of a number, finding a **cube root** is the opposite of cubing of a number. Finding a cube root answers the question "What number times itself three times makes the cube?" Because 2 cubed $= 2 \cdot 2 \cdot 2 = 2^3 = 8$, the cube root of 8 is 2. The symbol for cube root is $\sqrt[3]{\ }$. Therefore, $\sqrt[3]{8} = 2$.

EXAMPLE Finding the Cube Root of a Number

Find $\sqrt[3]{216}$.

- *Think:* What number times itself three times will make 216?

  $6 \cdot 6 \cdot 6 = 216$
- Find the cube root.

So, $\sqrt[3]{216} = 6$.

Check It Out

Find the cube root of each number.

13. $\sqrt[3]{64}$
14. $\sqrt[3]{343}$
15. $\sqrt[3]{1000}$
16. $\sqrt[3]{125}$

3•2 Exercises

Find each square root.

1. $\sqrt{9}$
2. $\sqrt{64}$
3. $\sqrt{121}$
4. $\sqrt{25}$
5. $\sqrt{196}$

6. $\sqrt{30}$ is between which two numbers?
   A. 3 and 4
   B. 5 and 6
   C. 29 and 31
   D. None of these
7. $\sqrt{84}$ is between which two numbers?
   A. 4 and 5
   B. 8 and 9
   C. 9 and 10
   D. 83 and 85

8. $\sqrt{21}$ is between what two consecutive numbers?
9. $\sqrt{65}$ is between what two consecutive numbers?
10. $\sqrt{106}$ is between what two consecutive numbers?

Estimate each square root to the nearest thousandth.

11. $\sqrt{3}$
12. $\sqrt{10}$
13. $\sqrt{47}$
14. $\sqrt{86}$
15. $\sqrt{102}$

Find the cube root of each number.

16. $\sqrt[3]{27}$
17. $\sqrt[3]{512}$
18. $\sqrt[3]{1331}$
19. $\sqrt[3]{1}$
20. $\sqrt[3]{8000}$

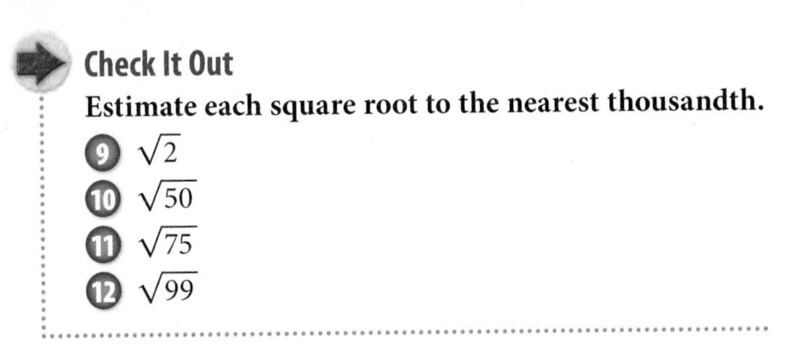

Check It Out

Estimate each square root to the nearest thousandth.

9. $\sqrt{2}$
10. $\sqrt{50}$
11. $\sqrt{75}$
12. $\sqrt{99}$

Cube Roots

In the same way that finding a square root is the opposite of squaring of a number, finding a **cube root** is the opposite of cubing of a number. Finding a cube root answers the question "What number times itself three times makes the cube?" Because 2 cubed $= 2 \cdot 2 \cdot 2 = 2^3 = 8$, the cube root of 8 is 2. The symbol for cube root is $\sqrt[3]{}$. Therefore, $\sqrt[3]{8} = 2$.

EXAMPLE Finding the Cube Root of a Number

Find $\sqrt[3]{216}$.

- *Think:* What number times itself three times will make 216?

  $6 \cdot 6 \cdot 6 = 216$

- Find the cube root.

So, $\sqrt[3]{216} = 6$.

Check It Out

Find the cube root of each number.

13. $\sqrt[3]{64}$
14. $\sqrt[3]{343}$
15. $\sqrt[3]{1000}$
16. $\sqrt[3]{125}$

3•2 Exercises

Find each square root.

1. $\sqrt{9}$
2. $\sqrt{64}$
3. $\sqrt{121}$
4. $\sqrt{25}$
5. $\sqrt{196}$

6. $\sqrt{30}$ is between which two numbers?
   - **A.** 3 and 4
   - **B.** 5 and 6
   - **C.** 29 and 31
   - **D.** None of these
7. $\sqrt{84}$ is between which two numbers?
   - **A.** 4 and 5
   - **B.** 8 and 9
   - **C.** 9 and 10
   - **D.** 83 and 85

8. $\sqrt{21}$ is between what two consecutive numbers?
9. $\sqrt{65}$ is between what two consecutive numbers?
10. $\sqrt{106}$ is between what two consecutive numbers?

Estimate each square root to the nearest thousandth.

11. $\sqrt{3}$
12. $\sqrt{10}$
13. $\sqrt{47}$
14. $\sqrt{86}$
15. $\sqrt{102}$

Find the cube root of each number.

16. $\sqrt[3]{27}$
17. $\sqrt[3]{512}$
18. $\sqrt[3]{1331}$
19. $\sqrt[3]{1}$
20. $\sqrt[3]{8000}$

3•3 Scientific Notation

Using Scientific Notation

Often, in science and in mathematics, numbers are used that are either very large or very small. Large numbers often have many zeros at the end. Small numbers often have many zeros in the beginning.

Large number: $\underbrace{450{,}000{,}000}_{\text{many zeros at the end}}$

Small number: $\underbrace{0.000000032}_{\text{many zeros at the beginning}}$

Check It Out

Identify each number as very large or very small.

1. 0.000015
2. 6,000,000
3. 0.00000901

Writing Large Numbers Using Scientific Notation

Scientific notation is a way of expressing large numbers as the product of a number between 1 and 10 and a power of 10. To write a number in scientific notation, first move the decimal point so that it is to the right of the first nonzero digit. Second, count the number of places you moved the decimal point. Finally, find the power of 10. If the absolute value of the original number is greater than 1, the exponent is positive.

$35{,}700 = 3.57 \cdot 10^4$

EXAMPLE Writing Large Numbers in Scientific Notation

Write 4,250,000,000 in scientific notation.

4.250000000.	• Move the decimal point so that only one digit is to the left of the decimal.
4.250000000. (9 places)	• Count the number of decimal places that the decimal has to be moved to the right.
$4.25 \cdot 10^9$	• Write the number without the ending zeros, and multiply by the correct power of 10.

So, 4,250,000,000 in scientific notation is $4.25 \cdot 10^9$.

Check It Out

Write each number in scientific notation.

4. 68,000
5. 7,000,000
6. 73,280,000
7. 30,500,000,000

APPLICATION Bugs

Insects are the most successful form of life on Earth. About one million have been classified and named. It is estimated that there are up to four million more. That's not total insects we are talking about; that's different *kinds* of insects!

Estimates are that there are 200,000,000 insects for each person on the planet. Given a world population of approximately 6,000,000,000, with about how many insects do we share Earth? Use a calculator to arrive at an estimate. Express the number in scientific notation. See **HotSolutions** for the answer.

Writing Small Numbers Using Scientific Notation

Scientific notation can also be used as a way of expressing small numbers as the product of a number between 1 and 10 and a power of 10. To write a number in scientific notation, first move the decimal point to the right of the first nonzero digit. Second, count the number of places you moved the decimal point. Finally, find the power of 10. If the absolute value of the original number is between 0 and 1, the exponent is negative.

$0.000357 = 3.57 \cdot 10^{-4}$

EXAMPLE Writing Small Numbers in Scientific Notation

Write 0.0000000425 in scientific notation.

0.00000004.25
- Move the decimal point so that only one nonzero digit is to the left of the decimal.

0.00000004.25 (8 places)
- Count the number of decimal places that the decimal has to be moved to the left.

$4.25 \cdot 10^{-8}$
- Write the number without the beginning zeros, and multiply by the correct power of 10. Use a negative exponent to move the decimal to the left.

So, 0.0000000425 in scientific notation is $4.25 \cdot 10^{-8}$.

Check It Out

Write each number in scientific notation.

8. 0.0038
9. 0.0000004
10. 0.0000000000603
11. 0.0007124

Converting from Scientific Notation to Standard Form

Converting to Standard Form When the Exponent Is Positive

When the power of 10 is positive, each factor of 10 moves the decimal point one place to the right. When the last digit of the number is reached, some factors of 10 may still remain. Add a zero at the end of the number for each remaining factor of 10.

EXAMPLE **Converting to Standard Form**

Write $7.035 \cdot 10^6$ in standard form.

The exponent is positive. • Study the exponent.

The decimal point moves to the right 6 places. • Move the decimal point the correct number of places to the right. Add necessary zeros at the end of the number to fill to the decimal point.

7.035000.

Move the decimal point to the right 6 places.

• Write the number in standard form.

So, $7.035 \cdot 10^6 = 7{,}035{,}000$.

Check It Out

Write each number in standard form.

12. $5.3 \cdot 10^4$
13. $9.24 \cdot 10^8$
14. $1.205 \cdot 10^5$
15. $8.84073 \cdot 10^{12}$

Converting to Standard Form When the Exponent Is Negative

When the power of 10 is negative, each factor of 10 moves the decimal point one place to the left. Because there is only one digit to the left of the decimal, you will have to add zeros at the beginning of the number.

EXAMPLE Converting to Standard Form

Write $4.16 \cdot 10^{-5}$ in standard form.

The exponent is negative.
- Study the exponent.

The decimal point moves 5 places to the left.

0.00004.16

Move the decimal point to the left 5 places.
- Move the decimal point the correct number of places to the left. Add zeros at the beginning of the number to fill to the decimal point.

So, $4.16 \cdot 10^{-5} = 0.0000416$.
- Write the number in standard form.

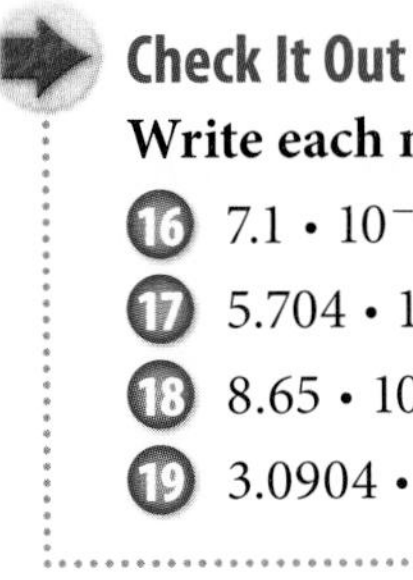

Check It Out

Write each number in standard form.

16. $7.1 \cdot 10^{-4}$
17. $5.704 \cdot 10^{-6}$
18. $8.65 \cdot 10^{-2}$
19. $3.0904 \cdot 10^{-11}$

3•3 Exercises

Identify each number as very large or very small.

1. 0.000034

2. 83,900,000

3. 0.000245

4. 302,000,000,000

Write each number in scientific notation.

5. 420,000

6. 804,000,000

7. 30,000,000

8. 13,060,000,000,000

9. 0.00037

10. 0.0000506

11. 0.002

12. 0.000000005507

Write each number in standard form.

13. $2.4 \cdot 10^{7}$

14. $7.15 \cdot 10^{4}$

15. $4.006 \cdot 10^{10}$

16. $8 \cdot 10^{8}$

17. $4.9 \cdot 10^{-7}$

18. $2.003 \cdot 10^{-3}$

19. $5 \cdot 10^{-5}$

20. $7.0601 \cdot 10^{-10}$

21. Which of the following expresses the number 5,030,000 in scientific notation?

A. $5 \cdot 10^{6}$ **B.** $5.03 \cdot 10^{6}$ **C.** $5.03 \cdot 10^{-6}$ **D.** $50.3 \cdot 10^{5}$

22. Which of the following expresses the number 0.0004 in scientific notation?

A. $4 \cdot 10^{4}$ **B.** $0.4 \cdot 10^{-3}$ **C.** $4 \cdot 10^{-4}$ **D.** $4 \cdot 10^{-3}$

23. Which of the following expresses the number $3.09 \cdot 10^{7}$ in standard form?

A. 30,000,000 **B.** 30,900,000

C. 0.000000309 **D.** 3,090,000,000

24. Which of the following expresses the number $5.2 \cdot 10^{-5}$ in standard form?

A. 0.000052 **B.** 0.0000052

C. 520,000 **D.** 5,200,000

25. When written in scientific notation, which of the following numbers will have the greatest power of 10?

A. 93,000 **B.** 408,000

C. 5,556,000 **D.** 100,000,000

3•4 Laws of Exponents

Revisiting Order of Operations

When evaluating expressions using the **order of operations**, you complete the operations in the following order.

First, complete the operations within grouping symbols.

Then, evaluate powers and roots.

Next, multiply and divide in order from left to right.

Finally, add and subtract in order from left to right.

EXAMPLE Evaluating Expressions with Exponents

Evaluate $3(6 - 2) + 4^3 \div 8 - 3^2$.

$= 3(4) + 4^3 \div 8 - 3^2$	• Complete the operations within parentheses first.
$= 3(4) + 64 \div 8 - 9$	• Evaluate the powers.
$= 12 + 8 - 9$	• Multiply and divide in order from left to right.
$= 11$	• Add and subtract in order from left to right.

So, $3(6 - 2) + 4^3 \div 8 - 3^2 = 11$.

Check It Out

Evaluate each expression.

1. $5^2 - 8 \div 4$
2. $(7 - 3)^2 + 16 \div 2^4$
3. $5 + (3^2 - 2 \cdot 4) + 12$
4. $16 - (4 \cdot 3 - 7) + 2^3$

Product Laws

You can make calculations with exponents much simpler by following the laws of exponents.

To multiply powers that have the same base, add the exponents.

$$a^b \cdot a^c = a^{b+c}$$

To multiply bases with the same power, multiply the bases.

$$a^c \cdot b^c = (ab)^c$$

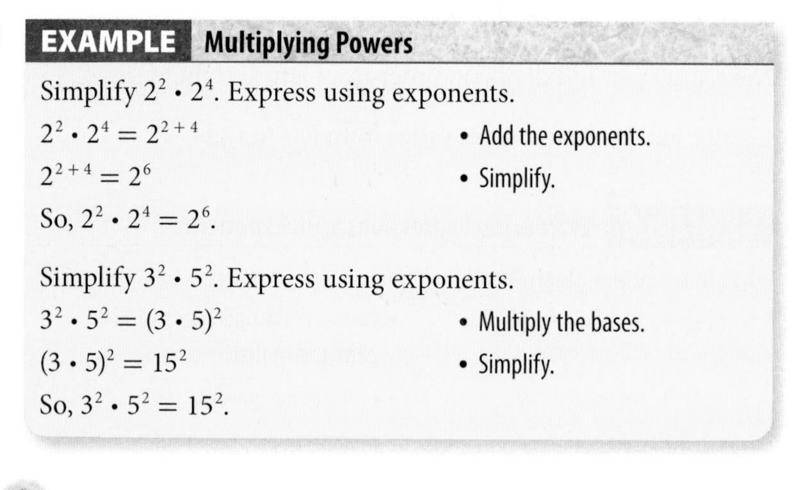

EXAMPLE **Multiplying Powers**

Simplify $2^2 \cdot 2^4$. Express using exponents.

$2^2 \cdot 2^4 = 2^{2+4}$ • Add the exponents.

$2^{2+4} = 2^6$ • Simplify.

So, $2^2 \cdot 2^4 = 2^6$.

Simplify $3^2 \cdot 5^2$. Express using exponents.

$3^2 \cdot 5^2 = (3 \cdot 5)^2$ • Multiply the bases.

$(3 \cdot 5)^2 = 15^2$ • Simplify.

So, $3^2 \cdot 5^2 = 15^2$.

Check It Out

Simplify. Express using exponents.

5. $3^4 \cdot 3^5$
6. $2^6 \cdot 2^{12}$
7. $2^3 \cdot 4^3$
8. $5^2 \cdot 6^2$

The Quotient Law

To divide two powers that have the same base, subtract the exponents.

$$\frac{a^b}{a^c} = a^{b-c};\ a \neq 0$$

To divide two bases with the same power, divide the bases.

$$\frac{a^c}{b^c} = \left(\frac{a}{b}\right)^c;\ b \neq 0$$

EXAMPLE **Dividing Powers**

Simplify $\frac{3^4}{3^2}$. Express using exponents.

$\frac{3^4}{3^2} = 3^{4-2}$ • Subtract the exponents.

$3^{4-2} = 3^2$ • Simplify.

So, $\frac{3^4}{3^2} = 3^2$.

Simplify $\frac{8^3}{4^3}$. Express using exponents.

$\frac{8^3}{4^3} = \left(\frac{8}{4}\right)^3$ • Divide the bases.

$\left(\frac{8}{4}\right)^3 = 2^3$ • Simplify.

So, $\frac{8^3}{4^3} = 2^3$.

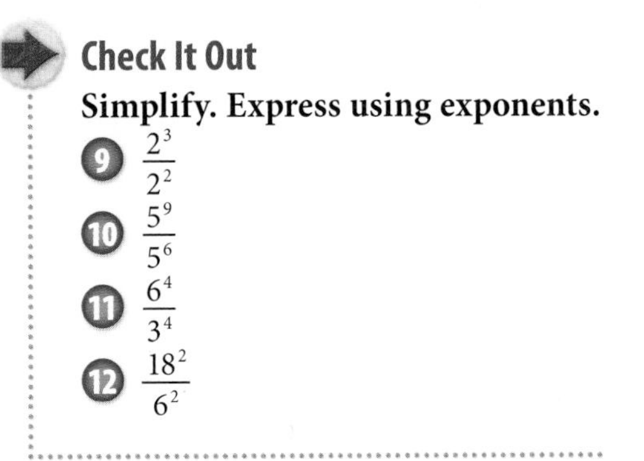

Check It Out

Simplify. Express using exponents.

9. $\frac{2^3}{2^2}$
10. $\frac{5^9}{5^6}$
11. $\frac{6^4}{3^4}$
12. $\frac{18^2}{6^2}$

Power of a Power Law

To find the power of a power, multiply the exponents.

$$(a^b)^c = a^{bc}$$

To find the power of a product, apply the exponent to each factor and multiply.

$$(ab)^c = a^c b^c$$

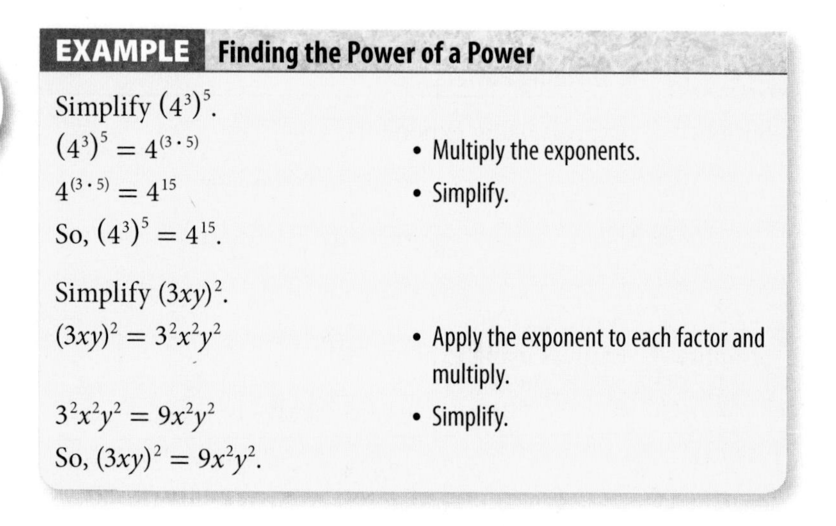

EXAMPLE **Finding the Power of a Power**

Simplify $(4^3)^5$.

$(4^3)^5 = 4^{(3 \cdot 5)}$ • Multiply the exponents.

$4^{(3 \cdot 5)} = 4^{15}$ • Simplify.

So, $(4^3)^5 = 4^{15}$.

Simplify $(3xy)^2$.

$(3xy)^2 = 3^2x^2y^2$ • Apply the exponent to each factor and multiply.

$3^2x^2y^2 = 9x^2y^2$ • Simplify.

So, $(3xy)^2 = 9x^2y^2$.

Check It Out

Simplify. Express using exponents.

13. $(3^2)^4$
14. $(3^6)^3$
15. $(4ab)^3$
16. $(3xy^6)^3$

3•4 Exercises

Evaluate each expression.

1. $4^2 \div 2^3$
2. $(5 - 3)^5 - 4 \cdot 5$
3. $7^2 - 3(5 + 3^2)$
4. $8^2 \div 4 \cdot 2$
5. $15 \div 3 + (10 - 7)^2 \cdot 2$
6. $7 \cdot 3 - (8 - 2 \cdot 3)^3 - 1$
7. $5^2 - 2 \cdot 3^2$
8. $2 \cdot 5 + 3^4 \div (4 + 5)$
9. $(7 - 3)^2 - (9 - 6)^3 \div 9$
10. $3 \cdot 4^2 \div 6 + 2(3^2 - 5)$

Simplify each expression.

11. $a^2 \cdot a^5$
12. $\frac{x^8}{x^5}$
13. $(m^3)^4$
14. $3^5 \cdot 3^7$
15. $2^4 \cdot 7^4$
16. $(6x^4)(8x^7)$
17. $\frac{3^7}{3^5}$
18. $\frac{15^3}{5^3}$
19. $(3^3)^5$
20. $[(4^2)^3]^2$
21. $(5ab)^4$

Powers and Roots

What have you learned?

You can use the problems and the list of words that follow to see what you learned in this chapter. You can find out more about a particular problem or word by referring to the topic number (*for example,* Lesson 3·2).

3

WHAT HAVE YOU LEARNED?

Problem Set

Write each product, using an exponent. (Lesson 3·1)

1. $7 \cdot 7 \cdot 7 \cdot 7 \cdot 7 \cdot 7 \cdot 7 \cdot 7 \cdot 7$

2. $n \cdot n \cdot n \cdot n$

Evaluate each square. (Lesson 3·1)

3. 3^2 **4.** 7^2 **5.** 12^2

6. $(-8)^2$ **7.** $\left(\frac{3}{4}\right)^2$

Evaluate each cube. (Lesson 3·1)

8. 4^3 **9.** 9^3 **10.** 5^3

11. $(-3)^3$ **12.** $\left(\frac{1}{2}\right)^3$

Evaluate each power. (Lesson 3·1)

13. 3^8 **14.** 7^4 **15.** 2^{11} **16.** $(-2)^5$

17. 5^0 **18.** 9^0 **19.** 4^{-3} **20.** 7^{-2}

Evaluate each power of 10. (Lesson 3·1)

21. 10^2 **22.** 10^5 **23.** 10^9

Evaluate each square root. (Lesson 3·2)

24. $\sqrt{9}$ **25.** $\sqrt{64}$ **26.** $\sqrt{169}$

Estimate each square root between two consecutive numbers. (Lesson 3·2)

27. $\sqrt{51}$ **28.** $\sqrt{18}$ **29.** $\sqrt{92}$

Estimate each square root to the nearest thousandth. (Lesson 3•2)

30. $\sqrt{23}$ **31.** $\sqrt{45}$

Evaluate each cube root. (Lesson 3•2)

32. $\sqrt[3]{27}$ **33.** $\sqrt[3]{125}$ **34.** $\sqrt[3]{729}$

Identify each number as very large or very small. (Lesson 3•3)

35. 0.000063

36. 8,600,000

Write each number in scientific notation. (Lesson 3•3)

37. 9,300,000

38. 800,000,000

39. 0.000054

40. 0.0605

Write each number in standard form. (Lesson 3•3)

41. $3.4 \cdot 10^4$

42. $7.001 \cdot 10^{10}$

43. $9 \cdot 10^6$

44. $5.3 \cdot 10^{-3}$

45. $6.02 \cdot 10^{-9}$

46. $4 \cdot 10^{-4}$

Evaluate each expression. (Lesson 3•4)

47. $3 \cdot 5^2 - 4^2 \cdot 2$

48. $6^2 - (8^2 \div 2^5 + 3 \cdot 5)$

49. $(1 + 2 \cdot 3)^2 - (2^3 - 4 \div 2^2)$

HotWords

Write definitions for the following words.

base (Lesson 3•1)
cube (Lesson 3•1)
cube root (Lesson 3•2)
exponent (Lesson 3•1)
order of operations (Lesson 3•4)
perfect squares (Lesson 3•2)
power (Lesson 3•1)
scientific notation (Lesson 3•3)
square (Lesson 3•1)
square root (Lesson 3•2)

HotTopic 4

Data, Statistics, and Probability

What do you know?

You can use the problems and the list of words that follow to see what you already know about this chapter. The answers to the problems are in **HotSolutions** at the back of the book, and the definitions of the words are in **HotWords** at the front of the book. You can find out more about a particular problem or word by referring to the topic number (*for example,* Lesson 4•2).

Problem Set

Use the following for Exercises 1–3. A student asked others riding on the school bus about their favorite Physical Education time. The answers are shown below. (Lesson 4•1)

Favorite Physical Education Time			
	6th Graders	7th Graders	8th Graders
Early morning	|||	卌	
Late morning	||||	卌 卌	||||
Early afternoon	卌	卌		
Late afternoon	||			卌	

1. When is the favorite Physical Education time among all students who gave answers?
2. Which grade had the most responses?
3. Is this a random sample?
4. In a class election, tally marks were used to count votes. What is the graph called that is made from these marks? (Lesson 4•2)

5. In a scatter plot, the line of best fit rises from left to right. What kind of correlation is illustrated? **(Lesson 4•3)**
6. In Mr. Dahl's class of 27 students, the lowest test grade is 58%, the highest is 92%, and the most common is 84%. What is the range of grades? **(Lesson 4•4)**
7. In Exercise 6, which can you find: the mean, median, or mode? **(Lesson 4•4)**
8. $C(7, 2) =$ ____ **(Lesson 4•5)**
9. A bag contains 10 chips—3 red, 4 blue, 1 green, and 2 black. A chip is drawn. A second chip is drawn without replacing the first one drawn. What is the probability that both are blue? **(Lesson 4•6)**

HotWords

biased sample (Lesson 4•1)
bimodal distribution (Lesson 4•3)
box plot (Lesson 4•2)
combination (Lesson 4•5)
correlation (Lesson 4•3)
dependent events (Lesson 4•6)
event (Lesson 4•5)
experimental probability (Lesson 4•6)
factorial (Lesson 4•5)
flat distribution (Lesson 4•3)
histogram (Lesson 4•2)
independent events (Lesson 4•6)
interquartile range (Lesson 4•4)
line of best fit (Lesson 4•3)
lower quartile (Lesson 4•4)
mean (Lesson 4•4)
measures of variation (Lesson 4•4)
median (Lesson 4•4)
mode (Lesson 4•4)
normal distribution (Lesson 4•3)
outcome (Lesson 4•5)
outlier (Lesson 4•4)
permutation (Lesson 4•5)
population (Lesson 4•1)
probability (Lesson 4•6)
quartiles (Lesson 4•4)
random sample (Lesson 4•1)
range (Lesson 4•4)
sample (Lesson 4•1)
sampling with replacement (Lesson 4•6)
scatter plot (Lesson 4•3)
skewed distribution (Lesson 4•3)
stem-and-leaf plot (Lesson 4•2)
theoretical probability (Lesson 4•6)
tree diagram (Lesson 4•5)
upper quartile (Lesson 4•4)
weighted average (Lesson 4•4)

4•1 Collecting Data

Surveys

Have you ever been asked to name your favorite movie? Have you been asked what kind of pizza you like? These kinds of questions are often asked in *surveys.* A statistician studies a group of people or objects, called a **population**. They usually get information from a small part of the population, called a **sample**.

In a survey, eighth-grade students were chosen at random from three countries and asked if they spent three or more hours on a normal school day watching TV, hanging out with friends, playing sports, reading for fun, or studying. The following bar graph shows the percent of students who said yes in each category.

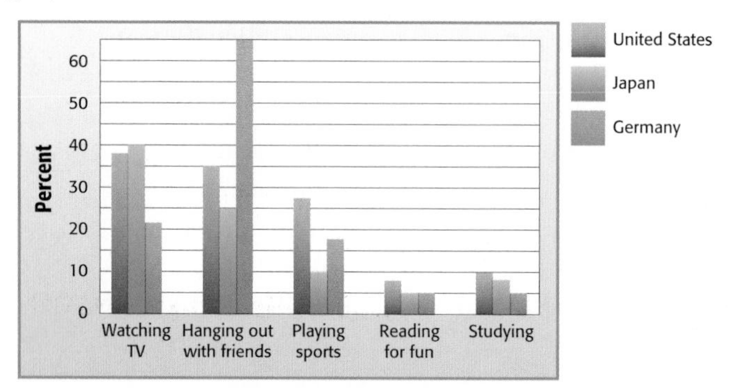

In this case, the population is all eighth-grade students in the United States, Japan, and Germany. The sample is the students who were actually surveyed.

In any survey,

- The population consists of the people or objects about which information is desired.
- The sample consists of the people or objects in the population that are actually studied.

Check It Out

Identify the population and the size of the sample.

1. In a survey, 150,000 adults over the age of 45 were asked if they listened to radio station KROK.
2. two hundred elk in Roosevelt National Forest
3. In a 2007 survey, 500 motor vehicle drivers in the state of California were asked their age.

Random Samples

When you choose a sample to survey for data, be sure that the sample is representative of the population. You also want to be sure it is a **random sample**, where each person in the population has an equal chance of being included.

Mr. Singh wants to determine whether his students want pizza, chicken fingers, ice cream, or bagels at a class party. He picks a sample by writing the names of his students on cards and drawing ten cards from a bag.

EXAMPLE Determining Whether a Sample is Random

Determine whether Mr. Singh's sample (above) is random.

The population is Mr. Singh's class. • Determine the population.

The sample consists of ten students. • Determine the sample.

Every student in Mr. Singh's class has the same chance of being chosen. • Determine if the sample is random.

So, the sample is random.

Check It Out

4. A student asked 20 of her parents' friends who they planned to vote for. Is the sample random?
5. A student assigns numbers to his 24 classmates and then uses a spinner divided into 24 equal parts to pick ten numbers. He asks those ten students to identify their favorite movie. Is the sample random?

Biased Samples

A **biased sample** is determined in such a way that one or more parts of the population are favored over others. A biased sample may be used to obtain a specific outcome or conclusion.

A sample may be biased if you are sampling from a population that is convenient or only from those who want to participate in the survey. For example, a survey may intend to represent all middle school students, but if only one class is asked to participate, the sample is biased because not everyone in the total population has an equal chance of being surveyed.

EXAMPLE Determining Whether a Sample is Biased

Determine if the sample is biased. Explain your answer.

During Mr. Thompson's English class, the students who had finished their homework in class were able to participate in an eighth grade survey. Sixty out of the 112 students who finished their homework were asked to name their favorite school subject.

- Determine the population.

  The population is Mr. Thompson's English class.
- Determine the sample.

  The sample consists of sixty students.
- Determine if the sample is biased.

  Not all eighth-grade students had an equal chance of participating in the survey.

So, the sample is biased.

Check It Out

6. A country music radio station surveyed 1,400,000 members of its audience to determine what kind of music people like best. Out of 100 listeners, 80 stated they liked country music the best. Is the sample biased? Explain.
7. To determine what type of game 360 middle-school students want to play on reward day, a teacher randomly picks 100 out of 360 responses out of a box. Is the sample biased? Explain.

Questionnaires

When you write questions for a survey, it is important to be sure that the questions are not biased. That is, the questions should not make assumptions or influence the responses. The following two questionnaires are designed to find out what kind of food your classmates like and what they do after school. The first questionnaire uses biased questions. The second questionnaire uses questions that are not biased.

Questionnaire 1:

A. What kind of pizza do you like?

B. What is your favorite afternoon TV program?

Questionnaire 2:

A. What is your favorite food?

B. What do you like to do after school?

When you are developing a questionnaire,

- Decide the topic you want to ask about.
- Define a population and decide how to select an unbiased sample from that population.
- Develop questions that are not biased.

Check It Out

8. Why is question A in Questionnaire 1 biased?
9. Why is question B in Questionnaire 2 better than question B in Survey 1?
10. Write a question that asks the same thing as the following question but is not biased. Are you a caring citizen who recycles newspapers?

Compiling Data

After Mr. Singh collected the data from his students, he had to decide how to show the results. As he asked students their food preference, he used tally marks to tally the answers in a table. The following frequency table shows their answers.

Food Preferred in Mr. Singh's Class

Preferred Food	Number of Students
Pizza	𝍸 𝍸 II
Chicken fingers	𝍸 I
Ice cream	𝍸 IIII
Bagels	III

Follow this procedure when you are making a table to compile data.

- List the categories or questions in the first column or row.
- Tally the responses in the second column or row.

Check It Out

11. How many students chose chicken fingers?
12. What was the food least preferred by the students surveyed?
13. If Mr. Singh uses the survey to pick food to serve at the class party, what should he serve? Explain.

4-1 Exercises

1. Norma chose businesses to survey by obtaining a list of businesses in the city and writing each name on a slip of paper. She placed the slips of paper in a bag and drew 50 names. Is the sample random?
2. Jonah knocked on 25 doors in his neighborhood. He asked the residents who answered if they were in favor of the idea of the city building a swimming pool. Is the sample random?

Are the following questions biased? Explain.

3. Are you happy about the ugly building being built in your neighborhood?
4. How many hours do you watch TV each week?

Write unbiased questions to replace the following questions.

5. Do you prefer cute, cuddly kittens as pets, or do you like dogs better?
6. Are you thoughtful about not playing your stereo after 10 P.M.?

Ms. Chow asked her students which type of book they prefer to read and tallied the following data.

Book Preferences of Ms. Chow's Students		
Type of Book	Number of Seventh Graders	Number of Eighth Graders
Biography	卌 卌	卌 卌 II
Mystery	卌 I	III
Fiction	卌 卌 II	卌 卌
Science fiction	卌 II	卌 I
Nonfiction	III	卌 I

7. Which type of book was most popular? How many students preferred that type?
8. Which type of book was preferred by 13 students?
9. How many students were surveyed?

4•2 Displaying Data

Interpret and Create a Table

You know that statisticians collect data about people or objects. One way to show the data is to use a table. Here are the number of letters in the words of the first two sentences in *Black Beauty.*

3 5 5 4 1 3 4 8 3 1 5 8 6 4 1 4 2 5 5 2 2 4 5 5 6 4 2 3 6 3 11 4 2 3 4 3

EXAMPLE **Making a Table**

Make a table to organize the data about letters in the words.

- Name the first row or column what you are counting.

  Label the first row *Number of Letters.*

- Tally the amounts for each category in the second row or column.

Number of Letters	1	2	3	4	5	6	7	8	more than 8
Number of Words	III	~~IIII~~	~~IIII~~ II	~~IIII~~ III	~~IIII~~ II	III		II	I

- Count the tallies and record the number in the second row or column.

Number of Letters	1	2	3	4	5	6	7	8	more than 8
Number of Words	3	5	7	8	7	3	0	2	1

The most common number of letters in a word is 4. Three words have 1 letter.

Check It Out

1. What information is lost by using the category "more than 8"?
2. Use the data below to make a table to show the number of gold medals won by each country in the 1994 Winter Olympics.
   10 9 11 7 6 3 3 2 4 0 1 0 0 2 1 0 0 1 0 0 1 0

Interpret a Box Plot

A **box plot** shows data by using the middle value of the data and the *quartiles* (p. 208), or 25% divisions of the data. The box plot below shows exam scores on a math test for a class of eighth graders.

On a box plot, 50% of the scores are above the median, and 50% are below it. The first quartile is the median score of the bottom half of the scores. The third quartile is the median score of the top half of the scores.

Exam Scores

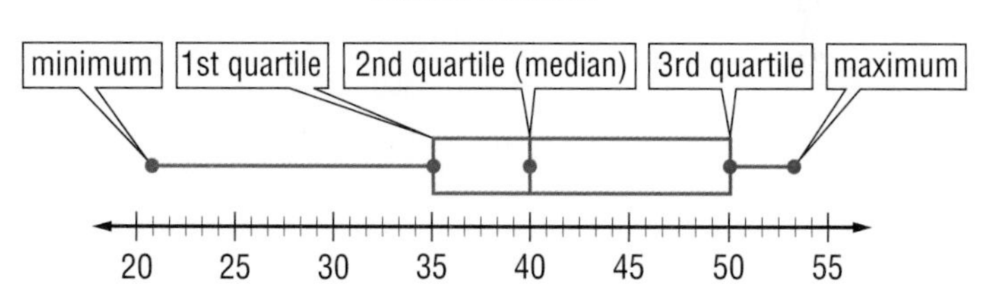

This is what we can tell about the exam scores.

- The high score is 53. The low score is 21.
- The median score is 40. The first quartile score is 35, and the third quartile score is 50.
- 50% of the scores are between 35 and 50.

Check It Out

Use the following box plot for Exercises 3–5.

Grams of Fat in Typical Fast-Food Milkshake

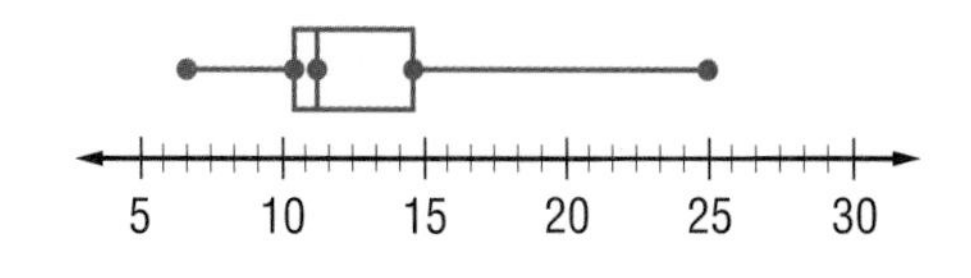

3. What is the greatest amount of fat in a fast-food milkshake?
4. What is the median amount of fat in a fast-food milkshake?
5. What percent of the milkshakes contain between 7 and 11.5 grams of fat?

Interpret and Create a Circle Graph

Another way to show data is to use a *circle graph*. A circle graph can be used to show parts of a whole.

Arturo conducted a survey to find out what kind of solid waste was thrown away. Arturo wants to make a circle graph to show the following data.

Solid Waste		
Type	**Percent**	**Central Angle Measure**
Paper	39%	$360° \cdot 0.39 = 140.4°$
Glass	6%	$360° \cdot 0.06 = 21.6°$
Metals	8%	$360° \cdot 0.08 = 28.8°$
Plastic	9%	$360° \cdot 0.09 = 32.4°$
Wood	7%	$360° \cdot 0.07 = 25.2°$
Food	7%	$360° \cdot 0.07 = 25.2°$
Yard Waste	15%	$360° \cdot 0.15 = 54°$
Miscellaneous Waste	9%	$360° \cdot 0.09 = 32.4°$

To make a circle graph,

- write each part of the data as a percent of the whole.
- find the degree measure of each part of the circle by multiplying the percent by 360°, the total number of degrees in a circle.
- draw a circle, measure each central angle, and complete the graph. Be sure to label the graph and include a title.

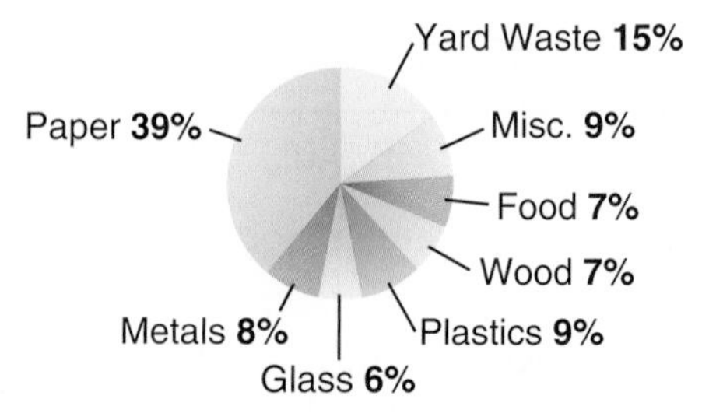

From the graph, you can see that more than half of the solid waste is made up of paper and yard waste. Equal amounts of food and wood are discarded.

Check It Out

Use the circle graph at the right to answer Exercises 6 and 7.

6. About what fraction of people buy used cars from a dealership?

7. About what fraction of people buy used cars from private owners?

Where We Buy Used Cars

Dealership
Other
Family
Used Car Lot
Private Owner

8. Make a circle graph to show the results of students' earnings.

Car wash: $355 Bake sale: $128
Recycling: $155 Book sale: $342

Interpret and Create a Line Plot

A *line plot,* sometimes called a frequency graph, displays data on a number line by using Xs to show the frequency of the data. Suppose that you collect the following data about the times your friends get up on a school day.

5:30, 6, 5:30, 8, 7:30, 8, 7:30, 9, 8, 8, 6, 6:30, 6, 8

To make a line plot:

- Draw a number line showing the numbers in your data set.
- Place an X to represent each result above the number line for each piece of data you have.
- Title the graph.

Your line plot should look like this:

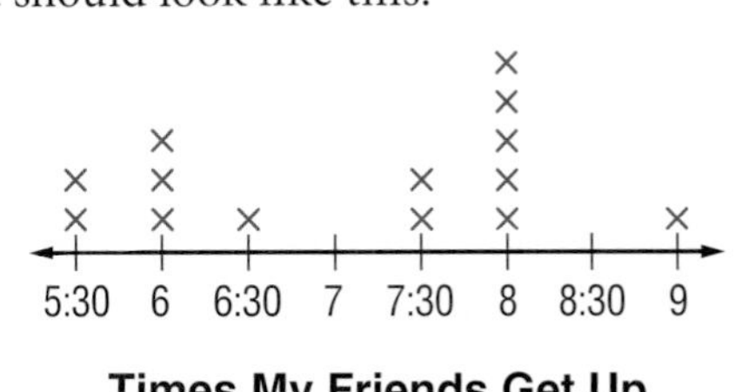

Times My Friends Get Up

You can tell from the line plot that your friends get up anywhere between 5:30 and 9:00 on school days.

Check It Out

9. What is the most common time for your friends to get up?
10. How many friends get up before 7:00 A.M.?
11. Make a line plot to show the number of letters in the words of the first two sentences in *Black Beauty* (p. 182).

Interpret a Line Graph

You know that a *line graph* can be used to show changes in data over time. The following line graph compares the monthly average vault scores of two gymnasts.

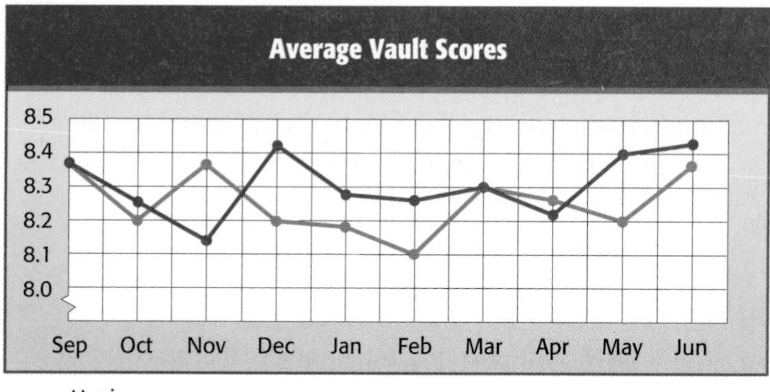

From the graph, you can see that Hani and Gabe had the same average scores in two months, September and March.

Check It Out

12. In December, which gymnast had better scores?
13. Which gymnast typically scores higher on the vault?

4•2 DISPLAYING DATA

Interpret a Stem-and-Leaf Plot

The following numbers show the ages of the students in a T'ai Chi class.

8 12 78 34 38 15 18 9 45 24 39 28 20 66 68 75 45 52 18 56

It is difficult to analyze the data when displayed as a list. You know that you could make a table, a box chart, or a line graph to show this information. Another way to show the information is to make a **stem-and-leaf plot**. The stem-and-leaf plot at the right shows the ages of the students.

Ages of Students

Stem	Leaf
0	8 9
1	2 5 8 8
2	0 4 8
3	4 8 9
4	5 5
5	2 6
6	6 8
7	5 8

1 | 2 = 12 years old

Notice that the tens digits appear in the left-hand column. These are called *stems*. Each digit on the right is called a *leaf*. From looking at the plot, you can tell that more students are in their teens than in their twenties or thirties and that two students are younger than ten.

Check It Out

The stem-and-leaf plot shows the average points per game of high-scoring players over several years.

14. How many players scored an average number of points between 30 and 31?
15. What was the highest average number of points scored? the lowest?

Average Points Per Game

Stem	Leaf
27	2
28	4
29	3 6 8
30	1 3 4 6 6 7 8
31	1 1 5
32	3 5 6 9
33	1 6
34	0 5
35	0
36	
37	1

30 | 1 = 30.1 points

Interpret and Create a Bar Graph

Another type of graph you can use to display data is called a *bar graph*. In this graph, either horizontal or vertical bars are used to show data. Consider the data showing Kirti's earnings from mowing lawns.

May	June	July	August	September
$78	$92	$104	$102	$66

You can make a bar graph to show Kirti's earnings.

To make a bar graph:

- Choose a vertical scale and decide what to place along the horizontal scale.
- For each item on the horizontal scale, draw a bar of the appropriate height.
- Write a title for the graph.

A bar graph of Kirti's earnings is shown below.

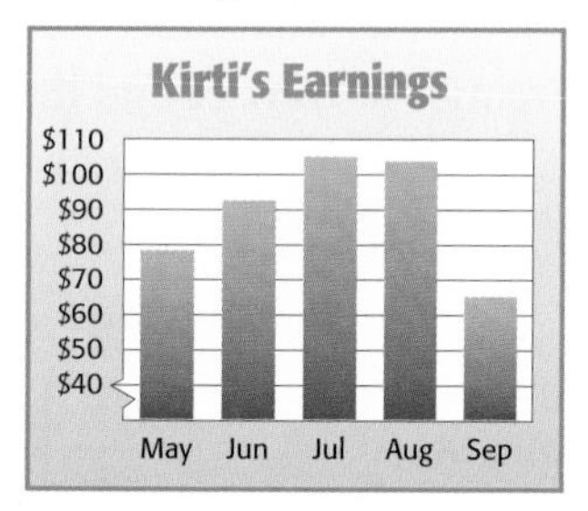

From the graph, you can see that his earnings were highest in July.

Check It Out

16. During which month were Kirti's earnings the lowest?
17. Write a sentence describing Kirti's earnings.
18. Use the data to make a bar graph to show the number of middle-school students on the honor role.

Sixth Grade	144
Seventh Grade	182
Eighth Grade	176

Interpret a Double-Bar Graph

If you want to show information about two or more things, you can use a *double-bar graph*. The following graph shows the sources of revenue for public schools for the past few years.

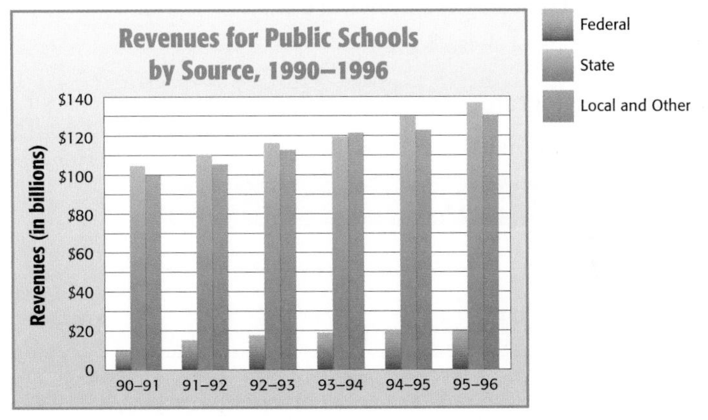

Source: National Education Association

You can see from the graph that the states usually contribute more toward public schools than do local and other sources. Note that the amounts are given in billions. That means $20 on the graph represents $20,000,000,000.

Check It Out

19. About how much did states contribute to public schools in 1993–94?

20. Write a sentence that describes the federal contribution during the years shown.

Interpret and Create a Histogram

A **histogram** is a special kind of bar graph that shows the frequency of data. Suppose that you ask several classmates how many hours, to the nearest hour, that they talk on the phone each week and collect the following data.

4 3 2 3 1 2 0 2 1 3 4 2 1 0 1 6

To create a histogram:

- Make a table showing frequencies.

Hours	Tally	Frequency
0	//	2
1	////	4
2	////	4
3	///	3
4	//	2
5		0
6	/	1

- Make a histogram showing the frequencies.
- Title the graph.

In this case, you might call it "Hours Spent on the Phone."

Your histogram might look like this.

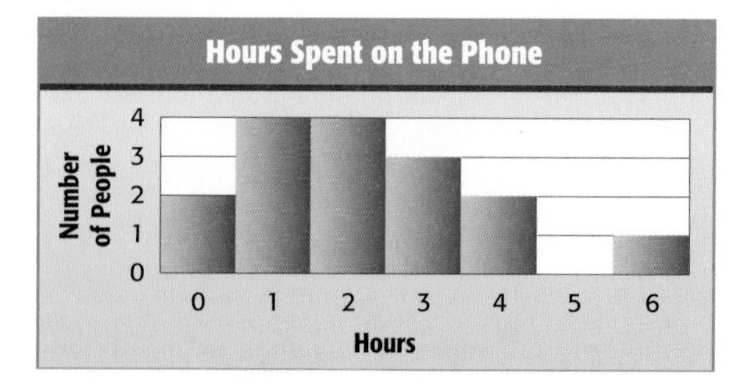

You can see from the diagram that as many students spent 1 hour on the phone as spent 2 hours.

Check It Out

21 Using the bar graph "Hours Spent on the Phone," determine how many classmates were surveyed.

22 Make a histogram from the data about *Black Beauty* (p. 182). How many words have 5 or more letters?

4•2 Exercises

1. Make a table and a histogram to show the following data.

   Hours Spent Each Week Reading for Pleasure

   3 2 5 4 3 1 5 0 2 3 1 4 3 5 1 7 0 3 0 2

2. Which was the most common amount of time spent each week reading for pleasure?
3. Make a line plot to show the data in Exercise 1.
4. Use your line plot to describe the hours spent reading for pleasure.

Use the following graphs for Exercises 5 and 6. These two circle graphs show whether cars made a right turn, a left turn, or drove straight ahead at an intersection near school. (Lesson 4•2)

5. Between 8 A.M. and 9 A.M., what percent of the cars turn?

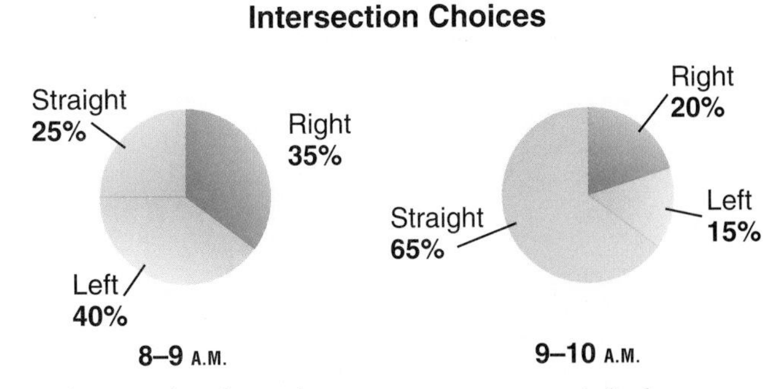

6. Do the graphs show that more cars go straight between 9 A.M. and 10 A.M. than between 8 A.M. and 9 A.M.?

7. Of the first ten presidents, two were born in Massachusetts, one in New York, one in South Carolina, and six in Virginia. Make a circle graph to show this information and write a sentence about your graph.

8. The stem-and-leaf plot shows the heights of 19 girls.

Girls' Heights

Stem	Leaf
5	3 4 4 4 6 8
6	0 0 3 4 4 4 5 6 8 8
7	0 1 2

5 | 3 = 53 inches

What can you say about the height of most of the girls?

9. The eighth-grade classes collected 56 pounds of aluminum in September, 73 pounds in October, 55 pounds in November, and 82 pounds in December. Make a bar graph to show the data.

10. The box plot shows the daily high temperatures in Seaside in July. What is the middle temperature? 50% of the temperatures are between 65° and what temperature?

Temperatures in Seaside in July

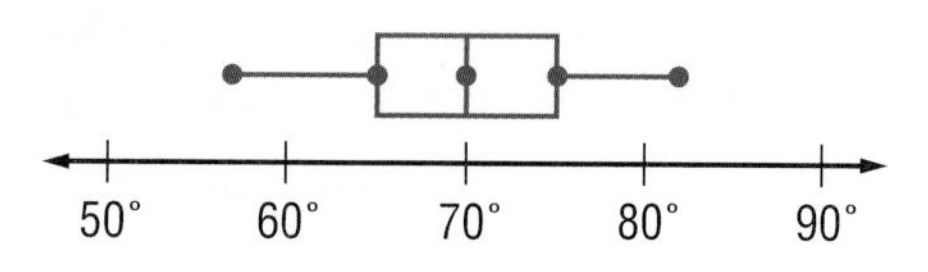

4-3 Analyzing Data

Scatter Plots

Once you have collected data, you may want to analyze and interpret it. You can plot points on a *coordinate graph* (p. 290) to make **scatter plots**. Then you can determine if the data are related.

Samuel collected information showing the number of candy boxes sold by each person in his soccer club and the number of years each person had been in the club.

Years in Club vs. Boxes Sold														
Years in Club	4	3	6	2	3	4	1	2	1	3	4	5	2	2
Boxes Sold	23	18	30	26	22	20	20	20	15	19	23	26	22	18

Make a scatter plot to determine whether there was any relationship between the two. First you write the data as ordered pairs, and then you graph the ordered pairs.

To make a scatter plot:

- Collect two sets of data that you can graph as ordered pairs.
- Label the vertical and horizontal axes and graph the ordered pairs.

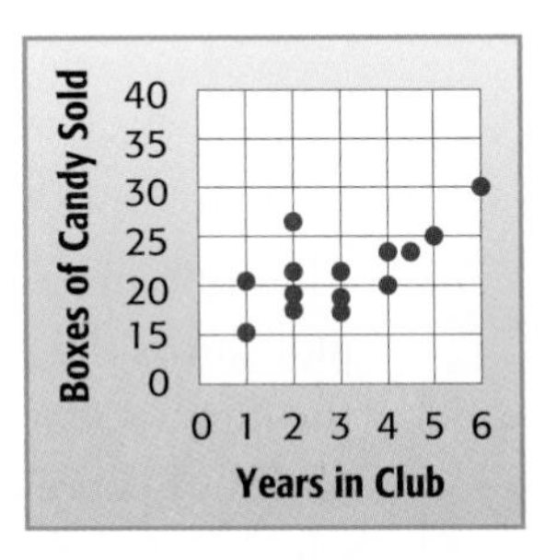

The scatter plot shows an upward trend in the data. You can say that the longer a person is in the soccer club, the more boxes of candy they tend to sell.

Check It Out

1. For the scatter plot below, determine whether the data are related. If they are, describe the relationship between the data.

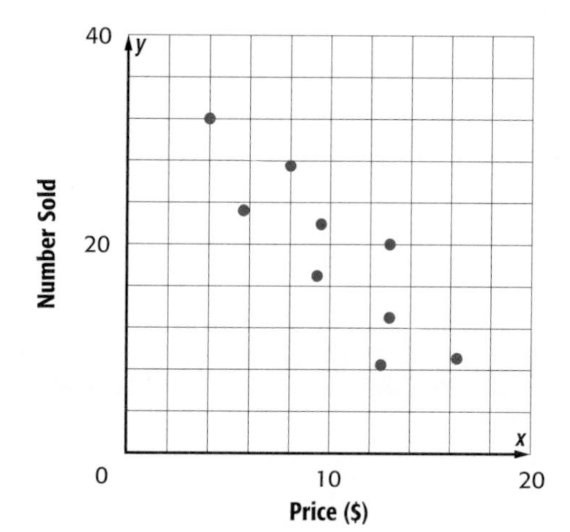

Make a scatter plot showing the following data.

2.

Winning Times for the Men's 100-meter Run Summer Olympics									
Year	1900	1912	1924	1936	1948	1960	1972	1984	1996
Time (in sec)	11.0	10.8	10.6	10.3	10.3	10.2	10.1	9.99	9.84

Correlation

The following scatter plots have slightly different appearances.

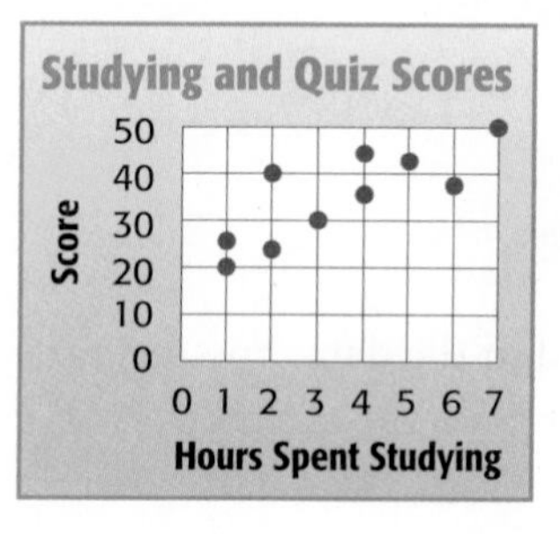

The studying and quiz scores scatter plot shows the relationship between the hours spent studying and quiz scores. A **correlation** is the way in which a change in one variable corresponds to a change in another. There is an upward trend in the scores. You call this a *positive correlation.*

This scatter plot shows the relationship between hours spent watching TV and quiz scores. There is a downward trend in the scores. You call this a *negative correlation.*

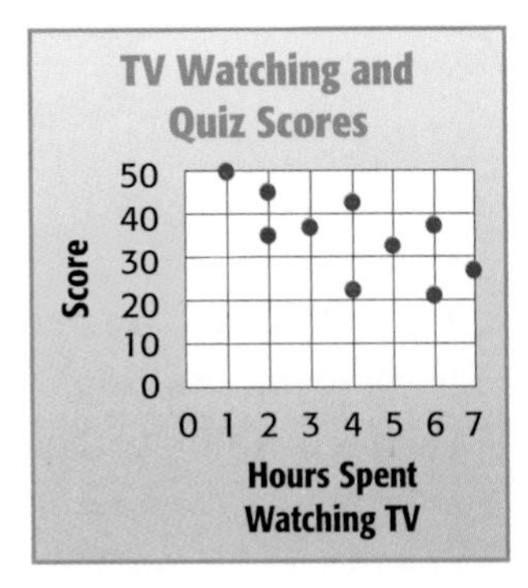

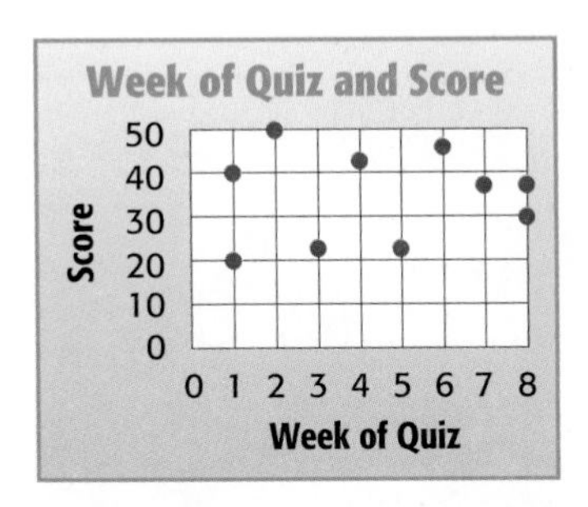

The third scatter plot shows the relationship between the week a quiz was taken and the score. There does not appear to be any relationship. You call this *no correlation.*

Check It Out

3. Which of the following scatter plots shows no relationship?

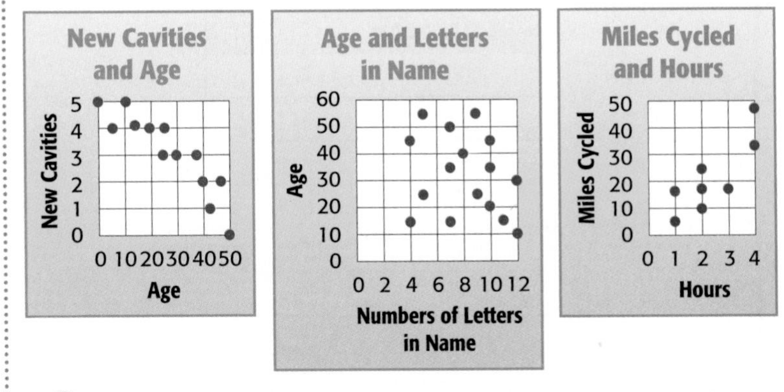

4. Describe the correlation in the scatter plot showing the relationship between age and number of cavities.
5. Which scatter plot shows a positive correlation?

APPLICATION How Risky Is It?

We are bombarded with statistics about risk. We are told that we are more likely to die as a result of Earth's collision with an asteroid than as a result of a tornado; more likely to come in contact with germs by handling paper money than by visiting someone in the hospital. We know the odds of finding radon in our houses (1 in 15) and how much one bad sunburn increases the risk of skin cancer (up to 50 percent).

How risky is modern life? Consider these statistics on life expectancy.

Year	Life Expectancy
1900	47.3
1920	54.1
1940	62.9
1960	69.7
1980	73.7
1990	75.4

Make a line graph of the data. What does the graph show about life expectancy? To what might you attribute this trend? See **HotSolutions** for the answers.

Line of Best Fit

When the points on a scatter plot have either a positive or negative correlation, you can sometimes draw a **line of best fit**. Consider the graph showing the relationship between age and number of cavities.

To draw a line of best fit:

- Decide if the points on the scatter plot show a trend.

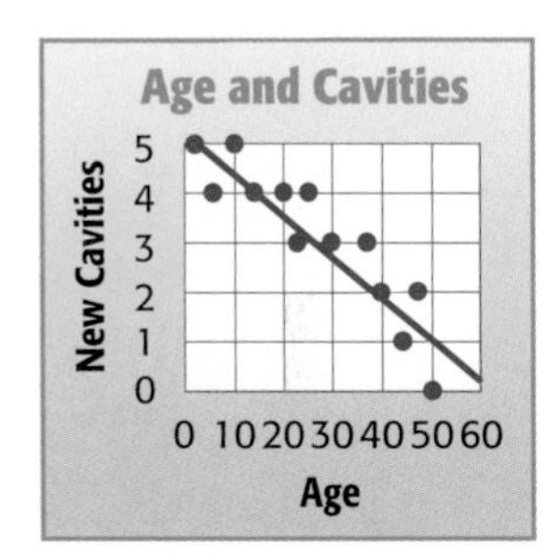

The points on this graph show a negative correlation.

- Draw a line that seems to run through the center of the group of points.

You can use the line to make predictions. From the line it appears that people at the age of 60 would be expected to have fewer than one new cavity, and people at the age of 70 would have no new cavities.

You use the line to help you predict, but the line can show data that is not possible. Always think about whether your prediction is reasonable. For example, people at the age of 60 would not get $\frac{1}{4}$ of a cavity. You would probably predict 1 or 0 cavities.

Check It Out

6. Use the following data to make a scatter plot and draw a line of best fit.

Latitude (°N)	35	34	39	42	35	42	33	42	21
Mean April Temperature (°F)	55	62	54	49	61	49	66	47	76

7. Predict the mean April temperature of a city, which has a latitude of 28°N.

Distribution of Data

A veterinarian measured the weights of 25 cats to the nearest pound and recorded the data on the following histogram. Notice the symmetry of the histogram. If you draw a curve over the histogram, the curve illustrates a **normal distribution**.

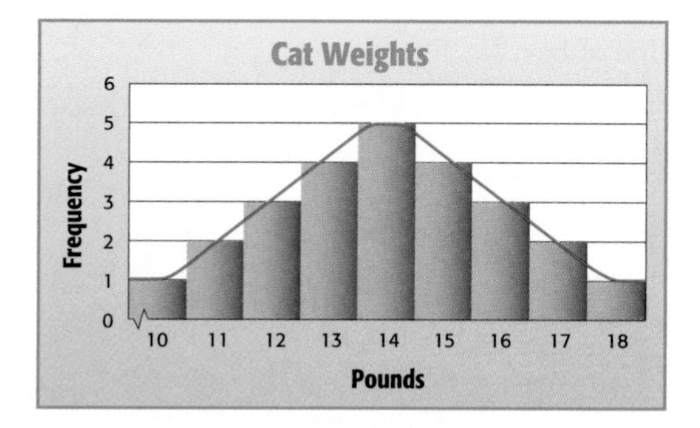

Often a histogram has a **skewed distribution**. These two histograms show the heights of students on a gymnastic team and a basketball team. Again, you can draw a curve to show the shape of the histogram. The graph showing the gymnastic team heights is skewed to the left. The graph showing the basketball team heights is skewed to the right.

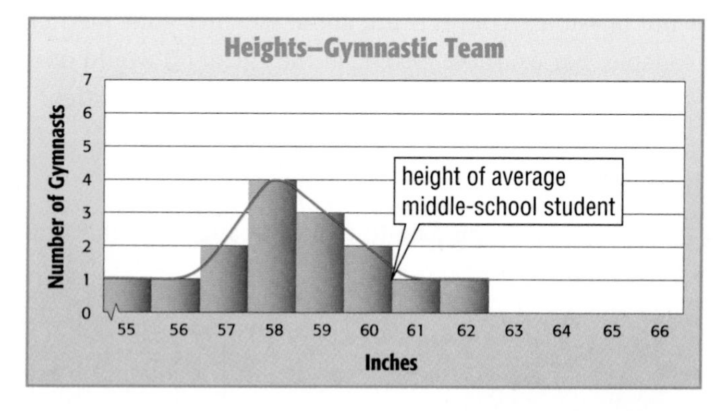

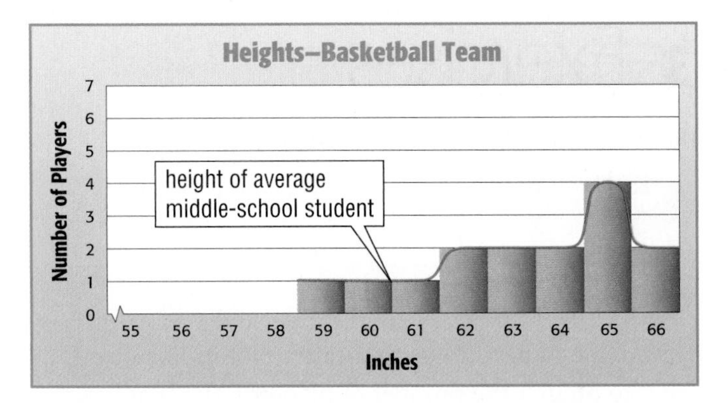

The graph on the left below illustrates heights of adults. This curve has two peaks, one for female heights and one for male heights. This kind of distribution is called a **bimodal distribution**. The one on the right shows the number of dogs boarded each week at a pet kennel. It is called a **flat distribution**.

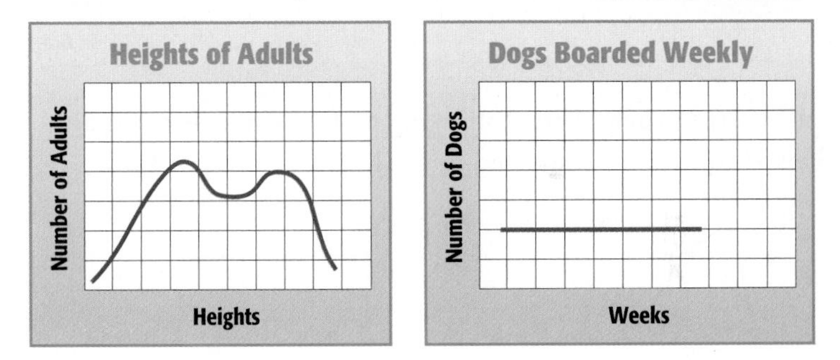

Check It Out

Identify each type of distribution as normal, skewed to the right, skewed to the left, bimodal, or flat.

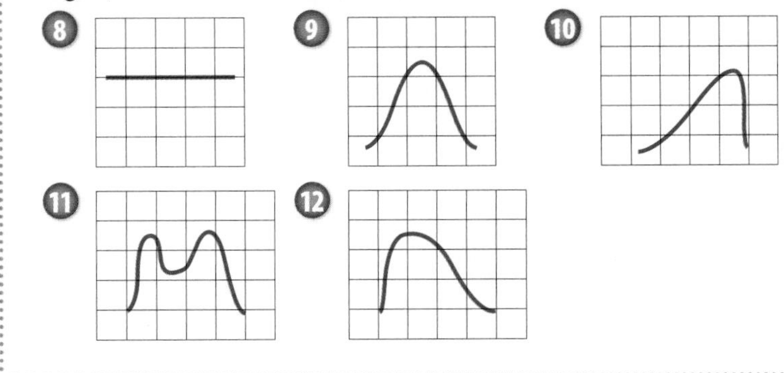

4•3 Exercises

1. Make a scatter plot of the following data.

Times at Bat	5	2	4	1	5	6	1	3	2	6
Hits	4	0	2	0	2	4	1	1	2	3

2. Describe the correlation in the scatter plot in Exercise 1.
3. Draw a line of best fit for the scatter plot in Exercise 1. Use it to predict the number of hits for 8 times at bat.

Describe the correlation in each of the following scatter plots.

4. **5.** **6.**

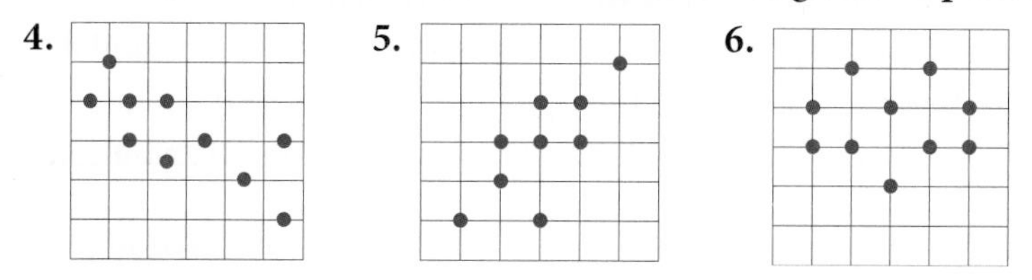

Tell whether each of the following distributions is normal, skewed to the right, skewed to the left, bimodal, or flat.

7.

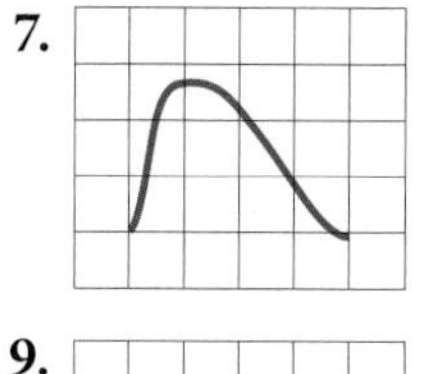

8.

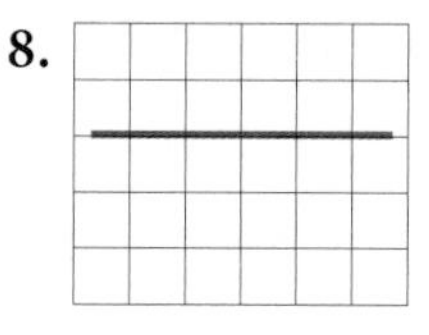

9.

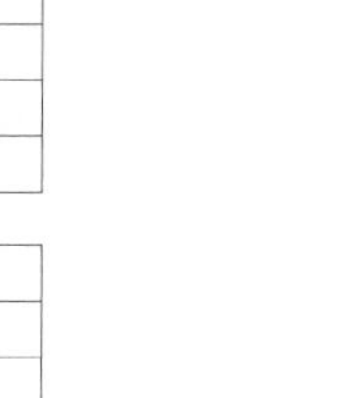

10.

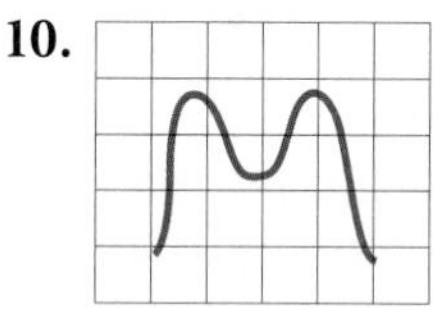

4•4 Statistics

Laila collected the following data about the amount her classmates spend on CDs each month.

$15, $15, $15, $15, $15, $15, $15
$25, $25, $25, $25, $25
$30, $30, $30
$45
$145

Laila said her classmates typically spend $15 per month, but Jacy disagreed. He said the typical amount was $25. A third classmate, Maria, said they were both wrong—the typical expenditure was $30. Each was correct because each was using a different common measure to describe the central tendency of the data.

Mean

One measure of central tendency of a set of data is the **mean**. To find the mean, or *average*, add the amounts the students spend and divide by the total number of amounts.

EXAMPLE Finding the Mean

Find the mean amount of money spent monthly on CDs by students in Laila's class.

$15 + $15 + $15 + $15 + $15 + $15 + $15 + $25 + $25 + $25 + $25 + $25 + $30 + $30 + $30 + $45 + $145 = $510

• Add the amounts.

In this case, there are 17 amounts.
$510 ÷ 17 = $30

• Divide the sum by the total number of amounts.

So, the mean amount each student spends on CDs is $30. Maria used the mean to describe the amounts when she said that each student typically spent $30.

Check It Out

Find the mean.

1. 15, 12, 6, 4.5, 12, 2, 11.5, 1, 8
2. 100, 79, 88, 100, 45, 92
3. The low temperatures in Pinetop the first week in February were 38°, 25°, 34°, 28°, 25°, 15°, and 24°. Find the mean temperature.
4. Ling averaged 86 points on five tests. What would she have to score on the sixth test to bring her average up one point?

Median

Another important measure of central tendency is the median. The **median** is the middle number in the data when the numbers are arranged in order from least to greatest. Recall the amounts spent on CDs.

$15, $15, $15, $15, $15, $15, $15
$25, $25, $25, $25, $25
$30, $30, $30
$45
$145

EXAMPLE Finding the Median

Find the median of amounts spent on CDs.

- Arrange the data in numerical order from least to greatest or greatest to least.

  Looking at the amounts spent on CDs, we can see that they are already arranged in order.

- Find the middle number.

  There are 17 numbers. The middle number is $25 because there are eight numbers above $25 and eight below it.

So, the median amount each student spends on CDs is $25.

Jacy was using the median when he said that the typical amount spent on CDs was $25.

When the number of amounts is even, you can find the median by finding the mean of the two middle numbers. To find the median of the numbers 1, 6, 4, 2, 5, and 8, you must find the two numbers in the middle.

EXAMPLE Finding the Median of an Even Number of Data

Find the median of the data 1 6 4 2 5 8.

1 2 4 5 6 8 or 8 6 5 4 2 1	• Arrange the numbers in order from least to greatest or greatest to least.
The two middle numbers are 4 and 5. $(4 + 5) \div 2 = 4.5$	• Find the mean of the two middle numbers.

So, the median is 4.5. Half the numbers are greater than 4.5 and half the numbers are less than 4.5.

Check It Out

Find the median.

5. 11, 15, 10, 7, 16, 18, 9
6. 1.4, 2.8, 5.7, 0.6
7. 11, 27, 16, 48, 25, 10, 18
8. The top ten scoring totals in the NBA are: 24,489; 31,419; 23,149; 25,192; 20,880; 20,708; 23,343; 25,389; 26,710; and 14,260 points. Find the median scoring total.

Mode

Another way to describe the central tendency of a set of numbers is the *mode.* The **mode** is the number in the set that occurs most frequently. Recall the amounts spent on CDs.

$15, $15, $15, $15, $15, $15, $15
$25, $25, $25, $25, $25
$30, $30, $30
$45
$145

To find the mode, group like numbers together and look for the one that appears most frequently.

EXAMPLE Finding the Mode

Find the mode of amounts spent on CDs.

Amount	Frequency
$ 15	7
$ 25	5
$ 30	3
$ 45	1
$145	1

- Arrange the numbers in order or make a frequency table of the numbers.

The most frequent amount spent is $15.

- Select the number that appears most frequently.

So, the mode of the amount each student spends on CDs is $15.

So Laila was using the mode when she said $15 was the typical amount students spent on CDs.

A group of numbers may have no mode or more than one mode. Data that have two modes is called *bimodal.*

Check It Out

Find the mode.

9. 1, 3, 3, 9, 7, 2, 7, 7, 4, 4
10. 1.6, 2.7, 5.3, 1.8, 1.6, 1.8, 2.7, 1.6
11. 2, 10, 8, 10, 4, 2, 8, 10, 6
12. The top 25 home run hitters of 1961 hit the following numbers of home runs in one season: 61, 49, 54, 49, 49, 60, 52, 50, 49, 52, 59, 54, 51, 49, 58, 54, 56, 54, 51, 52, 51, 49, 51, 58, 49. Find the mode.

APPLICATION Olympic Decimals

In Olympic gymnastics, the competitors perform a set of specific events. Scoring is based on a 10-point scale, where 10 is a perfect score.

For some of the events, gymnasts are judged on their technical merit and for composition and style.

	Technical Merit	Composition and Style
USA	9.4	9.8
China	9.6	9.7
France	9.3	9.9
Germany	9.5	9.6
Australia	9.6	9.7
Canada	9.5	9.6
Japan	9.7	9.8
Russia	9.6	9.5
Sweden	9.4	9.7
England	9.6	9.7

Marks for technical merit are based on the difficulty and variety of the routines and the skills of the gymnasts. Marks for composition and style are based on the originality and artistry of the routines.

Use these marks to determine the mean scores for technical merit and for composition and style. See **HotSolutions** for the answer.

Weighted Averages

When analyzing data where the numbers appear more than one time, the mean can be calculated using a weighted average. Consider the amounts spent on CDs by Laila's classmates:

$15, $25, $30, $45, $145

Since more people spent $15 on a CD than $45, a weighted average might give you a more accurate picture of the mean amount spent on CDs. A **weighted average** is where a data set is given different "weights."

EXAMPLE Finding the Weighted Average

Find the weighted average of amounts spent on CDs.

- Determine each amount and the number of times it occurs in the set.

  $15—7 times
  $25—5 times
  $30—3 times
  $45—1 time
  $145—1 time

- Multiply each amount by the number of times it occurs.

  $\$15 \cdot 7 = \$105$
  $\$25 \cdot 5 = \$125$
  $\$30 \cdot 3 = \$90$
  $\$45 \cdot 1 = \$45$
  $\$145 \cdot 1 = \$145$

- Add the products and divide by the total of the weights.

  $(\$105 + \$125 + \$90 + \$45 + \$145) \div (7 + 5 + 3 + 1 + 1)$
  $= \$510 \div 17 = \$30$

So, the weighted average spent on CDs was $30.

Check It Out

Find the weighted average.

13. 45 occurs 5 times, 36 occurs 10 times, and 35 occurs 15 times

14. The average number of checkout lanes in a Well-made department store is 8, and the average number in a Cost-easy store is 5. If there are 12 Well-made stores and 8 Cost-easy stores, find the average number of checkout lanes.

Measures of Variation

Measures of variation are used to describe the distribution or spread of a set of data.

Range

A measure of variation is the range. The **range** is the difference between the greatest and least numbers in a set. Consider the following miles of coastline along the Pacific Coast in the United States.

State	Miles of Coastline
California	1,200
Oregon	363
Washington	157
Hawaii	750
Alaska	6,640

To find the range, you subtract the least number of miles from the greatest.

EXAMPLE Finding the Range

Find the range of miles of Pacific coastline.

The greatest value is 5,580 mi and the least value is 157 mi. • Find the greatest and least values.

5,580 mi − 157 mi = 5,423 mi • Subtract.

So, the range is 5,423 miles.

Check It Out

Find the range.

15. 1.4, 2.8, 5.7, 0.6
16. 56°, 43°, 18°, 29°, 25°, 70°
17. The winning scores for the Candlelights basketball team are 78, 83, 83, 72, 83, 61, 75, 91, 95, and 72. Find the range in the scores.

Quartiles

Sometimes it is easier to summarize a set of data if you divide the set into equal-size groups. **Quartiles** are values that divide a set of data into four equal parts. A data set has three quartiles: the lower quartile, the median, and the upper quartile.

Remember that the *median* (p. 202) is the middle value of a set of data; therefore, there is an equal number of data points above and below the median. The **lower quartile** is the median of the lower half of the data set. The **upper quartile** is the median of the upper half of the data set. Suppose that you want to buy a cell phone. You can use quartiles to compare the prices of cell phones.

Prices of Cell Phones		
$30	$80	$250
$100	$40	$300
$120	$130	$350
$20	$150	$180

To find the quartiles, arrange the data in ascending order and divide the data into four equal parts. Then, separate the data into two equal parts by finding the median. Remember that if the data is an even amount of numbers, you find the mean of the two middle numbers. Then find the median of the lower half of the data. This is the *lower quartile.* Then find the median of the upper half of the data. This is the *upper quartile.*

EXAMPLE Finding Quartiles

Find the lower quartile of prices of cell phones.

- Arrange the data in numerical order from least to greatest and find the median of the set of data.

\$20 \$30 \$40 \$80 \$100 \$120 \$130 \$150 \$180 \$250 \$300 \$350

median

- Find the median of the lower half of the data.

\$20 \$30 \$40 \$80 \$100 \$120 \$130 \$150 \$180 \$250 \$300 \$350

$\frac{40 + 80}{2} = \$60$

The median of the lower half of the data is 60. So, the lower quartile is 60. Therefore, one fourth of the phones are priced at or below \$60.

Find the upper quartile of prices of cell phones.

- Find the median of the upper half of the data.

\$20 \$30 \$40 \$80 \$100 \$120 \$130 \$150 \$180 \$250 \$300 \$350

$\frac{180 + 250}{2} = \$215$

The median of the upper half of the data is 215. So, the upper quartile is 215. Therefore, one fourth of the phones are priced at or above \$215.

Check It Out

Find the lower quartile and upper quartile for each set of data.

18. 240, 253, 255, 270, 311
19. 73, 70, 66, 61, 60, 58, 58, 58, 57
20. 3.35, 3.38, 3.32, 3.12, 3.12, 3.13, 3.07, 3.07

Interquartile Range

Another measure of variation is the *interquartile range*. The **interquartile range** is the range of the middle half of the data. The interquartile range is a more stable measure than the range because the range depends on the greatest and least value. Also, the interquartile range is not affected by extremely large or small values.

To find the interquartile range, find the difference between the upper quartile and lower quartile. Recall the prices of cell phones.

EXAMPLE Finding the Interquartile Range

Find the interquartile range of prices of cell phones.

Interquartile range =
upper quartile − lower quartile
$155 = $215 − $60

- Subtract the lower quartile from the upper quartile.

So, the interquartile range is $155.

The price of cell phones range from approximately $60 to $215. Therefore, there is a difference of approximately $155 between the prices of cell phones.

Check It Out

Find the interquartile range for each set of data.

21. 240, 253, 255, 270, 311
22. 73, 70, 66, 61, 60, 58, 58, 58, 57
23. 3.35, 3.38, 3.32, 3.12, 3.12, 3.13, 3.07, 3.07

Outliers

The interquartile range can also be used to tell when data values are "too far" from the median. An **outlier** is a data value that is either much larger or much smaller than the median, which can affect measures used to interpret the data. The data value is considered an outlier if it is more than 1.5 times the interquartile range beyond either quartile.

Suppose in science class you are constructing a model bridge to measure how much weight it can hold. You want to determine which measure of variation, the mean or median, best describes the set of data. The following data is the weight held by the toothpick bridges.

2.3, 4.5, 5.6, 5.8, 6.4, 6.5, 7.2, 7.6, 7.8, 12.1

EXAMPLE **Finding the Outliers**

Find any outliers of weights held by toothpick bridges.

upper quartile = 7.6 lower quartile = 5.6	• Find the upper and lower quartiles.
$7.6 - 5.6 = 2$	• Find the interquartile range by subtracting the lower quartile from the upper quartile.
$2 \cdot 1.5 = 3$	• Multiply the interquartile range by 1.5.
$5.6 - 3 = 2.6$ $7.6 + 3 = 10.6$	• Find the outliers by subtracting 3 from the lower quartile and adding 3 to the upper quartile.
2.3 and 12.1	• Identify the data that falls below 2.6 and above 10.6.

So, the outliers of the data set are 2.3 and 12.1 because they are more than 1.5 times the interquartile range.

In this example, it is best to use the median as the measure of variance, since there are two outliers that skew the mean.

Check It Out

Find the outliers for each set of data.

24. 8, 12, 14, 16, 20, 2, 13, 13, 17, 17, 17, 18, 18
25. 42, 18, 17, 14, 12, 12, 8
26. 36.1, 9.0, 7.6, 6.4, 5.2, 4.0, 4.0, 2.9

4•4 Exercises

Find the mean, median, mode, and range.

1. 2, 2, 4, 4, 6, 6, 8, 8, 8, 8, 10, 10, 12, 14, 18
2. 5, 5, 5, 5, 5, 5, 5, 5, 5
3. 50, 80, 90, 50, 40, 30, 50, 80, 70, 10
4. 271, 221, 234, 240, 271, 234, 213, 253, 196
5. Are any of the sets of data above bimodal? Explain.
6. Find the weighted average: 15 occurs 3 times, 18 occurs 1 time, 20 occurs 5 times, and 80 occurs 1 time.
7. Kelly had 85, 83, 92, 88, and 69 on her first five math tests. She needs an average of 85 to get a B. What score must she get on her last test to get a B?
8. Which measure—the mean, median, or mode—must be a member of the set of data?
9. The following times represent the lengths of phone calls, in minutes, made by an eighth grader one weekend.

   10 2 16 8 55 2 18 11 9 5 4 7

   Find the mean, median, and mode of the calls. Which measure best represents the data? Explain.
10. The price of a house is higher than half of the other houses in the area. Would you use the mean, median, mode, or range to describe it?

For Exercises 11–14, use the data in the table at the right.

11. What is the range of the data?
12. Find the median, lower quartile, upper quartile, and the interquartile range for the data.
13. Identify any outliers.
14. Use the measures of variation to describe the data in the table.

Population of U.S. Cities	
Detroit, MI	918,849
San Francisco, CA	744,041
Columbus, OH	733,203
Austin, TX	709,893
Providence, RI	175,255

4•5 Combinations and Permutations

Tree Diagrams

A **tree diagram** is a diagram used to show the total number of possible outcomes in a probability experiment.

An **outcome** is any one of the possible results of an action. For example, there are 6 possible outcomes when a standard number cube is rolled. An **event** is an outcome or a collection of outcomes. An organized list of all possible outcomes is called a *sample space.*

You often need to count outcomes. For example, suppose you have two spinners. One spinner has equally-sized regions numbered 1 through 3 and the other spinner has equally-sized regions numbered 1 and 2. Suppose that you want to find the number of different two-digit numbers you can make by spinning the first spinner and then the second one. You can make a *tree diagram.*

To make a tree diagram, list the possible outcomes of the first spinner.

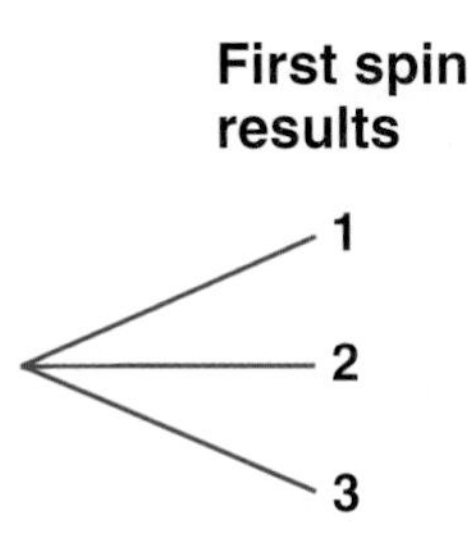

Then, to the right of each result, list the possible outcomes of the second spinner.

First spin results	Second spin results	Possible different numbers
1	1	1, 1
	2	1, 2
2	1	2, 1
	2	2, 2
3	1	3, 1
	2	3, 2

After listing all of the possibilities, you can count to see that there are six possible number combinations.

EXAMPLE Making a Tree Diagram

Make a tree diagram to find out how many possible ways three coins can land if you toss them into the air one at a time.

- List what happens with the first trial.

The first coin can come up heads or tails.

Head
Tail

- List what happens with the second and third (and so on) trials. List the results.

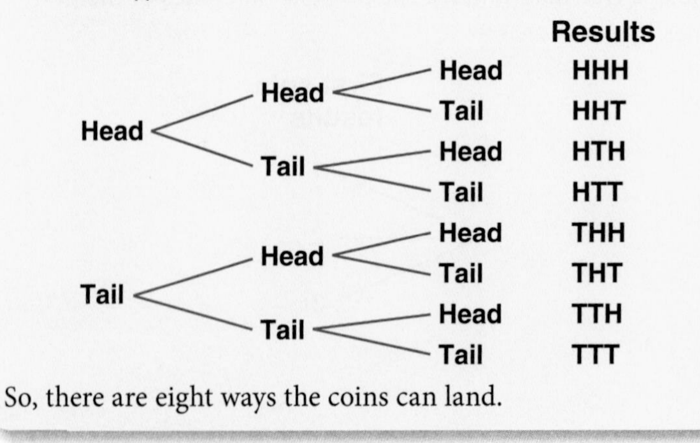

So, there are eight ways the coins can land.

You can also find the number of possibilities by multiplying the number of choices at each step. For the three coins problem, $2 \cdot 2 \cdot 2 = 8$ represents two possibilities for coin one, two possibilities for coin two, and two possibilities for coin three.

Check It Out

In Exercises 1–3, use multiplication to solve. Use a tree diagram if helpful.

1. If you toss three number cubes, each showing the numbers 1–6, how many possible three-digit numbers can you form?
2. How many possible routes are there from Creekside to Mountainville?

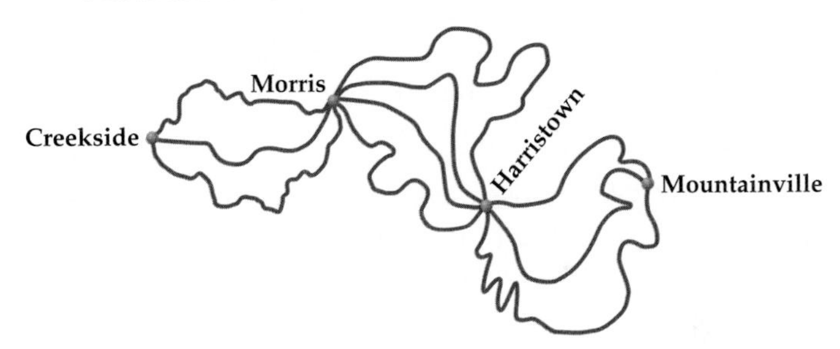

3. You are making cupcakes. Each cupcake is made with either chocolate or vanilla cake, and chocolate, vanilla, or strawberry frosting. Each cupcake also has either chopped nuts or sprinkles. How many different kinds of cupcakes can you make?

Permutations

You know that you can use a tree diagram to count all possible outcomes. A tree diagram also shows ways things can be arranged or listed. A listing in which the order is important is called a **permutation**. Suppose that you want to line up Rita, Jacob, and Zhao for a photograph. You can use a tree diagram to show all the different ways they could line up.

On left	In middle	On right	List
Rita	Jacob	Zhao	RJZ
	Zhao	Jacob	RZJ
Jacob	Rita	Zhao	JRZ
	Zhao	Rita	JZR
Zhao	Rita	Jacob	ZRJ
	Jacob	Rita	ZJR

There are 3 ways to choose the first person, 2 ways to choose the second, and 1 way to choose the third, so the total number of permutations is $3 \cdot 2 \cdot 1 = 6$. Remember that Rita, Jacob, Zhao is a different permutation from Zhao, Jacob, Rita.

$P(3, 3)$ represents the number of permutations of 3 things taken 3 at a time. Thus $P(3, 3) = 6$.

EXAMPLE **Finding Permutations**

Find $P(6, 5)$.

There are 6 choices for the first place, 5 for the second, 4 for the third, 3 for the fourth, and 2 for the fifth.
- Determine how many choices there are for each place.

$6 \cdot 5 \cdot 4 \cdot 3 \cdot 2 = 720$
- Find the product.

So, $P(6, 5) = 720$.

Factorial Notation

You saw that to find the number of permutations of 8 things, you find the product $8 \cdot 7 \cdot 6 \cdot 5 \cdot 4 \cdot 3 \cdot 2 \cdot 1$. The product $8 \cdot 7 \cdot 6 \cdot 5 \cdot 4 \cdot 3 \cdot 2 \cdot 1$ is called 8 **factorial**. The shorthand notation for a factorial is 8! So, $8! = 8 \cdot 7 \cdot 6 \cdot 5 \cdot 4 \cdot 3 \cdot 2 \cdot 1$.

Check It Out

Find each value.

4. $P(15, 2)$
5. $P(6, 6)$
6. The Grandview Middle School has a speech contest. There are 8 finalists. In how many different orders can the speeches be given?
7. One person from a class of 35 students is to be chosen as a delegate to Government Day, and another person is to be chosen as an alternate. In how many ways can a delegate and an alternate be chosen?

Find the value. Use a calculator if available.

8. 3!
9. 5!
10. 9!

Combinations

When you find the number of ways to select a delegate and an alternate from a class of 35, the order in which you select the students is important. Suppose, instead, that you simply pick two delegates. Then the order is not important. That is, choosing Elena and Rahshan is the same as choosing Rahshan and Elena when picking two delegates. An arrangement or listing in which order is not important is called a **combination**.

You can use the number of permutations to find the number of combinations. Say you want to select 2 students to be delegates from a group of 6 students (Elena, Rahshan, Felicia, Hani, Toshi, and Kelly).

First delegate	Second delegate	List
Elena	Rahshan	ER
	Felicia	EF
	Hani	EH
	Toshi	ET
	Kelly	EK
Rahshan	Elena	RE
	Felicia	RF
	Hani	RH
	Toshi	RT
	Kelly	RK
Felicia	Elena	FE
	Rahshan	FR
	Hani	FH
	Toshi	FT
	Kelly	FK
Hani	Elena	HE
	Rahshan	HR
	Felicia	HF
	Toshi	HT
	Kelly	HK
Toshi	Elena	TE
	Rahshan	TR
	Felicia	TF
	Hani	TH
	Kelly	TK
Kelly	Elena	KE
	Rahshan	KR
	Felicia	KF
	Hani	KH
	Toshi	KT

These repeat the same delegates since order does not matter. (ER, RE)

To find the number of combinations of six students taken two at a time, you start by finding the permutations. You have six ways to choose the first delegate and five ways to choose the second, so this is $6 \cdot 5 = 30$. But the order does not matter, so some combinations are counted more than once. Therefore, you need to divide by the number of different ways the two delegates can be arranged (2!).

$$C(6, 2) = \frac{P(6, 2)}{2!} = \frac{6 \cdot 5}{2 \cdot 1} = 15$$

EXAMPLE Finding Combinations

Find $C(6, 3)$.

$P(6, 3) = 6 \cdot 5 \cdot 4 = 120$ • Find the number of permutations.

$\frac{120}{3!} = \frac{120}{3 \cdot 2 \cdot 1} = 20$ • Divide by the number of ways the objects can be arranged.

So, $C(6, 3) = 20$.

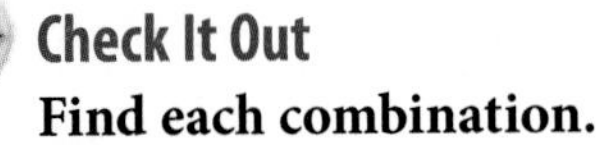

Check It Out

Find each combination.

11. $C(9, 6)$
12. $C(14, 2)$
13. How many different combinations of three plants can you choose from a dozen plants?
14. Are there more combinations or permutations of two books from a total of four? Explain.

4•5 Exercises

1. Make a tree diagram to show the results when you toss a coin and roll a number cube containing the numbers 1 through 6.
2. Write all the combinations of the digits 3, 5, and 7, using only two numbers at a time.

Find each value.

3. $P(7, 5)$
4. $C(8, 8)$
5. $P(9, 4)$
6. $C(7, 3)$
7. $5! \cdot 4!$
8. $P(8, 8)$
9. $P(4, 3)$

Solve.

10. Eight friends want to play enough games of tennis (singles) to be sure that everyone plays everyone else. How many games will they have to play?
11. At a chess tournament, trophies are given for first, second, third, and fourth places. Twenty students enter the tournament. How many different arrangements of four winning students are possible?
12. Determine whether the following is a permutation or a combination.
    a. choosing a team of 5 players from 20 people
    b. arranging 12 people in a line for a photograph
    c. choosing first, second, and third places from 20 show dogs

4•6 Probability

The **probability** of an event is a number from 0 to 1 that measures the chance that an event will occur.

Experimental Probability

One way to find the probability of an event is to conduct an experiment. Suppose that a pair of dice (one red, one blue) is cast 20 times, and on 6 of the occasions, the sum of the numbers facing up is 9. You compare the number of times the sum equals 9 to the number of times you cast the dice to find the probability. In this case, the **experimental probability** of the outcome 9 is $\frac{6}{20}$ *or* $\frac{3}{10}$.

EXAMPLE **Determining Experimental Probability**

Find the experimental probability of drawing a red marble from a bag of 10 colored marbles.

- Conduct an experiment. Record the number of trials and the result of each trial.

  Choose a marble from the bag, record its color, and replace it. Repeat 10 times. Suppose that you draw red, green, blue, green, red, blue, blue, red, green, blue.

- Compare the number of occurrences of one result to the number of trials. That is the probability for that result.

  Compare the number of red marbles to the total number of draws.

So, the experimental probability of drawing a red marble is $\frac{3}{10}$.

Check It Out

Three pennies are tossed 100 times. The results are shown on the circle graph.

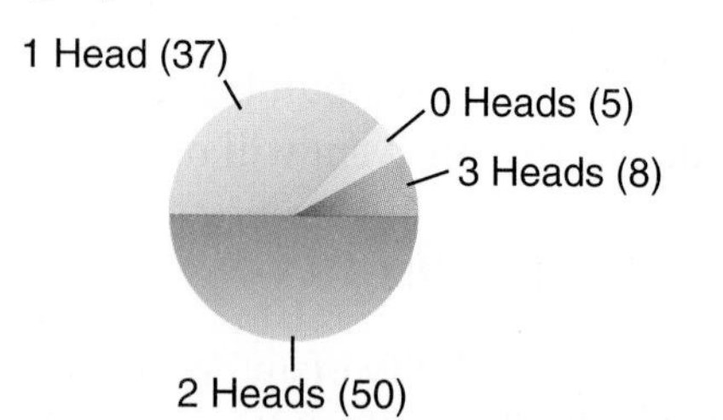

1. Find the experimental probability of getting two heads.
2. Find the experimental probability of getting no heads.
3. Drop a flat-head thumbtack 50 times and record the number of times it lands point up. Find the experimental probability of the tack landing point up. Compare your answers with other student answers.

Theoretical Probability

You know that you can find the experimental probability of tossing a head when you toss a coin by doing the experiment and recording the results. You can also find the **theoretical probability**, the probability based on known characteristics or facts.

$$P(\text{event}) = \frac{\text{number of ways an event occurs}}{\text{number of outcomes}}$$

For example, the outcomes when tossing a coin are head and tail. An event is a specific outcome, such as heads. So the probability of getting a head $P(H) = \frac{\text{number of heads}}{\text{number of outcomes}} = \frac{1}{2}$.

PROBABILITY 4·6

EXAMPLE Determining Theoretical Probability

Find the theoretical probability of drawing a red marble from a bag containing 5 red, 8 blue, and 7 white marbles.

- Determine the number of ways the event occurs.

  In this case, the event is drawing a red marble. There are 5 red marbles.

- Determine the total number of outcomes. Use a list, multiply, or make a *tree diagram* (p. 213).

  There are 20 marbles in the bag.

- Use the formula.

$$P(\text{event}) = \frac{\text{number of ways an event occurs}}{\text{number of outcomes}}$$

- Find the probability of the target event.

  In this case, drawing a red marble is represented by $P(\text{red})$.

  $P(\text{red}) = \frac{5}{20} = \frac{1}{4}$

So, the probability of drawing a red marble is $\frac{1}{4}$.

Check It Out

Find each probability. Use the spinner for Exercises 4 and 5.

4. $P(\text{even number})$
5. $P(\text{number greater than 10})$

6. $P(2)$ when tossing a number cube
7. The letters of the word *Mississippi* are written on identical slips of paper and placed in a box. If you draw a slip at random, what is the probability that it will be a vowel?

Expressing Probabilities

You can express a probability as a fraction, as shown before. But, just as you can write a fraction as a decimal, ratio, or percent, you can also write a probability in any of those forms (p. 125).

The probability of getting a head when you toss a coin is $\frac{1}{2}$. You can also express the probability as follows:

Fraction	Decimal	Ratio	Percent
$\frac{1}{2}$	0.5	1:2	50%

Check It Out

Express each of the following probabilities as a fraction, decimal, ratio, and percent.

8. the probability of drawing a red marble from a bag containing 4 red marbles and 12 green ones
9. the probability of getting an 8 when spinning a spinner divided into eight equal divisions numbered 1 through 8
10. the probability of getting a green gumball out of a machine containing 25 green, 50 red, 35 white, 20 black, 5 purple, 50 blue, and 15 orange gumballs
11. the probability of being chosen to do your oral report first if your teacher puts all 25 students' names in a bag and draws

APPLICATION Lottery Fever

You read the headline. You say to yourself, "Somebody's *bound* to win this time." But the truth is, you would be wrong! The chances of winning a Pick-6 lottery are always the same, and very, very, very small.

Start with the numbers from 1 to 7. There are always 7 different ways to choose 6 out of 7 things. (Try it for yourself.) So, your chances of winning a 6-out-of-7 lottery would be $\frac{1}{7}$ or about 14.3%. Suppose you try using 6 out of 10 numbers. There are 210 different ways you can do that, making the likelihood of winning a 6-out-of-10 lottery $\frac{1}{210}$ or 0.4%. For a 6-out-of-20 lottery, there are 38,760 possible ways to pick 6 numbers, and only 1 of these would be the winner. That's about a 0.003% chance of winning. Get the picture?

The chances of winning a 6-out-of-50 lottery are 1 in 15,890,700 or 1 in about 16 million. For comparison, think about the chances that you will get struck by lightning—a rare occurrence. It is estimated that in the U.S. roughly 260 people are struck by lightning each year. Suppose the population of the U.S. is about 260 million. Would you be more likely to win the lottery or be struck by lightning? See **HotSolutions** for the answer.

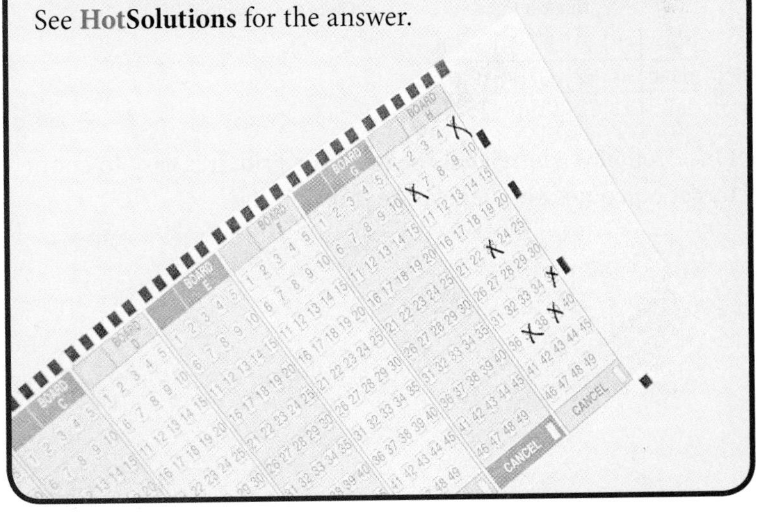

Outcome Grids

Another way to show the possible outcomes of an experiment is to use an *outcome grid*. The following outcome grid shows the outcomes when rolling two number cubes and observing the sum of the two numbers.

1st Number Cube \ 2nd Number Cube	1	2	3	4	5	6
1	2	3	4	5	6	7
2	3	4	5	6	7	8
3	4	5	6	7	8	9
4	5	6	7	8	9	10
5	6	7	8	9	10	11
6	7	8	9	10	11	12

You can use the grid to find the sum that occurs most often, which is 7.

EXAMPLE **Making Outcome Grids**

Make an outcome grid to show the results of tossing a coin and rolling a number cube.

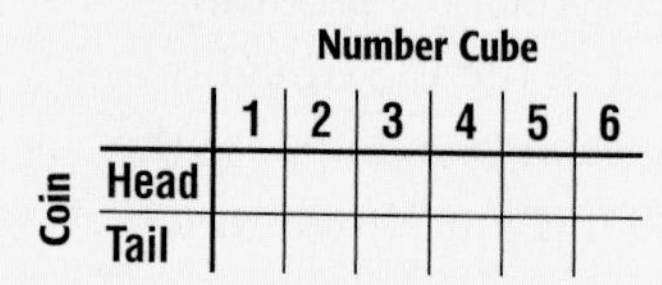

Coin \ Number Cube	1	2	3	4	5	6
Head						
Tail						

- List the outcomes of the first type down the side. List the outcomes of the second type across the top.

Coin \ Number Cube	1	2	3	4	5	6
Head	H1	H2	H3	H4	H5	H6
Tail	T1	T2	T3	T4	T5	T6

- Fill in the outcomes.

Once you have completed the outcome grid, it is easy to count target outcomes and determine probabilities.

Check It Out

12 Make an outcome grid to show the two-letter outcomes when spinning the spinner twice.

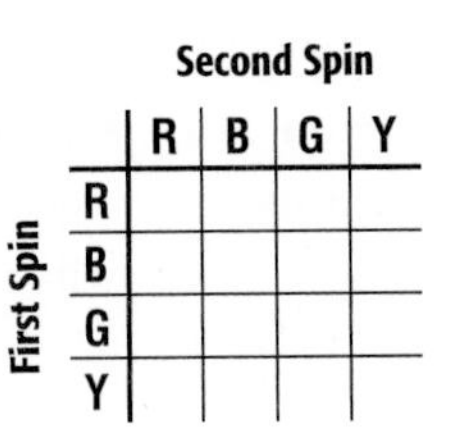

13 What is the probability of getting green as one color when you spin the spinner in Exercise 12 twice?

Probability Line

You know that the probability of an event is a number from 0 to 1. One way to show probabilities and how they relate to each other is to use a *probability line*. The following probability line shows the possible ranges of probability values.

Impossible Event	Equally Likely Events	Certain Event
0	$\frac{1}{2}$	1

The line shows that events which are certain have a probability of 1. Such an event is the probability of getting a number between 0 and 7 when rolling a standard number cube. An event that cannot happen has a probability of zero. The probability of getting an 8 when spinning a spinner that shows 0, 2, and 4 is 0. Events that are equally likely, such as getting a head or a tail when you toss a coin, have a probability of $\frac{1}{2}$.

EXAMPLE Showing Probability on a Probability Line

Suppose you roll two standard number cubes. Show the probabilities of rolling a sum of 4 and a sum of 7 on a probability line.

From the outcome grid on page 226, you can see that there are 3 sums of 4, and 6 sums of 7, out of 36 possible sums.

- Calculate the probabilities of the given events.

$P(\text{sum of 4}) = \frac{3}{36} = \frac{1}{12}$

$P(\text{sum of 7}) = \frac{6}{36} = \frac{1}{6}$

P(sum of 4) P(sum of 7)

0 $\frac{1}{12}$ $\frac{1}{6}$ $\frac{1}{2}$ 1

- Draw a number line and label it from 0 to 1. Plot the probabilities on the number line.

Check It Out

Draw a probability line for each event.

14. the probability of tossing a tail on one flip of a coin
15. the probability of rolling a 1 or a 2 on one roll of a die
16. the probability of being chosen if there are four people and an equal chance of any of them being chosen
17. the probability of getting a green gumball out of the machine if there are 25 each of green, yellow, red, and blue gumballs

Dependent and Independent Events

If you toss a coin and roll a number cube, the result of one does not affect the other. These are examples of *independent events.* For **independent events**, the outcome of one event does not affect the other event. To find the probability that we get a head and then a 5, you can find the probability of each event and then multiply. The probability of getting a head is $\frac{1}{2}$ and the probability of getting a 5 on a roll of the number cube is $\frac{1}{6}$. So the probability of getting a head and a 5 is $\frac{1}{2} \cdot \frac{1}{6} = \frac{1}{12}$.

Suppose that you have 4 oatmeal and 6 raisin cookies in a bag. The probability that you get an oatmeal cookie if you choose a cookie at random is $\frac{4}{10} = \frac{2}{5}$. However, once you have taken an oatmeal cookie out, there are only 9 cookies left, 3 of which are oatmeal. So the probability that a friend picks an oatmeal cookie once you have drawn is $\frac{3}{9} = \frac{1}{3}$. These events are called **dependent events** because the probability of one depends on the other.

In the case of dependent events, you still multiply to get the probability of both events. So the probability that your friend gets an oatmeal cookie after you have picked one is $\frac{2}{5} \cdot \frac{1}{3} = \frac{2}{15}$.

To find the probability of dependent and independent events:

- Find the probability of the first event.
- Find the probability of the second event.
- Find the product of the two probabilities.

Check It Out

Find the probability. Then determine whether the events are dependent or independent.

18. Find the probability of getting an even number and an odd number if you roll two number cubes. Are the events dependent or independent?
19. You draw two marbles from a bag containing six red marbles and fourteen white marbles. What is the probability that you get two white marbles? Are the events dependent or independent?

Sampling With and Without Replacement

If you draw a card from a deck of cards, the probability that it is an ace is $\frac{4}{52}$, or $\frac{1}{13}$. If you put the card back in the deck and draw another card, the probability that it is an ace is still $\frac{1}{13}$, and the events are independent. This is called **sampling with replacement**.

If you do not put the card back in, the probability of drawing an ace the second time depends on what you drew the first time. If you drew an ace, there will be only three aces left of 51 cards, so the probability of drawing a second ace will be $\frac{3}{51}$, or $\frac{1}{17}$. In sampling without replacement, the events are dependent.

Check It Out

Find the probability for each event.

20. You draw a card from a deck of cards and then put it back. Then you draw another card. What is the probability you get a spade and then a heart?
21. Answer the question again if you do not replace the card.
22. There are 8 balls in a box, 4 black and 4 white. If you draw a black ball out, what is the probability that the next ball will be black?
23. You have a bag of 5 yellow marbles, 6 blue marbles, and 4 red marbles. If two marbles are drawn one right after the other, and not replaced, what is the probability that each marble is either red or yellow?

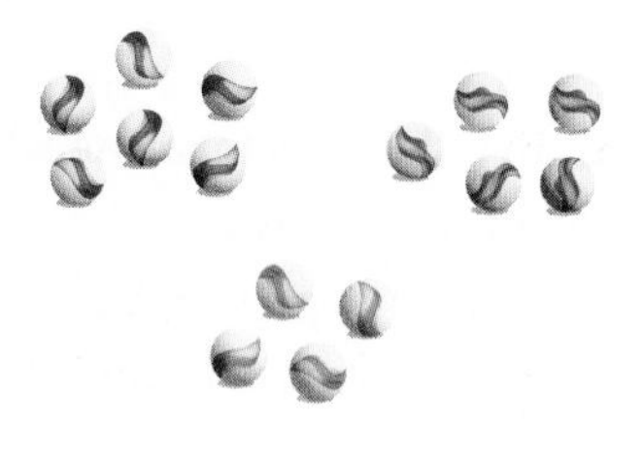

4•6 Exercises

Use the spinner shown for Exercises 1 and 2. Find each probability as a fraction, decimal, ratio, and percent.

1. $P(4)$
2. $P(\text{odd number})$
3. If you toss a coin 48 times and get 26 heads, what is the probability of getting a head? Is this experimental or theoretical probability?
4. If you roll a number cube, what is the probability of getting a 6? Is this experimental or theoretical probability?
5. Draw a probability line to show the probability of getting a number greater than 6 when rolling a number cube numbered 1 through 6.
6. Make an outcome grid to show the outcomes of spinning two spinners divided into four equal sections labeled 1 through 4.
7. Find the probability of drawing two red kings from a deck of cards if you replace the card between drawings.
8. Find the probability of drawing two red kings from a deck of cards if you do not replace the cards between drawings.
9. Look again at Exercises 7 and 8. In which exercise are the events dependent?
10. You want to choose a volleyball team from a combined group of 11 boys and 13 girls. The team consists of 6 players. What is the probability of picking a girl second if a boy was picked first?

Data, Statistics, and Probability

What have you learned?

You can use the problems and the list of words that follow to see what you learned in this chapter. You can find out more about a particular problem or word by referring to the topic number (*for example,* Lesson 4•2).

Problem Set

1. Taking a survey at the mall, Salvador asked, "What do you think of the beautiful new landscaping at the mall?" Was the question biased or unbiased? (Lesson 4•1)

Use the following box plot to answer Exercises 2–4. (Lesson 4•2)

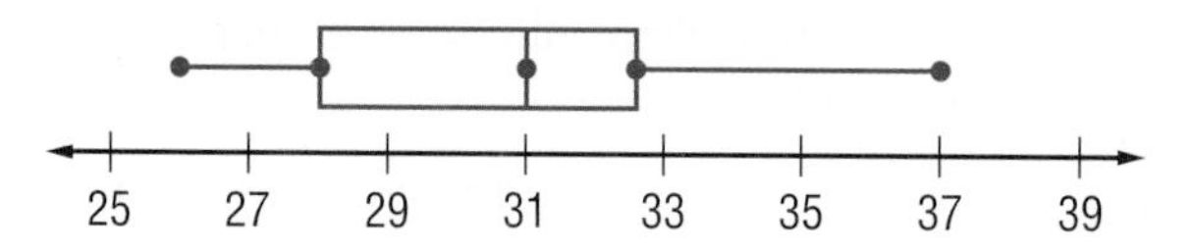

2. According to the box plot, what is the greatest miles per gallon on the highway you can expect from a vehicle?
3. What is the median highway miles per gallon?
4. What percent of vehicles get less than 31 miles per gallon on the highway?

Use this information for Exercises 5 and 6. A bookstore manager compared the prices of 100 new books to the number of pages in each book to see if there was a relationship between them. For each book, the manager made an ordered pair of the form (number of pages, price). (Lesson 4•3)

5. What kind of graph will these data make?
6. On the graph, many of these 100 points seem to lie on a straight line. What is this line called?

WHAT HAVE YOU LEARNED? 4

7. Find the mean, median, mode, and range of the numbers 42, 43, 19, 16, 16, 36, and 17. (Lesson 4•4)
8. $C(6, 3) =$ ____ (Lesson 4•5)

Use the following information to answer Exercises 9 and 10. A bag contains 4 red, 3 blue, 2 green, and 1 black marble. (Lesson 4•6)

9. One marble is drawn. What is the probability that it is red?
10. Three marbles are drawn. What is the probability that 2 are black and 1 is green?

HotWords

Write definitions for the following words.

biased sample (Lesson 4•1)
bimodal distribution (Lesson 4•3)
box plot (Lesson 4•2)
combination (Lesson 4•5)
correlation (Lesson 4•3)
dependent events (Lesson 4•6)
event (Lesson 4•5)
experimental probability (Lesson 4•6)
factorial (Lesson 4•5)
flat distribution (Lesson 4•3)
histogram (Lesson 4•2)
independent events (Lesson 4•6)
interquartile range (Lesson 4•4)
line of best fit (Lesson 4•3)
lower quartile (Lesson 4•4)
mean (Lesson 4•4)
measures of variation (Lesson 4•4)
median (Lesson 4•4)
mode (Lesson 4•4)
normal distribution (Lesson 4•3)
outcome (Lesson 4•5)
outlier (Lesson 4•4)
permutation (Lesson 4•5)
population (Lesson 4•1)
probability (Lesson 4•6)
quartiles (Lesson 4•4)
random sample (Lesson 4•1)
range (Lesson 4•4)
sample (Lesson 4•1)
sampling with replacement (Lesson 4•6)
scatter plot (Lesson 4•3)
skewed distribution (Lesson 4•3)
stem-and-leaf plot (Lesson 4•2)
theoretical probability (Lesson 4•6)
tree diagram (Lesson 4•5)
upper quartile (Lesson 4•4)
weighted average (Lesson 4•4)

HotTopic 5

Logic

What do you know?

You can use the problems and the list of words that follow to see what you already know about this chapter. The answers to the problems are in **HotSolutions** at the back of the book, and the definitions of the words are in **HotWords** at the front of the book. You can find out more about a particular problem or word by referring to the topic number (*for example,* Lesson 5·2).

5

LOGIC

Problem Set

Tell whether each statement is *true* or *false*.

1. You form the inverse of a conditional statement by switching the hypothesis and the conclusion. (Lesson 5·1)
2. If a conditional statement is true, then its related converse is always false. (Lesson 5·1)
3. Every set is a subset of itself. (Lesson 5·3)
4. A counterexample shows that a statement is false. (Lesson 5·2)
5. You form the union of two sets by combining all the elements in both sets. (Lesson 5·3)
6. The intersection of two sets can be the empty set. (Lesson 5·3)

Write each conditional in if/then form. (Lesson 5·1)

7. The jet flies to Belgium on Tuesday.
8. The bank is closed on Sunday.

Write the converse of each conditional statement. (Lesson 5·1)

9. If $x = 7$, then $x^2 = 49$.
10. If an angle has a measure less than 90°, then the angle is acute.

Write the negation of each statement. (Lesson 5•1)

11. The playground will close at sundown.

12. These two lines form an angle.

Write the inverse of the conditional statement. (Lesson 5•1)

13. If two lines intersect, then they form four angles.

Write the contrapositive of the conditional statement. (Lesson 5•1)

14. If a pentagon has five equal sides, then it is equilateral.

Find a counterexample that shows that each of these statements is false. (Lesson 5•2)

15. Tuesday is the only day of the week that begins with the letter T.

16. The legs of a trapezoid are equal.

Use the Venn diagram for Exercises 17–20. (Lesson 5•3)

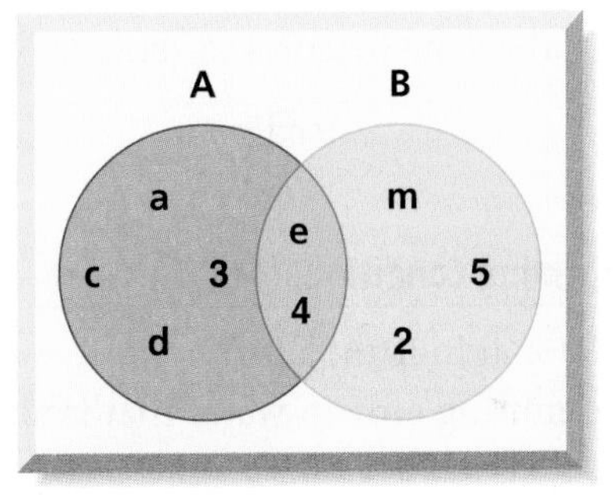

17. List the elements in set A.

18. List the elements in set B.

19. Find A ∪ B.

20. Find A ∩ B.

contraposititive (Lesson 5•1)
converse (Lesson 5•1)
counterexample (Lesson 5•2)
intersection (Lesson 5•3)
inverse (Lesson 5•1)
set (Lesson 5•3)
union (Lesson 5•3)
Venn diagram (Lesson 5•3)

5•1 If/Then Statements

Conditional Statements

A *conditional statement* is a statement that you can express in *if/then* form. The *if* part of a conditional is the *hypothesis,* and the *then* part is the *conclusion.* Often you can rewrite a statement that contains two related ideas as a conditional by making one of the ideas the hypothesis and the other the conclusion.

Statement: All members of the varsity swim team are seniors.

The conditional statement:

hypothesis

If a person is a varsity swim team member,
then the person is a senior.

conclusion

EXAMPLE Forming Conditional Statements

Write this conditional in if/then form:

Julie goes swimming only in water that is above 80°F.

- Find the two ideas.

  (1) Julie goes swimming. (2) Water is above 80°F.

- Decide which idea will be the hypothesis and which will be the conclusion.

  Hypothesis: Julie goes swimming.

  Conclusion: Water is above 80°F.

- Place the hypothesis in the *if* clause and the conclusion in the *then* clause. If necessary, add words so that your sentence makes sense.

  If Julie goes swimming, then the water is above 80°F.

Check It Out

Write each statement in if/then form.

1. Perpendicular lines meet to form right angles.
2. An integer that ends in 0 or 5 is a multiple of 5.
3. Runners participate in marathons.

Converse of a Conditional

When you switch the hypothesis and conclusion in a conditional statement, you form a new statement called the **converse**.

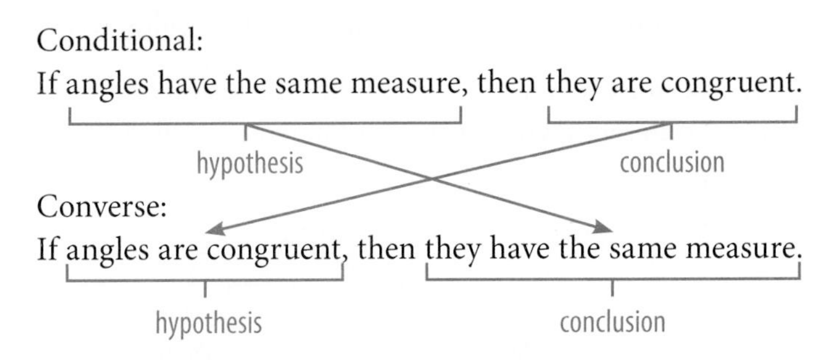

The converse of a conditional may or may not have the same *truth value* as the conditional on which it is based. In other words, the truth of the converse can be logically unrelated to the truth of the initial statement.

Check It Out

Write the converse of each conditional.

4. If an integer ends with 1, 3, 5, 7, or 9, then the integer is odd.
5. If Jacy is 15 years old, then he is too young to vote.
6. If it is raining, then you will see a cumulus cloud.

Negations and the Inverse of a Conditional

A *negation* of a given statement has the opposite truth value of the given statement. This means that if the given statement is true, the negation is false; if the given statement is false, the negation is true.

Statement: A square is a quadrilateral. (true)

Negation: A square is not a quadrilateral. (false)

Statement: A pentagon has four sides. (false)

Negation: A pentagon does not have four sides. (true)

When you negate the hypothesis and the conclusion of a conditional statement, you form a new statement called the **inverse**.

Conditional: If $3x = 6$, then $x = 2$.

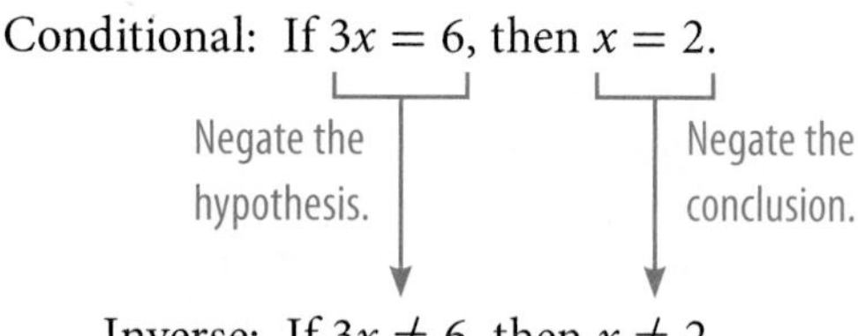

Inverse: If $3x \neq 6$, then $x \neq 2$.

The inverse of a conditional may or may not have the same truth value as the conditional.

Check It Out

Write the negation of each statement.

7. A rectangle has four sides.
8. The donuts were eaten before noon.

Write the inverse of each conditional.

9. If an integer ends with 0 or 5, then it is a multiple of 5.
10. If I am in Seattle, then I am in the state of Washington.

Contrapositive of a Conditional

You form the **contrapositive** of a conditional when you negate the hypothesis and conclusion and then interchange them.

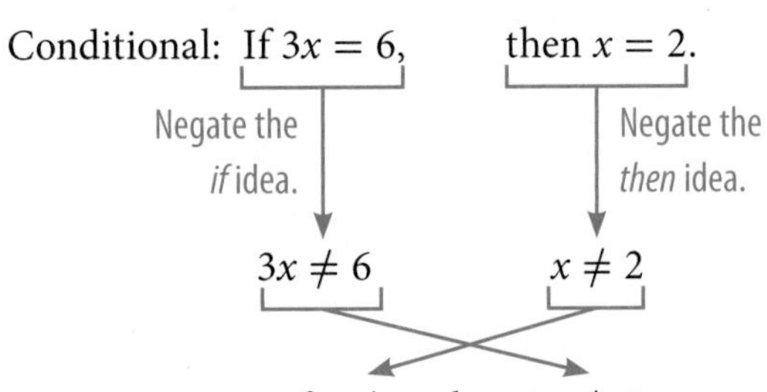

Contrapositive: If $x \neq 2$, then $3x \neq 6$.

The contrapositive of a conditional statement has the same truth value as the conditional.

Check It Out

Tell whether the statement is *true* or *false*.

11. If a conditional statement is true, then its related contrapositive is always true.

Write the contrapositive of each conditional.

12. If an angle has a measure of 90°, then the angle is a right angle.
13. If $x \neq 3$, then $2x \neq 6$.
14. If it snows, then school will be canceled.
15. If you are over 12 years old, then you buy an adult ticket.
16. If you bought your tickets in advance, then you paid less.

5•1 Exercises

Write each conditional in if/then form.

1. Perpendicular lines form right angles.
2. Positive integers are greater than zero.
3. Everyone in that town voted in the last election.
4. Equilateral triangles have three equal sides.
5. Numbers that end with 0, 2, 4, 6, or 8 are even numbers.
6. Elena visits her aunt every Friday.

Write the converse of each conditional.

7. If a triangle is equilateral, then it is isosceles.
8. If Chenelle is over 21, then she can vote.
9. If a number is a factor of 8, then it is a factor of 24.
10. If $x = 4$, then $3x = 12$.

Write the negation of each statement.

11. All the buildings are three stories tall.
12. x is a multiple of y.
13. The lines in the diagram intersect at point P.
14. A triangle has three sides.

Write the inverse of each conditional.

15. If $5x = 15$, then $x = 3$.
16. If the weather is good, then I will drive to work.

Write the contrapositive of each conditional.

17. If $x = 6$, then $x^2 = 36$.
18. If the perimeter of a square is 8 inches, then each side length is 2 inches.

For each conditional, write the converse, inverse, and contrapositive.

19. If a rectangle has a length of 4 feet and a width of 2 feet, then its perimeter is 12 feet.
20. If a triangle has three sides of different lengths, then it is scalene.

5•2 Counterexamples

Counterexamples

In the fields of logic and mathematics, if/then statements are either true or false. To show that a statement is false, find one example that agrees with the hypothesis but not with the conclusion. Such an example is called a **counterexample**.

When reading the conditional statement below, you may be tempted to think that it is true.

If a polygon has four equal sides, then it is a square. The statement is false, however, because there is a counterexample—the rhombus. A rhombus agrees with the hypothesis (it has four equal sides), but it does not agree with the conclusion (a rhombus is not a square).

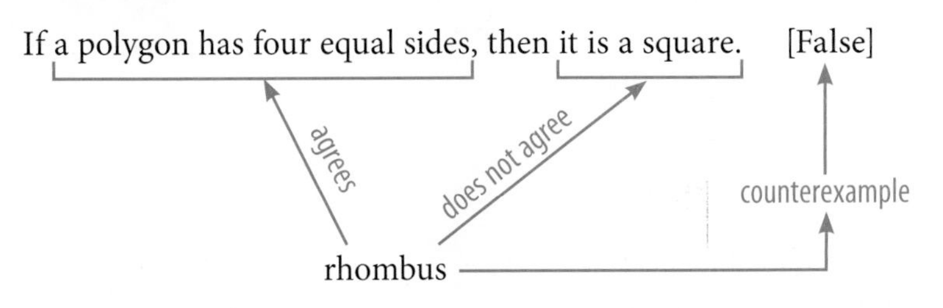

Check It Out

Tell whether each statement and its converse are *true* or *false*. If a statement is false, give a counterexample.

1. Statement: If two lines in the same plane are parallel, then they do not intersect.
   Converse: If two lines in the same plane do not intersect, then they are parallel.
2. Statement: If an angle has a measure of 90°, then it is a right angle.
   Converse: If an angle is a right angle, then it has a measure of 90°.

APPLICATION 150,000. . ., But Who's Counting

Do you think there are two people in the United States with exactly the same number of hairs on their head? In fact, you can prove that there are—not by counting the hairs on everyone's head, but by logic.

Consider these statements:

A. At the maximum, there are approximately 150,000 hairs on the human scalp.

B. The population of the United States is greater than 150,000.

Because statements A and B are both true, then there are two people in the United States with exactly the same number of hairs on their head.

Here's how to think about this. If you did count hairs, the first 150,000 people could each have a different number. Person 1 could have 1 hair; person 2, 2 hairs, and so on to 150,000. Person 150,001 would have to have a number of hairs between 1 and 150,000, and this would be a duplicate of one of the heads you have already counted.

Can you prove that there are two people in your town with the same number of hairs on their head?

See **HotSolutions** for the answer.

5•2 Exercises

Find a counterexample that shows that each statement is false.

1. If a number is a factor of 18, then it is a factor of 24.
2. If a figure is a quadrilateral, then it is a parallelogram.
3. If $x + y$ is an even number, then x and y are even numbers.

Tell whether each conditional is *true* or *false*. If false, give a counterexample.

4. If a number is prime, then it is an odd number.
5. If xy is an odd number, then both x and y are odd.
6. If you draw a line through a square, then you form two triangles.

Tell whether each statement and its converse are *true* or *false*. If false, give a counterexample.

7. Statement: If two angles have measures of 30°, then the angles are congruent.
   Converse: If two angles are congruent, then they have measures of 30°.
8. Statement: If $6x = 54$, then $x = 9$.
   Converse: If $x = 9$, then $6x = 54$.

Tell whether the statement and its inverse are *true* or *false*. If false, give a counterexample.

9. Statement: If an angle has a measure of 120°, then it is an obtuse angle.
   Inverse: If an angle does not have a measure of 120°, then it is not an obtuse angle.

Tell whether the statement and its contrapositive are *true* or *false*. If false, give a counterexample.

10. Statement: If a triangle is isosceles, then it is equilateral.
    Contrapositive: If a triangle is not equilateral, then it is not isosceles.

11. Write your own false conditional, and then give a counterexample that shows it is false.

5•3 Sets

Sets and Subsets

A **set** is a collection of objects. Each object is called a *member* or *element* of the set. Sets are often named with capital letters.

A = {1, 2, 3, 4} B = {a, b, c, d}

When a set has no elements, it is the *empty set.* You write { } or ∅ to indicate the empty set.

When all the elements of a set are also elements of another set, the first set is a *subset* of the other set.

{2, 4} ⊂ {1, 2, 3, 4} (⊂ is the subset symbol.)

Remember that every set is a subset of itself and that the empty set is a subset of every set.

Check It Out

Tell whether each statement is *true* or *false*.

1. {5} ⊂ {even numbers}
2. ∅ ⊂ {3, 5}
3. {2} ⊂ {2}

Find all the subsets of each set.

4. {1, 4}
5. {m}
6. {a, b, c}

Union of Sets

You find the **union** of two sets by creating a new set that contains all of the elements from the two sets.

J = {1, 3, 5, 7} L = {2, 4, 6, 8}

J ∪ L = {1, 2, 3, 4, 5, 6, 7, 8} (∪ is the union symbol.)

When the sets have elements in common, list the common elements only once in the intersection.

P = {r, s, t, v} Q = {a, k, r, t, w}

P ∪ Q = {a, k, r, s, t, v, w}

Check It Out

Find the union of each pair of sets.

7. {1, 2} ∪ {9, 10}
8. {m, a, t, h} ∪ {m, a, p}
9. {★, $, ♪, %, ▲, ∞} ∪ {∞, %, $, #}

Intersection of Sets

You find the **intersection** of two sets by creating a new set that contains all of the elements that are common to both sets.

A = {8, 12, 16, 20}

B = {4, 8, 12}

A ∩ B = {8, 12}   (∩ is the intersection symbol.)

If the sets have no elements in common, the intersection is the empty set ∅.

Check It Out

Find the intersection of each pair of sets.

10. {9} ∩ {9, 18}
11. {a, c, t} ∩ {b, d, u}
12. {★, $, ♪, %, ▲, ∞} ∩ {∞, %, $, #}

Venn Diagrams

A **Venn diagram** shows how the elements of two or more sets are related. When the circles in a Venn diagram overlap, the overlapping part contains the elements that are common to both sets.

When evaluating Venn diagrams, you have to look carefully to identify the overlapping parts to see which elements of the sets are in those parts. The white part of the diagram shows where all three sets overlap one another.

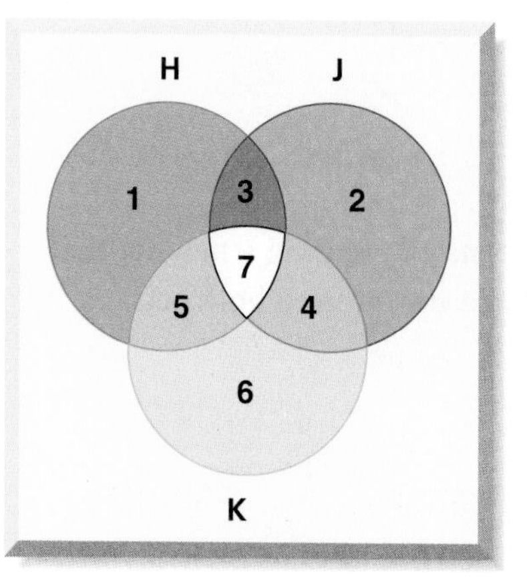

H = {1, 3, 5, 7}
J = {2, 3, 4, 7}
K = {4, 5, 6, 7}

H ∪ J = {1, 2, 3, 4, 5, 7}
H ∪ K = {1, 3, 4, 5, 6, 7}
J ∪ K = {2, 3, 4, 5, 6, 7}
H ∪ J ∪ K = {1, 2, 3, 4, 5, 6, 7}

H ∩ J = {3, 7}
H ∩ K = {5, 7}
J ∩ K = {7, 4}
H ∩ J ∩ K = {7}

Check It Out

Use the Venn diagram below for Exercises 13–16. List the elements in the following sets.

13. X
14. X ∪ Z
15. Y ∩ Z
16. X ∩ Y ∩ Z

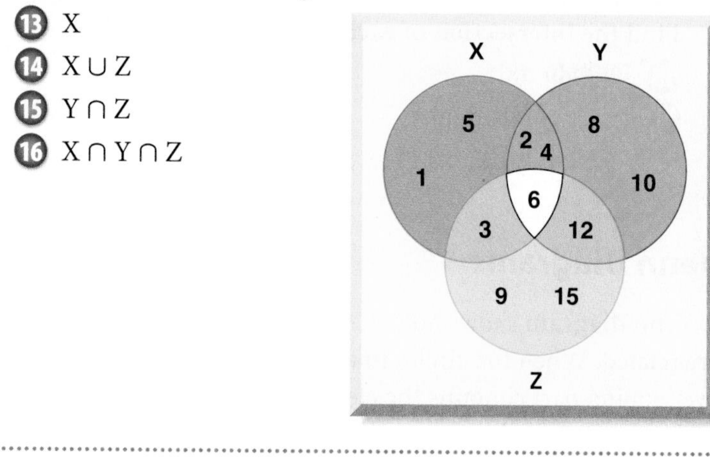

5•3 Exercises

Tell whether each statement is *true* or *false*.

1. $\{1, 2, 3\} \subset$ {counting numbers}

2. $\{1, 2, 3\} \subset \{1, 2\}$

3. $\{1, 2, 3\} \subset$ {even numbers}

4. $\varnothing \subset \{1, 2, 3\}$

Find the union of each pair of sets.

5. $\{2, 3\} \cup \{4, 5\}$

6. $\{x, y\} \cup \{y, z\}$

7. $\{r, o, y, a, l\} \cup \{m, o, a, t\}$

8. $\{2, 5, 7, 10\} \cup \{2, 7\}$

Find the intersection of each pair of sets.

9. $\{1, 3, 5, 7\} \cap \{6, 7, 8\}$

10. $\{6, 8, 10\} \cap \{7, 9, 11\}$

11. $\{r, o, y, a, l\} \cap \{m, o, a, t\}$

12. $\varnothing \cap \{4, 5\}$

Use the Venn diagram at the right for Exercises 13–16.

13. List the elements of set T.

14. List the elements of set R.

15. Find $T \cup R$.

16. Find $T \cap R$.

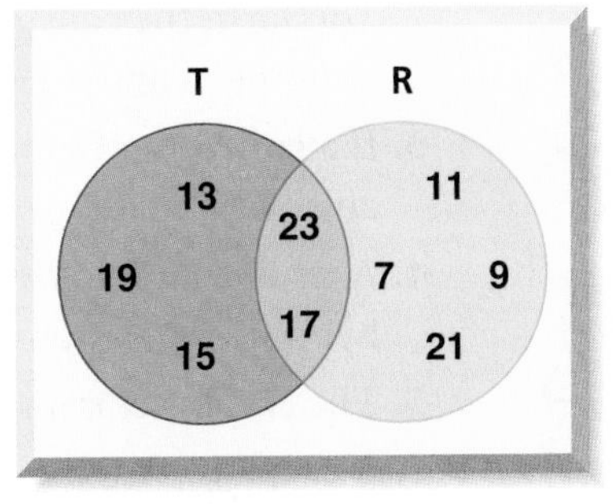

Use the Venn diagram at the right for Exercises 17–25.

17. List the elements of set M.

18. List the elements of set N.

19. Find P.

20. Find $M \cup N$.

21. Find $N \cup P$.

22. Find $M \cup P$.

23. Find $M \cap N$.

24. Find $P \cap N$.

25. Find $M \cap N \cap P$.

Logic

What have you learned?

You can use the problems and the list of words that follow to see what you learned in this chapter. You can find out more about a particular problem or word by referring to the topic number (*for example,* Lesson 5·2).

Problem Set

Tell whether each statement is *true* or *false*.

1. A conditional statement is always true. (Lesson 5·1)
2. You form the converse of a conditional by interchanging the hypothesis and the conclusion. (Lesson 5·1)
3. If a conditional statement is true, then its related inverse is always true. (Lesson 5·2)
4. A counterexample of a conditional agrees with the hypothesis but not with the conclusion. (Lesson 5·2)
5. The empty set is a subset of every set. (Lesson 5·3)
6. One counterexample is enough to show that a statement is false. (Lesson 5·2)

Write each conditional in if/then form. (Lesson 5·1)

7. A square is a quadrilateral with four equal sides and four equal angles.
8. A right angle has a measure of 90°.

Write the converse of each conditional statement. (Lesson 5·1)

9. If $y = 9$, then $y^2 = 81$.
10. If an angle has a measure greater than 90° and less than 180°, then the angle is obtuse.

Write the negation of each statement. (Lesson 5•1)

11. I am glad it's Friday!

12. These two lines are perpendicular.

Write the inverse of the conditional statement. (Lesson 5•1)

13. If the weather is warm, then we will go for a walk.

Write the contrapositive of the conditional statement. (Lesson 5•1)

14. If a quadrilateral has two pairs of parallel sides, then it is a parallelogram.

Find a counterexample that shows that the statement is false. (Lesson 5•2)

15. The number 24 has only even factors.

16. Find all the subsets of {7, 8, 9}.

Use the Venn diagram for Exercises 17–21. (Lesson 5•3)

17. List the elements in set A.

18. List the elements in set C.

19. Find A ∪ B.

20. Find B ∩ C.

21. Find A ∩ B ∩ C.

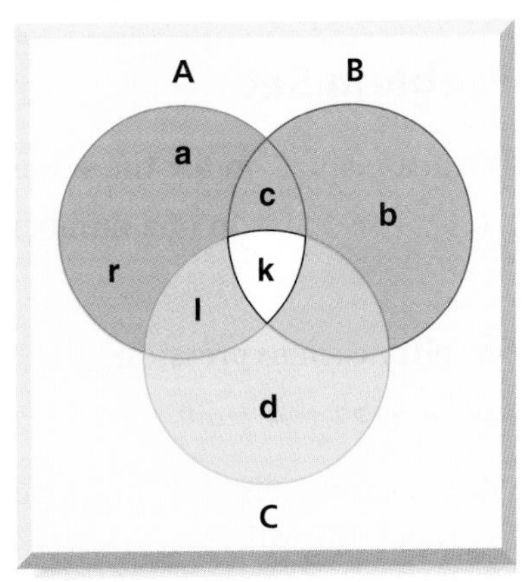

Write definitions for the following words.

contraposititive (Lesson 5•1)
converse (Lesson 5•1)
counterexample (Lesson 5•2)
intersection (Lesson 5•3)
inverse (Lesson 5•1)
set (Lesson 5•3)
union (Lesson 5•3)
Venn diagram (Lesson 5•3)

HotTopic 6

Algebra

You can use the problems and the list of words that follow to see what you already know about this chapter. The answers to the problems are in **HotSolutions** at the back of the book, and the definitions of the words are in **HotWords** at the front of the book. You can find out more about a particular problem or word by referring to the topic number (*for example,* Lesson 6•2).

Problem Set

Write an equation for the sentence. (Lesson 6•1)

1. 4 times the sum of a number and 2 is 4 less than twice the number.

Simplify each expression. (Lesson 6•2)

2. $2a + 5b - a - 2b$
3. $4(3n - 1) - (n + 6)$

4. If a car travels at 55 miles per hour, how long does it take to travel 165 miles? (Lesson 6•3)

Solve each equation. Check your solution. (Lesson 6•4)

5. $\frac{y}{6} - 1 = 8$
6. $5(2n - 3) = 4n + 9$

Use a proportion to solve Exercise 7. (Lesson 6•5)

7. In a class, the ratio of boys to girls is $\frac{3}{4}$. If there are 12 boys in the class, how many girls are there?

Solve each inequality. Graph the solution. (Lesson 6•6)

8. $x + 7 < 5$
9. $3x + 5 \geq 17$
10. $-4n + 11 < 3$

Locate each point on the coordinate plane and identify in which quadrant or on which axis it lies. (Lesson 6•7)

11. $A(4, -3)$ **12.** $B(-2, -1)$ **13.** $C(0, 1)$ **14.** $D(-3, 0)$

Write the equation of the line that contains the points. (Lesson 6•8)

15. $(-2, -6)$ and $(5, 1)$

HotWords

Addition Property of Equality (Lesson 6•4)
additive inverse (Lesson 6•4)
arithmetic sequences (Lesson 6•7)
Associative Property (Lesson 6•2)
coefficient (Lesson 6•2)
Commutative Property (Lesson 6•2)
constant of variation (Lesson 6•9)
cross product (Lesson 6•5)
direct variation (Lesson 6•9)
Distributive Property (Lesson 6•2)
Division Property of Equality (Lesson 6•4)
domain (Lesson 6•7)
equation (Lesson 6•1)
equivalent (Lesson 6•1)
equivalent expression (Lesson 6•2)
expression (Lesson 6•1)
function (Lesson 6•7)
Identity Property (Lesson 6•2)
inequality (Lesson 6•6)
like terms (Lesson 6•2)
Multiplication Property of Equality (Lesson 6•4)
ordered pair (Lesson 6•7)
origin (Lesson 6•7)
point (Lesson 6•7)
product (Lesson 6•1)
proportion (Lesson 6•5)
quadrant (Lesson 6•7)
quotient (Lesson 6•1)
range (Lesson 6•7)
rate (Lesson 6•5)
ratio (Lesson 6•5)
rise (Lesson 6•8)
run (Lesson 6•8)
slope (Lesson 6•8)
Subtraction Property of Equality (Lesson 6•4)
system of equations (Lesson 6•10)
term (Lesson 6•1)
unit rate (Lesson 6•5)
variable (Lesson 6•1)
x-axis (Lesson 6•7)
y-axis (Lesson 6•7)
y-intercept (Lesson 6•8)

6•1 Writing Expressions and Equations

Expressions

In mathematics, the value of a particular number may be unknown. A **variable** is a symbol, usually a letter, that is used to represent an unknown number. Some commonly used variables follow.

$$x \quad n \quad y \quad a$$

A **term** is a number, a variable, or a number and variable combined by multiplication or division. Some examples of terms follow.

$$w \quad 5 \quad 3x \quad \frac{y}{8}$$

An **expression** is a term or a collection of terms separated by addition or subtraction signs. Some expressions, with the number of terms, are listed below.

Expression	Number of Terms	Description
$4z$	1	a number multiplied by a variable
$5y + 3$	2	terms separated by a +
$2x + 8y - 6$	3	terms separated by a + or −
$\frac{7ac}{b}$	1	all multiplication and division

Check It Out

Count the number of terms in each expression.

1. $5x + 12$
2. $3abc$
3. $9xy - 3c - 8$
4. $3a^2b + 2ab$

Writing Expressions Involving Addition

To write an expression, you will often have to interpret a written phrase. For example, the phrase "three added to some number" could be written as the expression $x + 3$, where the variable x represents the unknown number.

Notice that the words "added to" indicate that the operation between three and the unknown number is addition. Other words and phrases that indicate addition are "more than," "plus," and "increased by." Another word that indicates addition is "sum." The sum is the result of adding terms together.

Some common addition phrases and their corresponding expressions are listed below.

Phrase	Expression
four more than some number	$y + 4$
a number increased by eight	$n + 8$
five plus some number	$5 + a$
the sum of a number and seven	$x + 7$

Check It Out

Write an expression for each phrase.

5. a number added to seven
6. the sum of a number and ten
7. a number increased by three
8. one more than some number

Writing Expressions Involving Subtraction

The phrase "five subtracted from some number" could be written as the expression $x - 5$, where the variable x represents the unknown number. Notice that the words "subtracted from" indicate that the operation between the unknown number and five is subtraction.

Some other words and phrases that indicate subtraction are "less than," "minus," and "decreased by." Another word that indicates subtraction is "difference." The difference is the result of subtracting two terms.

In a subtraction expression, the order of the terms is very important. You have to know which term is being subtracted and which is being subtracted from. To help interpret the phrase "six less than a number," replace "a number" with a numerical example like 10. What is 6 less than 10? The answer is 4, which is expressed $10 - 6$, not $6 - 10$. Therefore, the phrase "six less than a number" translates to the expression $x - 6$.

Below are some common subtraction phrases and their corresponding expressions.

Phrase	Expression
two less than some number	$z - 2$
a number decreased by six	$a - 6$
nine minus some number	$9 - n$
the difference between a number and three	$x - 3$

Check It Out

Write an expression for each phrase.

9. a number subtracted from 14
10. the difference between a number and two
11. some number decreased by eight
12. nine less than some number

Writing Expressions Involving Multiplication

The phrase "six multiplied by some number" could be written as the expression $6x$, where the variable x represents the unknown number. Notice that the words "multiplied by" indicate that the operation between the unknown number and six is multiplication.

Some other words and phrases that indicate multiplication are "times," "twice," and "of." "Twice" is used to mean "two times." "Of" is used primarily with fractions and percents. Another word that indicates multiplication is **product**. The product is the result of multiplying terms.

Here are some common multiplication phrases and their corresponding expressions.

Phrase	Expression
seven times some number	$7x$
twice a number	$2y$
one-third of some number	$\frac{1}{3}n$
the product of a number and four	$4a$

Check It Out

Write an expression for each phrase.

13. a number multiplied by three
14. the product of a number and seven
15. 35% of some number
16. 12 times some number

Writing Expressions Involving Division

The phrase "seven divided by some number" could be written as the expression $\frac{7}{x}$, where the variable x represents the unknown number. Notice that the words "divided by" indicate that the operation between the unknown number and seven is division.

Some other words and phrases that indicate division are "ratio of" and "divide." Another word that indicates division is **quotient**. The quotient is the result of dividing two terms.

Some common division phrases and their corresponding expressions are listed below.

Phrase	Expression
the quotient of 18 and some number	$\frac{18}{n}$
a number divided by 4	$\frac{x}{4}$
the ratio of 12 and some number	$\frac{12}{y}$
the quotient of a number and 9	$\frac{n}{9}$

Check It Out

Write an expression for each phrase.

17 a number divided by 7

18 the quotient of 16 and a number

19 the ratio of 40 and some number

20 the quotient of some number and 11

Writing Expressions Involving Two Operations

To write the phrase "four added to the product of five and some number" as an expression, first notice that "four added to" means "something plus four." That "something" is "the product of five and some number," which can be expressed as $5x$, since "product" indicates multiplication. Therefore, the mathematical expression is $5x + 4$. Here are a few more examples.

Phrase	Expression	Think
two less than the quotient of a number and five	$\frac{x}{5} - 2$	"two less than" means "something − 2"; "quotient" indicates division.
three times the sum of a number and four	$3(x + 4)$	Write the sum inside parentheses so that the entire sum is multiplied by three.
five more than six times a number	$6x + 5$	"five more than" means "something + 5"; "times" indicates multiplication.

Check It Out

Write each phrase as an expression.

21. 12 less than the product of eight and a number
22. 1 subtracted from the quotient of four and a number
23. twice the difference between a number and six

Writing Equations

An expression is a phrase; an **equation** is a mathematical sentence. An equation indicates that two expressions are **equivalent**, or equal. The symbol used in an equation is the equal sign =.

To write the following sentence as an equation, first identify the word or phrase that indicates equals. "Two less than the product of a number and five is the same as six more than the number." In this sentence, equals is indicated by the phrase "is the same as." In other sentences, equals may be indicated by "is," "the result is," "you get," or "equals."

Once you have identified the location of the equal sign, you can translate and write the expressions that go on the left and the right sides of the equation.

2 less than the product of a number and 5	is the same as	6 more than the number
$5x - 2$	$=$	$x + 6$

Check It Out

Write an equation for each sentence.

24. Eight subtracted from a number is the same as the product of five and the number.
25. Five less than four times a number is four more than twice the number.
26. When one is added to the quotient of a number and six, the result is nine less than the number.

6•1 Exercises

Count the number of terms in each expression.

1. $4x + 2y - 3z$
2. $4n - 20$

Write an expression for each phrase.

3. eight more than a number
4. the sum of a number and nine
5. five less than a number
6. the difference between a number and four
7. one-half of some number
8. twice a number
9. the product of a number and six
10. a number divided by eight
11. the ratio of ten and some number
12. the quotient of a number and five
13. four more than the product of a number and three
14. five less than twice a number
15. twice the sum of eight and a number

Write an equation for each sentence.

16. Eight more than the quotient of a number and six is the same as two less than the number.
17. If nine is subtracted from twice a number, the result is eleven.
18. Three times the sum of a number and five is four more than twice the number.

19. Which of the following words is used to indicate multiplication?
    A. sum
    B. difference
    C. product
    D. quotient
20. Which of the following does not indicate subtraction?
    A. less than
    B. difference
    C. decreased by
    D. ratio of

6•2 Simplifying Expressions

Terms

A term that contains a number only is called a *constant* term. Compare the terms 7 and $5x$. Notice that the value of 7 never changes—it remains constant. The value of $5x$ will change as the value of x changes. If $x = 2$, then $5x = 5(2) = 10$, and if $x = 3$, then $5x = 5(3) = 15$.

The numerical factor of a term that contains a variable is called the **coefficient** of the variable. In the term $5x$, the 5 is the coefficient of x.

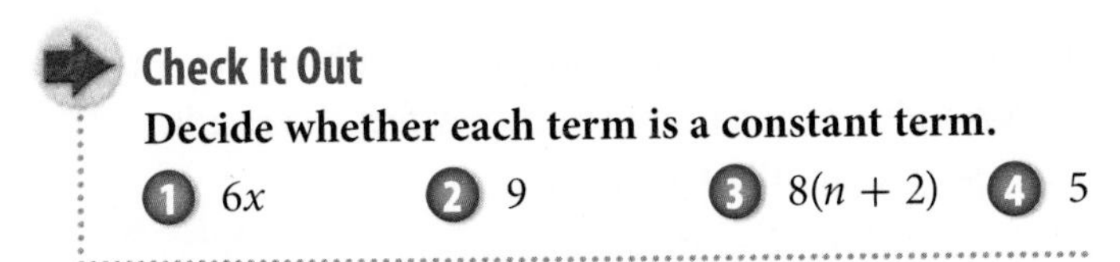

Check It Out

Decide whether each term is a constant term.

1. $6x$
2. 9
3. $8(n + 2)$
4. 5

The Commutative Property of Addition and Multiplication

The **Commutative Property** of Addition states that the order of terms being added may be switched without changing the sum.

$4 + 9 = 9 + 4$ and $x + 5 = 5 + x$.

The Commutative Property of Multiplication states that the order of terms being multiplied may be switched without changing the product.

$2(8) = 8(2)$ and $y \cdot 6 = 6y$.

The Commutative Property does not hold true for subtraction or division. The order of the terms does affect the result.

$7 - 3 = 4$ but $3 - 7 = -4$.
$6 \div 2 = 3$ but $2 \div 6 = \frac{1}{3}$.

Check It Out

Rewrite each expression using the Commutative Property of Addition or Multiplication.

5. $2x + 5$
6. $n \cdot 7$
7. $9 + 4y$
8. $5 \cdot 6$

The Associative Property of Addition and Multiplication

The **Associative Property** of Addition states that the grouping of terms being added does not affect the sum.

$(2 + 3) + 4 = 2 + (3 + 4)$ and $(y + 5) + 9 = y + (5 + 9)$.

The Associative Property of Multiplication states that the grouping of terms being multiplied does not affect the product.

$(3 \cdot 4) \cdot 5 = 3 \cdot (4 \cdot 5)$ and $6 \cdot (4y) = (6 \cdot 4)y$.

The Associative Property does not hold true for subtraction or division. The grouping of the numbers does affect the result.

$(9 - 7) - 5 = -3$ but $9 - (7 - 4) = 6$.

$(12 \div 6) \div 2 = 1$ but $12 \div (6 \div 2) = 4$.

Check It Out

Rewrite each expression using the Associative Property of Addition or Multiplication.

9. $(4 + 8) + 11$
10. $(5 \cdot 2) \cdot 9$
11. $(2x + 5y) + 4$
12. $7 \cdot (8n)$

The Distributive Property

The **Distributive Property** of Multiplication over Addition states that multiplying a sum by a number is the same as multiplying each addend by that number and then adding the products.

So, $4(3 + 5) = (4 \cdot 3) + (4 \cdot 5)$.

How would you multiply $6 \cdot 99$ in your head? You might think $600 - 6 = 594$. If you did, you used the Distributive Property.

$6(100 - 1) = 6(100 + (-1))$ • Rewrite $100 - 1$ as $100 + (-1)$.

$6(100 + (-1))$ • Distribute the factor of 6 to each term inside the parentheses.

$= 6 \cdot 100 - 6 \cdot 1$
$= 600 - 6 = 594$ • Simplify, using the order of operations.

The Distributive Property does not hold for division.

$$4 \div (3 + 5) \neq (4 \div 3) + (4 \div 5)$$

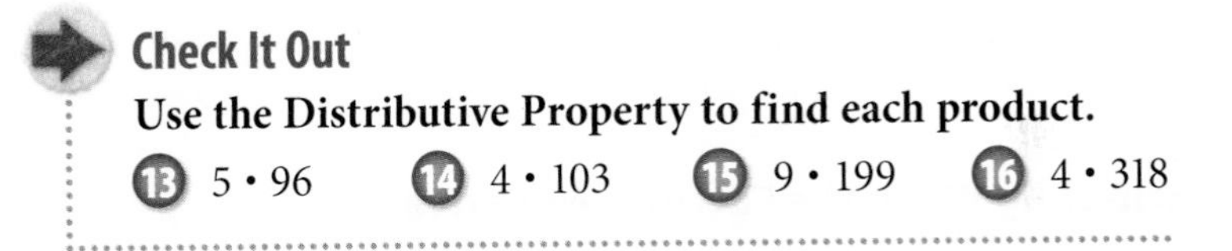

Check It Out

Use the Distributive Property to find each product.

13. $5 \cdot 96$
14. $4 \cdot 103$
15. $9 \cdot 199$
16. $4 \cdot 318$

Properties of Zero and One

The **Identity Property** of Addition states that the sum of any number and 0 is that number. The additive identity is 0 because $a + 0 = a$ and $0 + a = a$. The Identity Property of Multiplication states that the product of any number and 1 is that number. The multiplicative identity is 1 because $a \cdot 1 = a$ and $1 \cdot a = a$. Subtraction and division do not have identity properties.

The Zero Property of Multiplication states the product of zero and any number is zero: $a \cdot 0 = 0$ and $0 \cdot a = 0$. Since you cannot divide a number by zero, any number divided by zero is undefined.

Check It Out

Name the property shown by each statement.

17. $\sqrt{9} \cdot 1 = \sqrt{9}$
18. $6x + 0 = 6x$
19. $325 \cdot 0 = 0$

Equivalent Expressions

The Distributive Property can be used to write an **equivalent expression**. Equivalent expressions are different ways of writing one expression.

EXAMPLE **Writing an Equivalent Expression**

Write an equivalent expression for $2(6x - 4)$.

$2(6x - 4)$
$2 \cdot 6x - 2 \cdot 4$
$12x - 8$

- Distribute the factor of 2 to each term inside the parentheses.
- Simplify.

So, $2(6x - 4) = 12x - 8$.

Distributing When the Factor is Negative

The Distributive Property is applied in the same way if the factor to be distributed is negative.

EXAMPLE **Writing an Equivalent Expression**

Write an equivalent expression for $-3(5x - 6)$.

$-3(5x - 6)$
$-3 \cdot 5x - (-3) \cdot 6$
$-15x + 18$

- Distribute the factor to each term inside the parentheses.
- Simplify. Remember: $(-3) \cdot 6 = -18$ and $-(-18) = +18$.

So, $-3(5x - 6) = -15x + 18$.

Check It Out

Write an equivalent expression by using the Distributive Property.

20. $2(7x + 4)$
21. $8(3n - 2)$
22. $-1(7y - 4)$
23. $-3(-3x + 5)$

The Distributive Property with Common Factors

Given the expression $6x + 9$, you can use the Distributive Property to factor out the greatest common factor and write an equivalent expression. Recognize that the greatest common factor of the two terms is 3.

Rewrite the expression as $3 \cdot 2x + 3 \cdot 3$. Then write the greatest common factor 3 in front of the parentheses and the remaining factors inside the parentheses: $3(2x + 3)$.

EXAMPLE Factoring Out the Common Factor

Factor out the greatest common factor from the expression $8n - 20$.

$4 \cdot 2n - 4 \cdot 5$	• Find a common factor and rewrite the expression.
$4 \cdot (2n - 5)$	• Use the Distributive Property.

So, $8n - 20 = 4 \cdot (2n - 5)$.

Check It Out

Factor out the greatest common factor in each expression.

24. $7x + 35$
25. $18n - 15$
26. $15c + 60$
27. $40a - 100$

Like Terms

Like terms are terms that contain the exact same variables. Constant terms are *like terms* because they do not contain variables. Here are some examples of like terms.

Like Terms	Reason
$4x$ and $5x$	Both contain the same variable.
5 and 13	Both are constant terms.
$3n^2$ and $9n^2$	Both contain the same variable with the same exponent.

Below are some examples of terms that are not like terms.

Not Like Terms	Reason
$7x$ and $9y$	Variables are different.
$2n$ and 6	One term has a variable; the other is constant.
$8x^2$ and $8x$	The variables are the same, but the exponents are different.

Like terms may be combined into one term by addition or subtraction. Consider the expression $3x + 5x$. Notice that the two terms have a common factor, x. Use the Distributive Property to write $x(3 + 5)$. The expression simplifies to $8x$, so $3x + 5x = 8x$.

EXAMPLE Combining Like Terms

Simplify $6n - 8n$.

$n(6 - 8)$ • Recognize that the variable is a common factor. Rewrite the expression using the Distributive Property.

$n(-2)$ • Simplify.

$-2n$ • Use the Commutative Property of Multiplication.

So, $6n - 8n = -2n$.

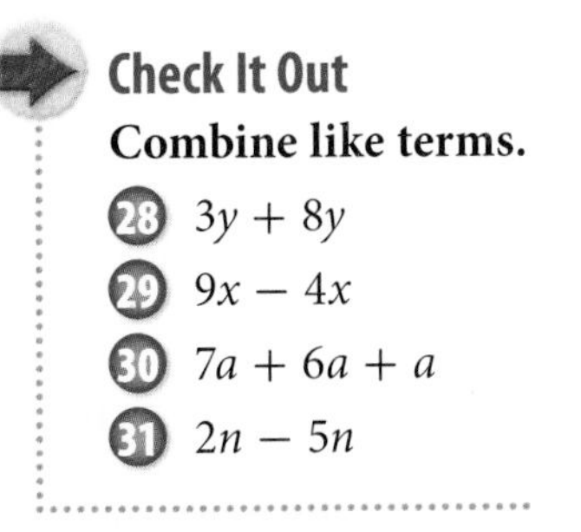

Check It Out

Combine like terms.

28. $3y + 8y$
29. $9x - 4x$
30. $7a + 6a + a$
31. $2n - 5n$

Simplifying Expressions

Expressions are simplified when all of the like terms have been combined. Terms that are not like terms cannot be combined. In the expression $5x - 7y + 8x$, there are three terms. Two of the terms are like terms, $5x$ and $8x$, which add to $13x$. The expression can be simplified to $13x - 7y$.

EXAMPLE Simplifying Expressions

Simplify the expression $2(4x - 5) - 12x + 17$.

$2 \cdot 4x - 2 \cdot 5 - 12x + 17$	• Use the Distributive Property.
$8x - 10 - 12x + 17$	• Simplify.
$-4x + 7$	• Combine like terms.

So, $2(4x - 5) - 12x + 17 = -4x + 7$.

Check It Out

Simplify each expression.

32. $4y + 5z - y + 3z$
33. $x + 4(3x - 5)$
34. $15a + 8 - 2(3a + 2)$
35. $2(5n - 3) - (n - 2)$

6•2 Exercises

Decide whether each term is a constant term.

1. $4n$

2. -9

3. Identify the coefficients in the expression $4x - 7 + 6x + 10$.

Rewrite each expression using the Commutative Property of Addition or Multiplication.

4. $9 + 5$

5. $n \cdot 4$

6. $8x + 11$

Rewrite each expression using the Associative Property of Addition or Multiplication.

7. $2 + (7 + 14)$

8. $(8 \cdot 5) \cdot 3$

9. $3 \cdot (6n)$

Use the Distributive Property to find each product.

10. $8 \cdot 99$

11. $6 \cdot 108$

Name the property shown by each statement.

12. $\sqrt{8} + 0 = \sqrt{8}$

13. $a^3 \cdot 1 = a^3$

Write an equivalent expression.

14. $-7(2n + 8)$

15. $12(3a - 10)$

Factor out the common factor in each expression.

16. $8x + 32$

17. $6n - 9$

18. $30a - 50$

Combine like terms.

19. $14x - 8x$

20. $7n + 8n - n$

21. $2a - 11a$

Simplify each expression.

22. $8a + b - 3a - 5b$

23. $3x + 2(6x - 5) + 8$

24. Which property is illustrated by $7(2x + 1) = 14x + 7$?

A. Commutative Property of Multiplication

B. Distributive Property

C. Associative Property of Addition

D. The example does not illustrate a property.

6•3 Evaluating Expressions and Formulas

Evaluating Expressions

You can *evaluate* an algebraic expression for different values of the variable. To evaluate $5x - 3$ for $x = 4$, *substitute* 4 in place of the x: $5(4) - 3$.

Use the order of operations to simplify: multiply first, then subtract. So, $5(4) - 3 = 20 - 3 = 17$.

EXAMPLE **Evaluating An Expression**

Evaluate the expression $3x^2 - 4x + 5$, when $x = -2$.

$3(-2)^2 - 4(-2) + 5$ • Substitute the numeric value for x.

$3 \cdot 4 - 4(-2) + 5$ • Use the order of operations to simplify. Evaluate the power.

$12 + 8 + 5 = 25$ • Multiply and then add from left to right.

So, when $x = -2$, then $3x^2 - 4x + 5 = 25$.

Check It Out

Evaluate each expression for the given value.

1. $9x - 14$ for $x = 4$
2. $5a + 7 + a^2$ for $a = -3$
3. $\frac{n}{3} + 2n - 5$ for $n = 12$
4. $2(y^2 - 2y + 1) + 4y$ for $y = 3$

Evaluating Formulas

Interest is the amount of money paid or earned for the use of money. If you invest money in a savings account, you earn interest from the bank. For a loan, you pay interest to the bank. The amount of money invested or borrowed is called the principal. The interest rate, I, is the percent charged or paid during a given period of time, t, in years. To solve problems involving simple interest (I), the formula $I = prt$ can be used.

EXAMPLE Finding Simple Interest

Find the simple interest for $10,000 at 4% for 2 years.

$I = prt$ • Write the simple interest formula.

$I = (10{,}000)(0.04)(2)$ • Substitute the values of p, r, and t into the formula for simple interest.

$I = \$800$ • Simplify.

So, the simple interest is $800.

Check It Out

Find the simple interest to the nearest cent.

5. $5,000 at 6% for 10 years
6. $135,000 at 6.25% for 30 years
7. $26,000 at 6.86% for 4 years

The Formula for Distance Traveled

The distance traveled by a person, a vehicle, or an object depends on the rate and the amount of time traveled. The formula $d = rt$ can be used to find the distance traveled, d, if the rate, r, and the amount of time, t, are known.

EXAMPLE Finding the Distance Traveled

Find the distance traveled by a train that averages 60 miles per hour for $3\frac{1}{4}$ hours.

$d = (60) \cdot \left(3\frac{1}{4}\right)$ • Substitute values into the distance formula ($d = rt$).

$60\left(\frac{13}{4}\right) = 195$ • Multiply.

So, the train traveled 195 miles.

Check It Out

Calculate the distance traveled.

8. A person rides 12 miles per hour for 3 hours.
9. A plane flies 750 kilometers per hour for $2\frac{1}{2}$ hours.
10. A person drives a car 55 miles per hour for 8 hours.
11. A snail moves 2.3 feet per hour for 4 hours.

APPLICATION Maglev

Maglev (short for *magnetic levitation*) trains fly above the tracks. Magnetic forces lift and propel the trains. Without the friction of the tracks, the maglevs run at speeds of 150 to 300 miles per hour. Are they the trains of the future? At a speed of 200 miles per hour with no stops, how long would it take to travel the distance between these cities? Round to the nearest quarter of an hour.

235 miles from Boston, MA, to New York, NY
440 miles from Los Angeles, CA, to San Francisco, CA
750 miles from Mobile, AL, to Miami, FL
See **HotSolutions** for the answers.

6•3 Exercises

Evaluate each expression for the given value.

1. $6x - 11$ for $x = 5$
2. $5a^2 + 7 - 3a$ for $a = 4$
3. $\frac{n}{6} - 3n + 10$ for $n = -6$
4. $3(4y - 1) - \frac{12}{y} + 8$ for $y = 2$

Use the formula $I = prt$ to calculate the simple interest.

5. Johnny borrowed $12,000 to buy a car. If the yearly simple interest rate was 18%, how much interest will she pay on a 5-year loan?
6. Jamil invested $500 in a savings account for 3 years. Find the total amount in his account if it earns a yearly simple interest of 3.25%.
7. Susan borrowed $5,000 at a yearly simple interest rate of 4.5%. How much interest will she pay on a 2-year loan?

Use the formula $d = rt$ to find the distance traveled.

8. Find the distance traveled by a jogger who jogs at 6 miles per hour for $1\frac{1}{2}$ hours.
9. A race car driver averaged 180 miles per hour. If the driver completed the race in $2\frac{1}{2}$ hours, how many miles was the race?
10. The speed of light is approximately 186,000 miles per second. About how far does light travel in 5 seconds?

6•4 Solving Linear Equations

Additive Inverses

The sum of any term and its **additive inverse** is 0. The additive inverse of 7 is -7, because $7 + (-7) = 0$, and the additive inverse of $-8n$ is $8n$, because $-8n + 8n = 0$.

Check It Out

Give the additive inverse of each term.

1. 4
2. $-x$
3. -35
4. $10y$

Solving Addition and Subtraction Equations

In order to solve an equation, the variable needs to be by itself or isolated on one side of the equal sign. The **Subtraction Property of Equality** states that if you subtract the same number from each side of an equation, the two sides remain equal. Consider the equation $x + 3 = 11$. In order to get x by itself in this equation, subtract 3 from each side.

EXAMPLE Solving Addition Equations

Solve $x + 5 = 9$.

$x + 5 - 5 = 9 - 5$ • Isolate x by subtracting 5 from each side.

$x = 4$ • Simplify.

$(4) + 5 \stackrel{?}{=} 9$ • Check by substituting the solution into the original equation.

$9 = 9$

So, $x = 4$.

The **Addition Property of Equality** states that if you add the same number to each side of an equation, the two sides remain equal.

EXAMPLE **Solving Subtraction Equations**

Solve $n - 8 = 7$.

$n - 8 + 8 = 7 + 8$	• Isolate *n* by adding 8 to each side.
$n = 15$	• Simplify.
$(15) - 8 \stackrel{?}{=} 7$ $7 = 7$	• Check by substituting the solution into the original equation.

So, $n = 15$.

Check It Out

Solve each equation. Check your solution.

5. $x + 4 = 13$
6. $n - 5 = 11$
7. $y + 10 = 3$

Solving Equations by Multiplication or Division

The **Division Property of Equality** states that if you divide each side of the equation by the same nonzero number, the two sides remain equal.

EXAMPLE **Solving Multiplication Equations Using the Division Property of Equality**

Solve $3x = 15$.

$\frac{3x}{3} = \frac{15}{3}$	• Isolate *x* by dividing by 3 on each side.
$x = 5$	• Simplify.
$3(5) \stackrel{?}{=} 15$ $15 = 15$	• Check by substituting the solution into the original equation.

So, $x = 5$.

The **Multiplication Property of Equality** states that if you multiply each side of the equation by the same number, the two sides remain equal. In order to get x by itself in the equation $\frac{x}{7} = 4$, multiply each side by 7.

EXAMPLE Solving Division Equations Using the Multiplication Property of Equality

Solve the equation $\frac{n}{6} = 3$.

$\frac{n}{6} \cdot 6 = 3 \cdot 6$ • Isolate n by multiplying 6 on each side.

$n = 18$ • Simplify.

$\frac{(18)}{6} \stackrel{?}{=} 3$ • Check by substituting the solution into the original equation.

$3 = 3$

So, $n = 18$.

Check It Out

Solve each equation. Check your solution.

8. $5x = 35$
9. $\frac{y}{8} = 4$
10. $9n = -27$

APPLICATION Prime Time

One week the top five prime-time TV shows were rated like this:

Rating (%)	Program
23.3	Drama
21.6	Comedy
20.5	Movie
17.0	Cartoon
17.6	Sitcom

Let a equal the number of families watching TV. If the number of families watching the comedy program was 35 million, how many total families were watching TV? See **HotSolutions** for the answer.

Solving Two-Step Equations

Consider the equation $4x - 7 = 13$. This type of equation is sometimes referred to as a "two-step" equation because it contains two operations. To solve this type of equation, you undo each operation to isolate the term that contains the variable.

EXAMPLE Solving Two-Step Equations

Solve $4x - 7 = 13$.

$4x - 7 + 7 = 13 + 7$ • Isolate the term that contains the variable by adding 7 to each side.

$4x = 20$ • Simplify.

$\frac{4x}{4} = \frac{20}{4}$ • Divide by 4 on each side to isolate the variable.

$x = 5$

$4(5) - 7 \stackrel{?}{=} 13$ • Check by substituting the solution into the original equation.

$20 - 7 = 13$ • Simplify, using the order of operations.

So, $x = 5$.

Solve $\frac{n}{4} + 8 = 2$.

$\frac{n}{4} + 8 - 8 = 2 - 8$ • Isolate the term that contains the variable by subtracting 8 from each side.

$\frac{n}{4} = -6$ • Simplify.

$\frac{n}{4} \cdot 4 = -6 \cdot 4$ • Multiply each side by 4 to isolate the variable.

$n = -24$

$\frac{(-24)}{4} + 8 \stackrel{?}{=} 2$ • Check by substituting the solution into the original equation.

$-6 + 8 = 2$ • Simplify, using the order of operations.

So, $n = -24$.

Check It Out

Solve each equation. Check your solution.

11. $6x + 11 = 29$

12. $\frac{y}{5} - 3 = 7$

13. $2n + 15 = 1$

14. $\frac{a}{3} + 11 = 9$

Solving Equations with the Variable on Each Side

Consider the equation $5x + 4 = 8x - 5$. Notice that each side of the equation has a term that contains the variable. To solve this equation, you use the Addition or Subtraction Properties of Equality to write equivalent equations with the variables collected on one side of the equal sign and the constant terms on the other side of the equal sign. Then solve the equation.

EXAMPLE **Solving an Equation with Variables on Each Side**

Solve $5x + 4 = 8x - 5$.

$5x + 4 - 5x = 8x - 5 - 5x$
- Subtract $5x$ from each side to collect the terms that contain the variable on one side of the equal sign.

$4 = 3x - 5$
- Simplify by combining like terms.

$4 + 5 = 3x - 5 + 5$
- Add 5 to each side to collect constant terms on the other side of the equal sign.

$9 = 3x$
- Simplify.

$\frac{9}{3} = \frac{3x}{3}$
- Divide each side by 3 to isolate the variable.

$3 = x$
- Simplify.

$5(3) + 4 \stackrel{?}{=} 8(3) - 5$
- Check by substituting the solution into the original equation.

$15 + 4 \stackrel{?}{=} 24 - 5$
- Simplify, using order of operations.

So, $x = 3$.

Check It Out

Solve each equation. Check your solution.

15. $2m - 36 = 6m$
16. $9n - 4 = 6n + 8$
17. $12x + 9 = 2x - 11$
18. $3a + 24 = 9a - 12$

Equations Involving the Distributive Property

You may need to use the Distributive Property to solve an equation.

EXAMPLE **Solving Equations by Using the Distributive Property**

$3x - 4(2x + 5) = 3(x - 2) + 10$ • Apply the Distributive Property.

$3x - 8x - 20 = 3x - 6 + 10$ • Combine like terms.

$-5x - 20 = 3x + 4$

$-5x - 20 + 5x = 3x + 4 + 5x$ • Add $5x$ to each side to collect the x terms on one side of the equation.

$-20 = 8x + 4$ • Combine like terms.

$-20 - 4 = 8x + 4 - 4$ • Subtract 4 from each side to collect the constant terms on the other side of the equation.

$-24 = 8x$ • Combine like terms.

$-\frac{24}{8} = \frac{8x}{8}$ • Divide each side by 8 to isolate the variable.

$-3 = x$ • Simplify.

Check the solution.

$3(-3) - 4(2(-3) + 5) \stackrel{?}{=} 3((-3) - 2) + 10$ • Evaluate the original equation for $x = -3$.

$3(-3) - 4(-6 + 5) \stackrel{?}{=} 3(-5) + 10$

$3(-3) - 4(-1) \stackrel{?}{=} -15 + 10$

$-9 + 4 \stackrel{?}{=} -5$

$-5 = -5$ • The solution is correct.

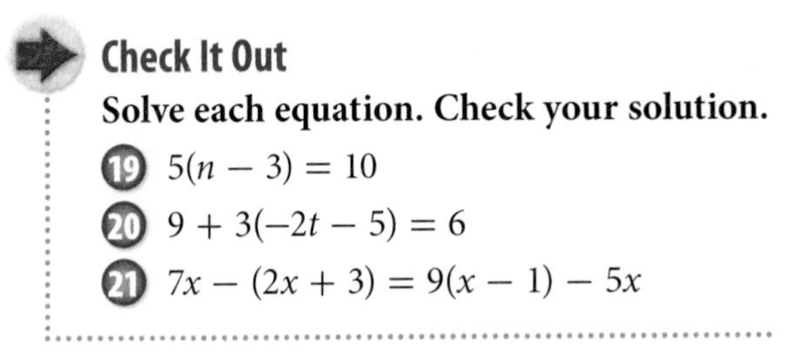

Check It Out

Solve each equation. Check your solution.

19. $5(n - 3) = 10$
20. $9 + 3(-2t - 5) = 6$
21. $7x - (2x + 3) = 9(x - 1) - 5x$

Solving for a Variable in a Formula

The formula $d = rt$ can be used to find the distance traveled d, by multiplying the rate r, by the time t. You can solve the formula for time t by dividing each side by r.

$d = rt$ • Divide by r on each side.

$\frac{d}{r} = \frac{rt}{r}$ • Simplify.

$\frac{d}{r} = t$

EXAMPLE Solving for a Variable

Solve $P = 2w + 2\ell$ for w.

$P = 2w + 2\ell$

$P - 2\ell = 2w + 2\ell - 2\ell$ • To isolate the term that contains w, subtract 2ℓ from each side.

$P - 2\ell = 2w$ • Combine like terms.

$\frac{P - 2\ell}{2} = \frac{2w}{2}$ • To isolate w, divide each side by 2.

$\frac{P - 2\ell}{2} = w$ • Simplify.

Check It Out

Solve for the indicated variable in each formula.

22. $A = \ell w$, solve for w
23. $2y - 3x = 8$, solve for y
24. $3a + 6b = 9$, solve for b

6•4 Exercises

Give the additive inverse of each term.

1. 8

2. $-6x$

Solve each equation. Check your solution.

3. $x + 8 = 15$

4. $n - 3 = 9$

5. $\frac{y}{5} = 9$

6. $4a = -28$

7. $x + 14 = 9$

8. $n - 12 = 4$

9. $7x = 63$

10. $\frac{a}{6} = -2$

11. $3x + 7 = 25$

12. $\frac{y}{9} - 2 = 5$

13. $4n + 11 = 7$

14. $\frac{a}{5} + 8 = 5$

15. $13n - 5 = 10n + 7$

16. $y + 8 = 3y - 6$

17. $7x + 9 = 2x - 1$

18. $6a + 4 = 7a - 3$

19. $8(2n - 5) = 4n + 8$

20. $9y - 5 - 3y = 4(y + 1) - 5$

21. $8x - 3(x - 1) = 4(x + 2)$

22. $14 - (6x - 5) = 5(2x - 1) - 4x$

Solve for the indicated variable in each formula.

23. $d = rt$, solve for r

24. $A = \ell w$, solve for ℓ

25. $4y - 5x = 12$, solve for y

26. $8y + 3x = 11$, solve for y

27. Which of the following equations can be solved by adding 6 to each side and dividing by 5 on each side?

A. $5x + 6 = 16$

B. $\frac{x}{5} + 6 = 16$

C. $5x - 6 = 14$

D. $\frac{x}{5} - 6 = 14$

28. Which equation does not have $x = 4$ as its solution?

A. $3x + 5 = 17$

B. $2(x + 2) = 10$

C. $\frac{x}{2} + 5 = 7$

D. $x + 2 = 2x - 2$

6•5 Ratio and Proportion

Ratio

A **ratio** is a comparison of two quantities. If there are 10 boys and 15 girls in a class, the ratio of the number of boys to the number of girls is 10 to 15, which can be expressed as the fraction $\frac{10}{15}$, which in simplest form is $\frac{2}{3}$. Look at some other ratios in the table below.

Comparison	Ratio	As a Fraction
Number of girls to number of boys	16 to 12	$\frac{16}{12} = \frac{4}{3}$
Number of boys to number of students	12 to 28	$\frac{12}{28} = \frac{3}{7}$
Number of students to number of girls	28 to 16	$\frac{28}{16} = \frac{7}{4}$

Check It Out

A coin bank contains 3 nickels and 9 dimes. Write each ratio in fraction form.

1. number of nickels to number of dimes
2. number of dimes to number of coins
3. number of coins to number of nickels

Rate

A **rate** is a ratio that compares quantities with different units. Some examples of rates are listed below.

$$\frac{100 \text{ mi}}{2 \text{ h}} \qquad \frac{\$400}{2 \text{ wks}} \qquad \frac{\$3}{2 \text{ lbs}}$$

A **unit rate** is a rate that has been simplified so that it has a denominator of 1.

$$\frac{50 \text{ mi}}{1 \text{ h}} \qquad \frac{\$200}{1 \text{ wk}} \qquad \frac{\$1.50}{1 \text{ lb}}$$

Proportions

When two ratios are equal, they form a **proportion**. For example, if a car averages 18 miles per 1 gallon of fuel, then the car averages $\frac{36 \text{ mi}}{2 \text{ gal}}$, $\frac{54 \text{ mi}}{3 \text{ gal}}$, and so on. The ratios are all equal because they are equal to $\frac{18}{1}$.

One way to determine whether two ratios form a proportion is to compare their **cross products**. Every proportion has two cross products: the numerator of one ratio multiplied by the denominator of the other ratio. If the cross products are equal, then the ratios form a proportion.

EXAMPLE Determining Whether Ratios Form a Proportion

Determine whether a proportion is formed.

$\frac{9}{12} \stackrel{?}{=} \frac{96}{117}$ $\qquad$ $\frac{14}{8} \stackrel{?}{=} \frac{63}{36}$

$\frac{9}{12} \stackrel{?}{=} \frac{96}{117}$ $\qquad$ $\frac{14}{8} \stackrel{?}{=} \frac{63}{36}$ $\qquad$ • Find the cross products.

$9 \cdot 117 \stackrel{?}{=} 96 \cdot 12$ $\qquad$ $14 \cdot 36 \stackrel{?}{=} 63 \cdot 8$

$1{,}053 \neq 1{,}152$ $\qquad$ $504 = 504$ $\qquad$ • If the cross products are equal, the ratios are proportional.

Because $\frac{9}{12} \neq \frac{96}{117}$, the ratios do not form a proportion.

Because $\frac{14}{8} = \frac{63}{36}$, the ratios form a proportion.

Check It Out

Determine whether a proportion is formed.

4. $\frac{6}{9} = \frac{18}{27}$
5. $\frac{7}{4} = \frac{40}{49}$
6. $\frac{3}{5} = \frac{21}{35}$

Using Proportions to Solve Problems

You can use proportions to solve problems.

Suppose that you can buy 2 CDs for \$25. At that rate, how much does it cost to buy 7 CDs? Let c represent the cost of 7 CDs.

If you express each ratio as $\frac{\text{CDs}}{\$}$, then one ratio is $\frac{2}{25}$ and the other is $\frac{7}{c}$. Write a proportion.

$$\frac{2}{25} = \frac{7}{c}$$

Now that you have written a proportion, you can use the cross products to solve for c.

$$2c = 175$$

To isolate the variable, divide each side by 2 and simplify.

$$\frac{2c}{2} = \frac{175}{2}$$

$$c = 87.5$$

So, 7 CDs cost \$87.50.

Check It Out

Use proportions to solve Exercises 7–10.

7. A car gets 22 miles per gallon of fuel. At this rate, how many gallons of fuel does the car need to travel 121 miles?
8. A worker earns \$100 every 8 hours. At this rate, how much will the worker earn in 36 hours?

Use the table below for Exercises 9 and 10.

Top-Five TV Shows in Prime-Time

Rating	Program
23.3	Drama
22.6	Comedy
20.9	Movie
19.0	Cartoon
18.6	Sitcom

9. The cartoon earns a 19.0 rating for 18,430,000 viewers. At this rate, how many viewers are needed for a 1.0 rating?
10. If 18,042,000 viewers watched the sitcom, how many viewers watched the drama?

6•5 Exercises

A basketball team has 20 wins and 10 losses. Write each ratio.

1. number of wins to number of losses
2. number of wins to number of games
3. number of losses to number of games

Express each rate as a unit rate.

4. 120 students for every 3 buses
5. \$3.28 for 10 pencils
6. \$274 for 40 hours of work

Find the unit rate.

7. \$8.95 for 3 lbs
8. \$570 for 660 feet of rope
9. 420 miles in 7 hours

Determine whether a proportion is formed.

10. $\frac{3}{8} = \frac{16}{42}$
11. $\frac{10}{4} = \frac{25}{10}$
12. $\frac{4}{6} = \frac{15}{22}$

Use a proportion to solve each exercise.

13. Clay jogs 2 miles in 17 minutes. At this rate, how far can he run in 30 minutes?
14. If the cost of gasoline is \$3.89 per gallon, how much does 14 gallons of gasoline cost?
15. Nick can text message 29 words in 45 seconds. At this rate, how many words can he text message in 3 minutes?
16. An online retailer sells each DVD for \$39.99. What is the cost of 5 DVDs?
17. A map is drawn using a scale of 8 miles to 1 centimeter. On the map, two cities are 7.5 centimeters apart. What is the actual distance between the two cities?

6•6 Inequalities

An **inequality** is a mathematical sentence comparing two quantities. The inequality symbols are shown in the chart below.

Symbol	Meaning	Example
$>$	Is greater than	$7 > 4$
$<$	Is less than	$4 < 7$
$\geq$	Is greater than or equal to	$x \geq 3$
$\leq$	Is less than or equal to	$-2 \leq x$

Graphing Inequalities

The equation $x = 5$ has one solution, 5. The inequality $x > 5$ has a solution of all real numbers greater than 5. Note that 5 is not a solution because the solution has to be greater than 5. Since you cannot list all of the solutions, you can show them on a number line. When graphing an inequality on a number line, a closed circle indicates that the endpoint is a solution. If the endpoint is an open circle, the number is not a solution of the inequality.

To show all the values that are greater than 5, but not including 5, use an open circle on 5 and shade the number line to the right.

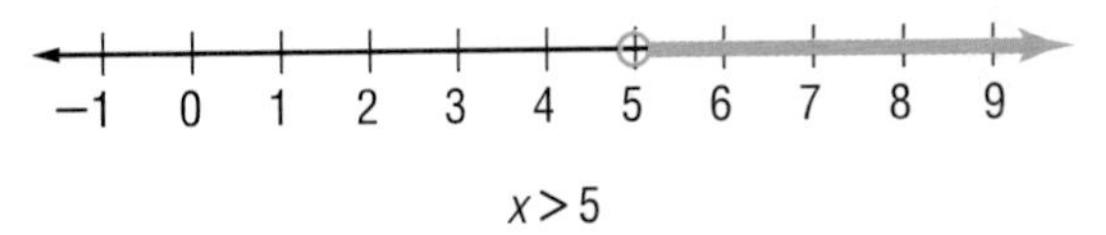

$x > 5$

The inequality $y \leq -1$ has a solution of all real numbers less than or equal to −1. In this case, −1 is also a solution, because −1 is less than *or* equal to −1.

To graph all the values that are less than or equal to −1, use a closed (filled-in) circle on −1 and shade the number line to the left. Be sure to include the arrow.

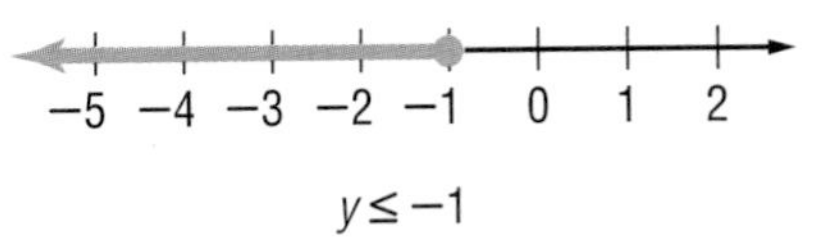

$y \leq -1$

Check It Out

Graph the solutions to each inequality on a number line.

1. $x \geq 2$
2. $y < -4$
3. $n > -3$
4. $x \leq 1$

Writing Inequalities

An inequality can be used to describe a range of values. You can translate a word problem into an inequality.

EXAMPLE Writing Inequalities

Define a variable and write an inequality for the sentence, 5 more than a number is greater than 7.

Let n = a number — • Define a variable.

$n + 5 > 7$ — • Write an inequality.

The sum of the variable and 5 is greater than 7.

Check It Out

Define a variable and write an inequality for each sentence.

5. You must be 16 years of age or older to drive.
6. A cell phone costs more than $19.99.
7. The average life span for a snapping turtle is 57 years.
8. The difference between a number and 9 is no more than 3.

Solving Inequalities by Addition and Subtraction

To solve an inequality, you must identify the values of the variable that make the inequality true. You can use addition and subtraction to isolate the variable in an inequality in the same way that you do to solve an equation. When you add or subtract the same number from each side of an inequality, the inequality remains true.

It is important to recognize that there are an infinite number of solutions to an inequality.

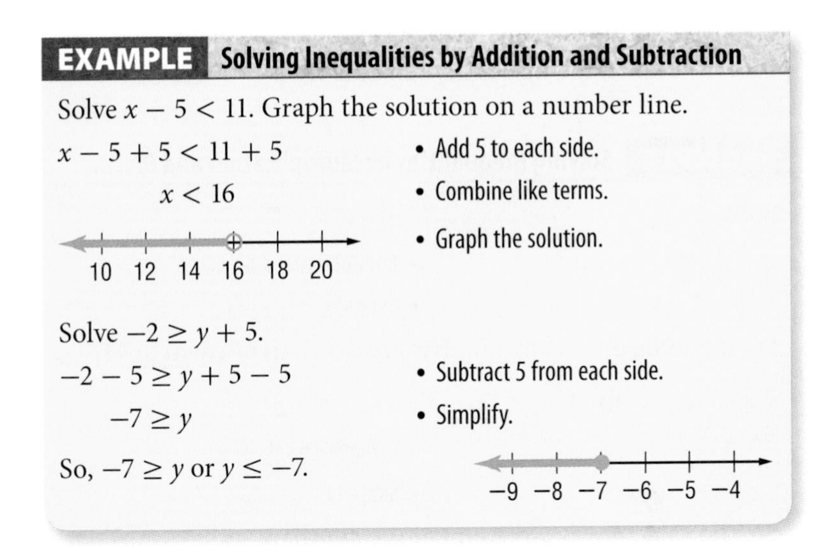

EXAMPLE Solving Inequalities by Addition and Subtraction

Solve $x - 5 < 11$. Graph the solution on a number line.

$x - 5 + 5 < 11 + 5$ • Add 5 to each side.

$x < 16$ • Combine like terms.

• Graph the solution.

Solve $-2 \geq y + 5$.

$-2 - 5 \geq y + 5 - 5$ • Subtract 5 from each side.

$-7 \geq y$ • Simplify.

So, $-7 \geq y$ or $y \leq -7$.

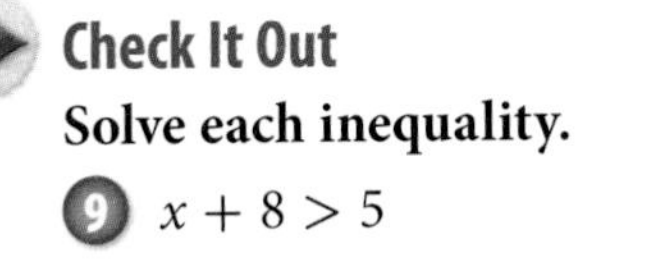

Check It Out

Solve each inequality.

9. $x + 8 > 5$
10. $n - 8 \leq 12$

APPLICATION Oops!

Seventeen-year-old Colin Rizzio took the SAT test and found a mistake in its math portion. One of the questions used the letter a to represent a number. The test makers assumed a was a positive number. But Colin Rizzio thought it could stand for any integer. Rizzio was right!

He notified the test makers by E-mail. They had to change the test scores of 45,000 students. Explain how $2 + a > 2$ changes if a can be positive, zero, or negative. See **HotSolutions** for the answer.

Solving Inequalities by Multiplication and Division

When you multiply or divide each side of an inequality by a positive number, the inequality remains true.

EXAMPLE **Solving Inequalities by Multiplication and Division**

Solve $\frac{1}{2}y \geq 7$.

$2\left(\frac{1}{2}y\right) \geq 2(7)$ • Multiply each side by 2.

$y \geq 14$ • Simplify.

So, the value of y is any number greater than or equal to 14.

Solve $6x > -30$.

$\frac{6x}{6} > \frac{-30}{6}$ • Divide each side by 6.

$x > -5$ • Simplify.

So, the value of x is any number greater than -5.

When you multiply or divide each side of an inequality by a negative number, the direction of the inequality symbol must be reversed for the inequality to remain true.

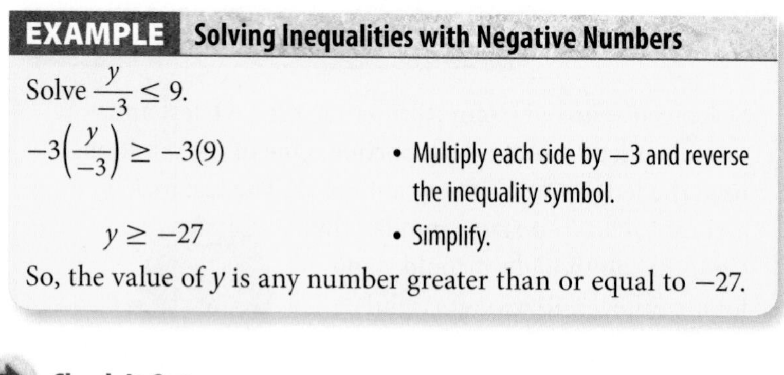

EXAMPLE **Solving Inequalities with Negative Numbers**

Solve $\frac{y}{-3} \leq 9$.

$-3\left(\frac{y}{-3}\right) \geq -3(9)$ • Multiply each side by -3 and reverse the inequality symbol.

$y \geq -27$ • Simplify.

So, the value of y is any number greater than or equal to -27.

Check It Out

Solve each inequality.

11. $3x < 12$
12. $5x \geq 35$
13. $-2x > 8$
14. $-4x \leq 12$

6•6 Exercises

Draw the number line showing the solutions to each inequality.

1. $x < -2$ **2.** $y \geq 0$ **3.** $n > -1$ **4.** $x \leq 7$

Write the inequality for each number line.

5. −3 −2 −1 0 1 2 3

6. −4 −3 −2 −1 0 1 2 3 4

Write an inequality for each sentence.

7. You must be 13 years of age or older to sit in the front seat.

8. The difference between a number and 12 is less than 7.

Solve each inequality.

9. $-3n + 7 \leq 1$

10. $8 - y > 5$

11. $-7 \leq \frac{a}{6}$

12. $-72 > -9y$

13. Which operation(s) requires that the inequality sign be reversed?

A. addition of -2

B. subtraction of -2

C. multiplication by -2

D. division by -2

14. If $x = -3$, is it true that $3(x - 4) \leq 2x$?

15. If $x = 6$, is it true that $2(x - 4) < 8$?

16. Which of the following statements is false?

A. $-7 \leq 2$ **B.** $0 \leq -4$ **C.** $6 \geq -6$ **D.** $3 \geq 3$

17. Which of the following inequalities does not have $x < 2$ as its solution?

A. $-4x < -8$

B. $x + 6 < 8$

C. $4x - 1 < 7$

D. $-x > -2$

6•7 Graphing on the Coordinate Plane

Axes and Quadrants

A coordinate plane is a horizontal number line and a vertical number line that intersect at their zero points. The **origin** is the point of intersection of the two number lines. The ***x*-axis** is the horizontal number line, and the ***y*-axis** is the vertical number line. There are four sections of the coordinate plane called **quadrants**. Each quadrant is named by a Roman numeral, as shown in the diagram.

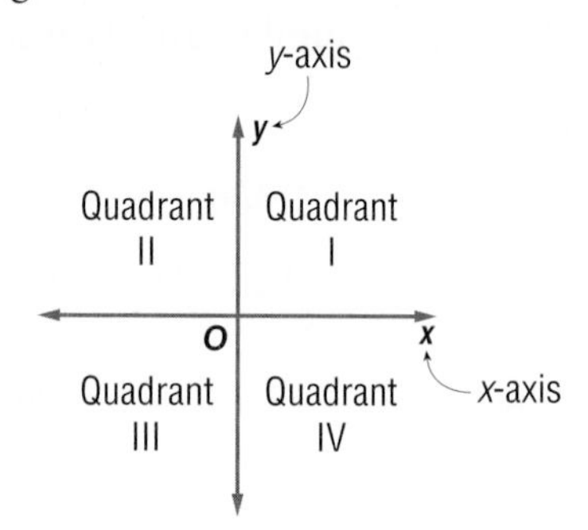

Check It Out

Fill in the blanks.

1. The vertical number line is called the ____.
2. The upper left section of the coordinate plane is called ____.
3. The lower right section of the coordinate plane is called ____.
4. The horizontal number line is called the ____.

Writing an Ordered Pair

A **point** on the coordinate plane is named by an *ordered pair.* An **ordered pair** is a set of numbers, or coordinates, written in the form (x, y). The first number in the ordered pair is the x-coordinate. The x-coordinate represents the horizontal placement of the point. The second number is the y-coordinate. The y-coordinate represents the vertical placement of the point.

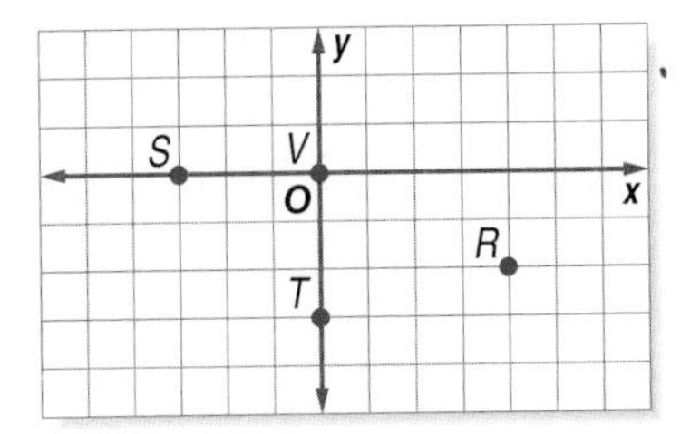

The x-coordinate of point R is 4 units to the right of the origin. The y-coordinate of point R is 2 units down. So, the ordered pair for point R is $(4, -2)$. Point S is 3 units to the left of the origin and 0 units up or down, so its ordered pair is $(-3, 0)$. Point T is 0 units to the left or right of the origin and 3 units down, so its ordered pair is $(0, -3)$. Point V is the origin, and its ordered pair is $(0, 0)$.

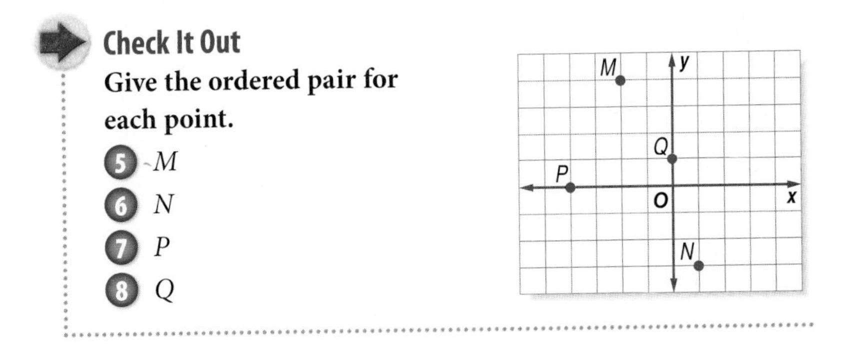

Check It Out

Give the ordered pair for each point.

5. M
6. N
7. P
8. Q

Locating Points on the Coordinate Plane

To locate point $A(2, -2)$, move 2 units to the right of the origin and then 2 units down. Point A lies in Quadrant IV. To locate point $B(-1, 0)$, move 1 unit to the left of the origin and 0 units up or down. Point B lies on the x-axis. Point $C(4, 3)$ is 4 units to the right and 3 units up from the origin. Point C lies in Quadrant I. Point $D(-2, -4)$ is 2 units to the left of the origin and 4 units down. Point D lies in Quadrant III.

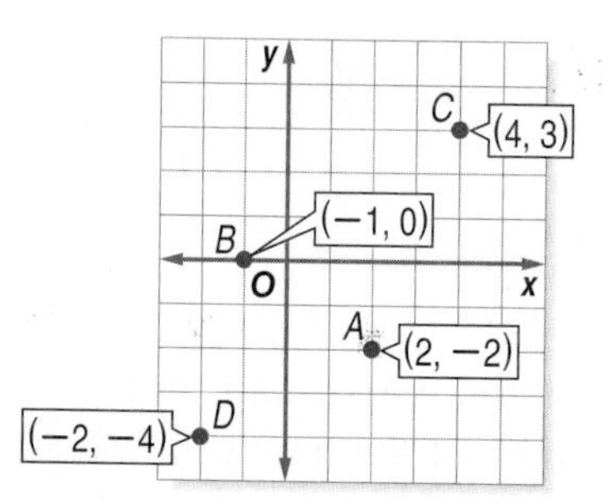

Check It Out

Plot and label each point on a coordinate plane. Then tell in which quadrant or on which axis it lies.

9. $H(-5, 2)$
10. $J(2, -5)$
11. $K(-3, -4)$
12. $L(-1, 0)$

Arithmetic Sequences

An **arithmetic sequence** is an ordered list of numbers in which the difference between any two consecutive terms is the same. The difference between the consecutive terms is called the common difference. You can write an algebraic expression to find the *n*th term of a sequence. First, use a table to examine the sequence, and then find how the term relates to the term number.

EXAMPLE **Arithmetic Sequences**

Write an expression to find the *n*th term of the arithmetic sequence 9, 18, 27, 36, . . . Then find the next three terms.

Term Number (n)	1	2	3	4
Term	9	18	27	36

- Use a table to examine the sequence.

The term is 9 times the term number (n).

- Find how the term relates to the term number.

The expression $9n$ can be used to find the *n*th term.

- Write an expression that can be used to find the *n*th term.

$9 \cdot 5 = 45; 9 \cdot 6 = 54; 9 \cdot 7 = 63$

- Find the next three terms.

So, the next three terms of the sequence are 45, 54, and 63.

Check It Out

Write an expression that can be used to find the *n*th term of each sequence. Then find the next three terms.

13. $-12, -18, -24, -30, \ldots$
14. $\frac{1}{1}, \frac{1}{4}, \frac{1}{9}, \frac{1}{16}, \ldots$
15. 4, 7, 10, 13, . . .

Linear Functions

A relationship that assigns exactly one output value for each input value is called a **function**. The set of input values in a function is called the **domain**. The set of output values is called the **range**.

All linear equations are functions since every x-value corresponds to exactly one y-value. For example, the linear equation $y = 2x - 1$ is a function because every value of x will result in a unique y-value. You can write this equation in function notation by replacing the y with the notation $f(x)$. So, $y = 2x - 1$ is written as $f(x) = 2x - 1$. In a function, x represents the elements of the domain, and $f(x)$ represents the elements of the range. Suppose that you want to find the value of the range that corresponds to $x = 2$ in the domain. This is written $f(2)$ and is read "f of 2." The value of $f(2)$ is found by substituting 2 for x in the equation.

You can organize the input, rule, and output into a function table.

Input	Rule	Output
x	$f(x) = 2x - 1$	$f(x)$
0	$f(x) = 2(0) - 1$	-1
1	$f(x) = 2(1) - 1$	1
-1	$f(x) = 2(-1) - 1$	-3

Recall that a function has exactly one output (y) for each input (x). Therefore, the solutions can be represented as ordered pairs (x, y). Four ordered pairs of the function $f(x) = 2x - 1$ are: $(0, -1)$, $(1, 1)$, $(-1, -3)$, and $(2, 3)$.

A function can also be represented with a graph. The equation $y = \frac{1}{3}x - 2$ represents a function. Choose values for the input x to find the output y. Graph the ordered pairs and draw a line that passes through each point.

EXAMPLE Graphing the Equation of a Line

Graph $f(x) = \frac{1}{3}x - 2$.

- Choose five values for x.

  Because the value of x is to be multiplied by $\frac{1}{3}$, choose values that are multiples of 3, such as −3, 0, 3, 6, and 9.

- Complete a function table.

Input x	Rule	Output $f(x)$
−3	$f(x) = \frac{1}{3}(-3) - 2$	−3
0	$f(x) = \frac{1}{3}(0) - 2$	−2
3	$f(x) = \frac{1}{3}(3) - 2$	−1
6	$f(x) = \frac{1}{3}(6) - 2$	0
9	$f(x) = \frac{1}{3}(9) - 2$	1

- Write the five solutions as ordered pairs.

  (−3, −3), (0, −2), (3, −1), (6, 0), and (9, 1)

- Plot the points on a coordinate plane, and draw the line.

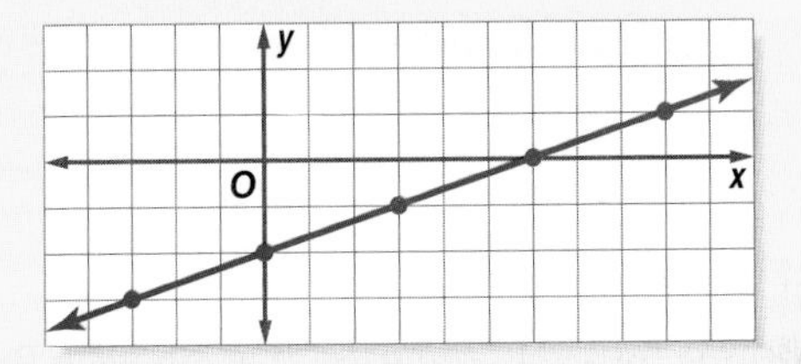

The ordered pair corresponding to any point on the line is a solution of the function $f(x) = \frac{1}{3}x - 2$. A function in which the graph of the solutions form a line is called a linear function.

Check It Out

Complete a function table of five values for each equation. Then graph the line.

16. $y = 3x - 2$
17. $y = 2x + 1$
18. $y = \frac{1}{2}x - 3$
19. $y = -2x + 3$

6•7 Exercises

Fill in the blanks.

1. The horizontal number line is called the ____.
2. The lower left region of the coordinate plane is called ____.
3. The upper right region of the coordinate plane is called ____.

Give the ordered pair for each point.

4. *A*
5. *B*
6. *C*
7. *D*

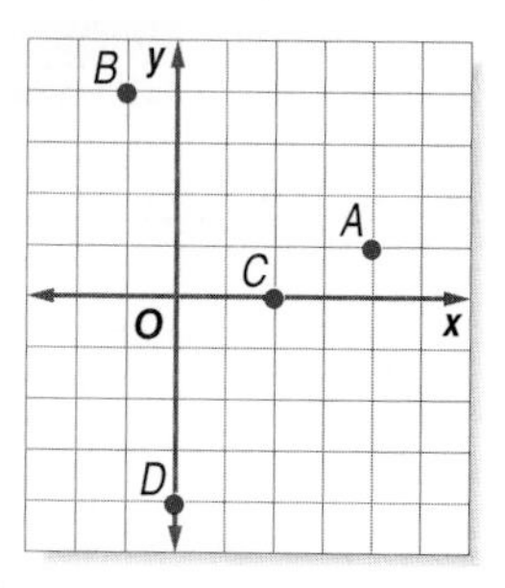

Plot each point on the coordinate plane and tell in which quadrant or on which axis it lies.

8. $H(2, 5)$ 9. $J(-1, -2)$ 10. $K(0, 3)$ 11. $L(-4, 0)$

Write an expression that can be used to find the *n*th term of the sequence. Then find the next three terms.

12. 1, 4, 9, 16, . . .
13. 5, 8, 11, 14, . . .
14. $1, \frac{3}{2}, 2, \frac{5}{2}, \ldots$

Find five solutions for each equation. Graph each line.

15. $y = 2x - 2$
16. $y = -3x + 3$
17. $y = \frac{1}{2}x - 1$

6•8 Slope and Intercept

Slope

One important characteristic of a line is its *slope*. **Slope** is a measure of a line's steepness. The slope, or rate of change, is given by the ratio of **rise** (vertical change) to **run** (horizontal change). The rise is the difference in the y-coordinates. The run is the difference in the x-coordinates.

$$\text{slope} = \frac{\text{rise (difference in the } y\text{-coordinates)}}{\text{run (difference in the } x\text{-coordinates)}}$$

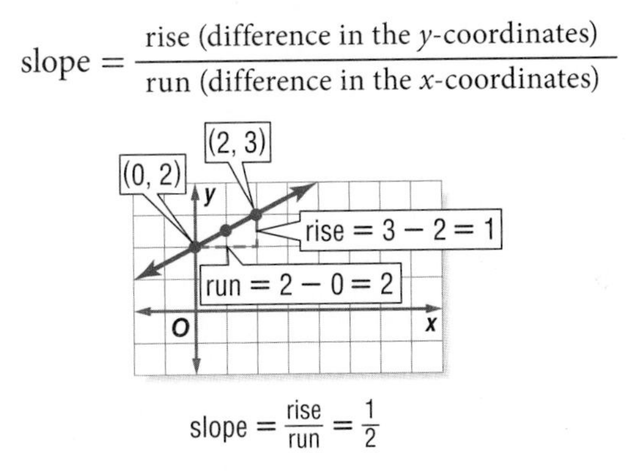

$$\text{slope} = \frac{\text{rise}}{\text{run}} = \frac{1}{2}$$

Notice that for line a, the rise between the two given points is $7 - 2$, or 5 units, and the run is $3 - 1$, or 2 units. Therefore, the slope of the line is $\frac{5}{2}$.

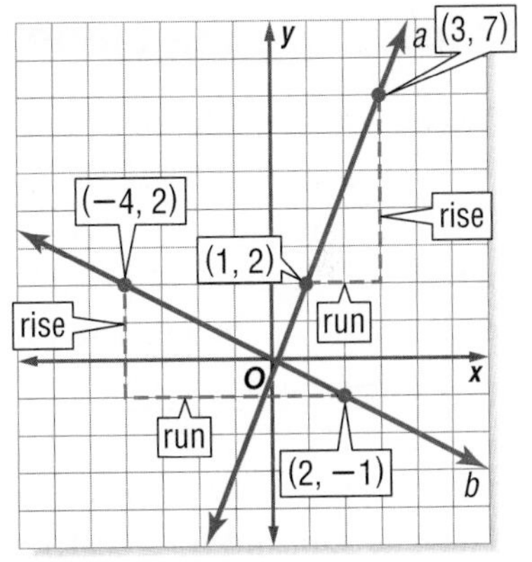

For line b, the rise between the two given points is -3 and the run is 6, so the slope of the line is $-\frac{3}{6} = -\frac{1}{2}$.

The slope of a straight line is constant. Therefore, the slope between any two points on line a will always equal $\frac{5}{2}$. Similarly, the slope between any two points on line b will equal $-\frac{1}{2}$.

Check It Out

Find the slope of each line.

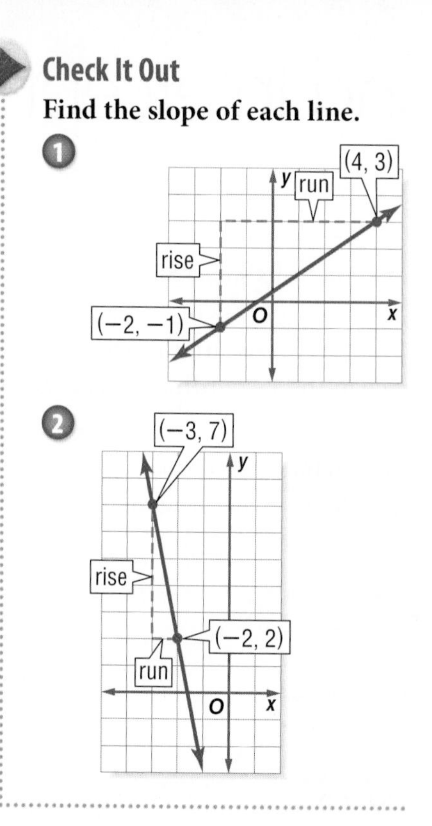

Calculating the Slope of a Line

You can calculate the slope of a line if you are given any two points on a line. The rise is the difference of the y-coordinates and the run is the difference of the x-coordinates. For the line that passes through the points (−2, 3) and (5, −6), the slope can be calculated as shown. The variable m is used to represent slope.

$$m = \frac{\text{rise}}{\text{run}} = \frac{3 - (-6)}{-2 - 5} = -\frac{9}{7}$$

The order in which you subtract the coordinates does not matter, as long as you find both differences in the same order.

$$m = \frac{\text{rise}}{\text{run}} = \frac{-6 - 3}{5 - (-2)} = -\frac{9}{7}$$

EXAMPLE Calculating the Slope of a Line

Find the slope of the line that contains the points (3, 1) and (−2, −3).

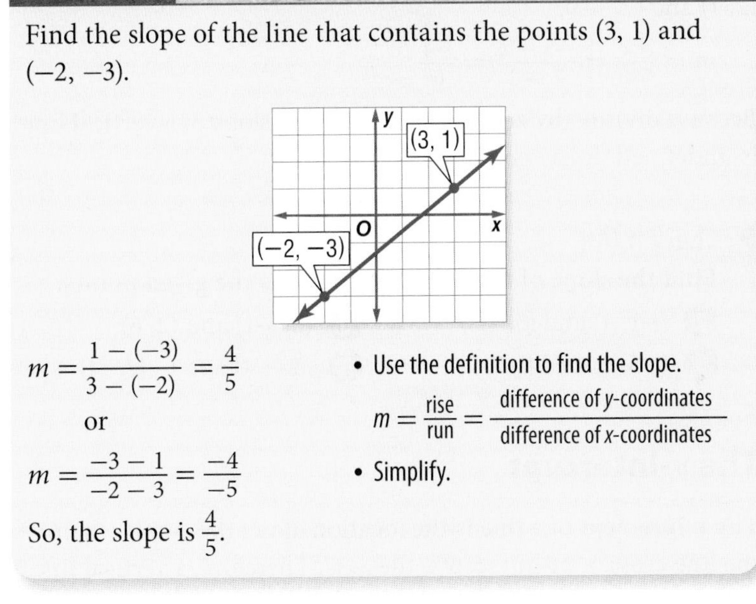

$m = \frac{1-(-3)}{3-(-2)} = \frac{4}{5}$

or

$m = \frac{-3-1}{-2-3} = \frac{-4}{-5}$

- Use the definition to find the slope.

  $m = \frac{\text{rise}}{\text{run}} = \frac{\text{difference of } y\text{-coordinates}}{\text{difference of } x\text{-coordinates}}$

- Simplify.

So, the slope is $\frac{4}{5}$.

Check It Out

Find the slope of the line that contains the given points.

3. (−1, 7) and (4, 2)
4. (−3, −4) and (1, 2)
5. (−2, 0) and (4, −3)
6. (0, −3) and (2, 7)

Slopes of Horizontal and Vertical Lines

Calculate the slope of a horizontal line that contains the points (−1, −3) and (2, −3).

$$m = \frac{\text{rise}}{\text{run}} = \frac{-3-(-3)}{2-(-1)} = \frac{0}{3} = 0$$

A horizontal line has no rise; its slope is 0.

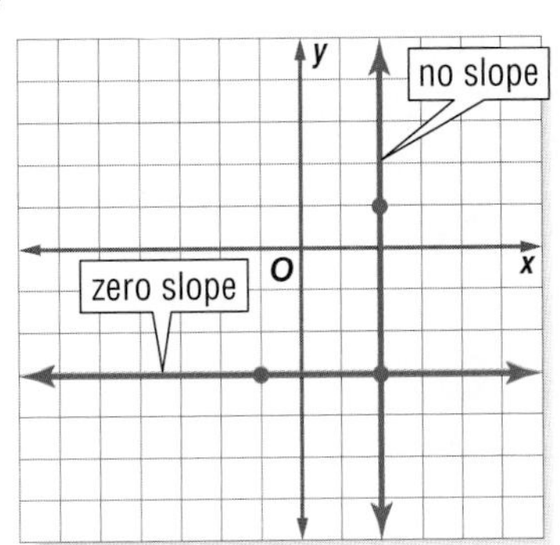

Calculate the slope of a vertical line that contains the points (2, 1) and (2, −3).

$$m = \frac{\text{rise}}{\text{run}} = \frac{-3 - 1}{2 - 2} = \frac{-4}{0}$$

Because division by zero is undefined, the slope of a vertical line is undefined. It has *no slope.*

Check It Out

Find the slope of the line that contains the given points.

7. (−1, 4) and (5, 4)
8. (2, −1) and (2, 6)
9. (−5, 0) and (−5, 7)
10. (4, −4) and (−1, −4)

The *y*-Intercept

The ***y*-intercept** of a line is the location along the *y*-axis where the line crosses, or intercepts, the axis. Therefore, a vertical line, with the exception of $x = 0$, does not have a *y*-intercept.

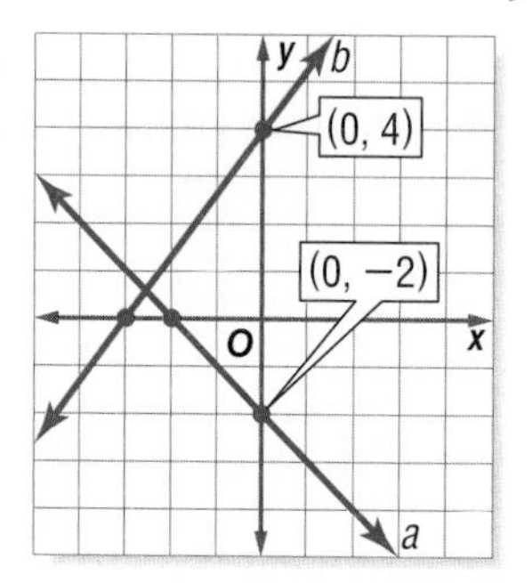

The *y*-intercept of line *a* is −2, and the *y*-intercept of line *b* is 4.

Check It Out

Identify the *y*-intercept of each line.

11. *c*
12. *d*
13. *e*
14. *f*

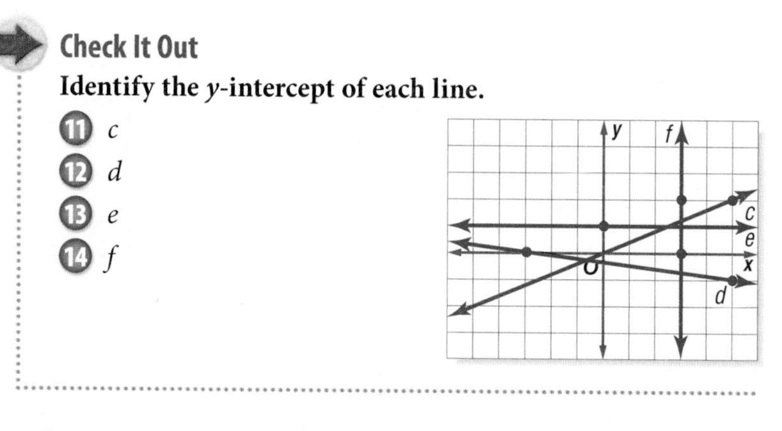

6•8 SLOPE AND INTERCEPT

Using the Slope and *y*-Intercept to Graph a Line

A line can be graphed by using the slope and the *y*-intercept. First, plot the *y*-intercept. Then, use the rise and run of the slope to locate a second point on the line. Connect the two points to graph the line.

EXAMPLE Graphing a Line by Using the Slope and *y*-Intercept

Graph the line with slope −2 and *y*-intercept 3.

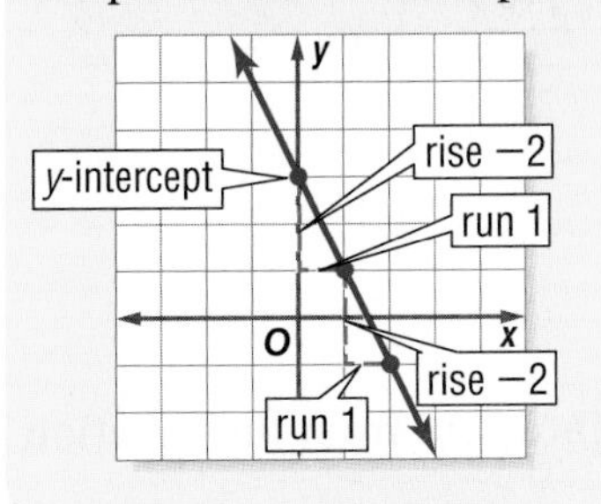

- Plot the *y*-intercept.
- Use the slope to locate other points on the line. If the slope is a whole number *a*, remember $a = \frac{a}{1}$, so rise is *a* and run is 1.
- Draw a line through the points.

Check It Out

Graph each line.

15. slope $= \frac{1}{3}$, *y*-intercept is −2.
16. slope $= -\frac{2}{5}$, *y*-intercept is 4.
17. slope = 3, *y*-intercept is −3.
18. slope = −2, *y*-intercept is 0.

Slope-Intercept Form

The equation $y = mx + b$ is the *slope-intercept form* of the equation of a line. When an equation is in this form, it is easy to identify the slope of the line and the y-intercept. The slope of the line is given by m and the y-intercept is b. The graph of the equation $y = \frac{2}{3}x - 2$ is a line that has a slope of $\frac{2}{3}$ and a y-intercept at (0, −2). The graph is shown below.

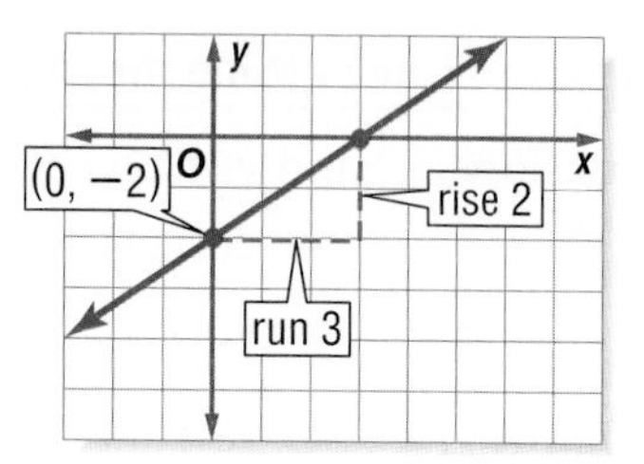

Check It Out

Determine the slope and the y-intercept from the equation of each line.

19. $y = -2x + 3$
20. $y = \frac{1}{5}x - 1$
21. $y = -\frac{3}{4}x$
22. $y = 4x - 3$

Writing Equations in Slope-Intercept Form

To change the equation $4x - 3y = 9$ from standard form to slope-intercept form, isolate the y on one side of the equation.

EXAMPLE **Writing the Equation of a Line in Slope-Intercept Form**

Write $4x - 3y = 9$ in slope-intercept form.

$4x - 3y - 4x = 9 - 4x$ • Isolate the term that contains y by subtracting $4x$ from each side.

$-3y = -4x + 9$ • Combine like terms.

$\frac{-3y}{-3} = \frac{-4x + 9}{-3}$ • Isolate y by dividing each side by -3.

$y = \frac{-4}{-3}x + \frac{9}{-3}$ • Simplify.

$y = \frac{4}{3}x - 3$

So, the slope-intercept form of the equation $4x - 3y = 9$ is $y = \frac{4}{3}x - 3$.

The slope of the line in the example above is $\frac{4}{3}$ and the y-intercept is located at -3. The graph of the line is shown.

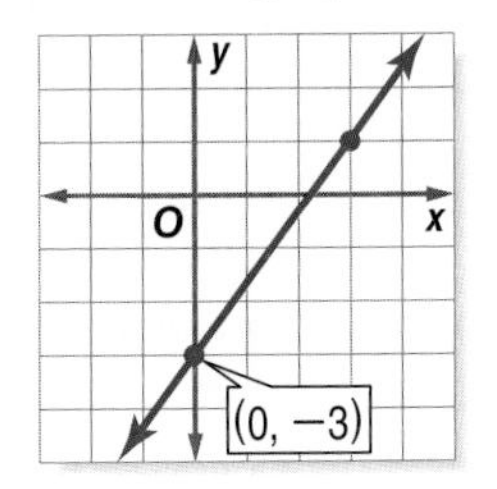

Check It Out

Write each equation in slope-intercept form. Graph the line.

23. $x + 2y = 6$
24. $2x - 3y = 9$
25. $4x - 2y = 4$
26. $7x + y = 8$

Slope-Intercept Form and Horizontal and Vertical Lines

The equation of a horizontal line is $y = b$, where b is the y-intercept of the line. In the graph below, the horizontal line has the equation $y = 2$. This equation is in slope-intercept form since the equation could be written $y = 0x + 2$. The slope is 0, and the y-intercept is 2.

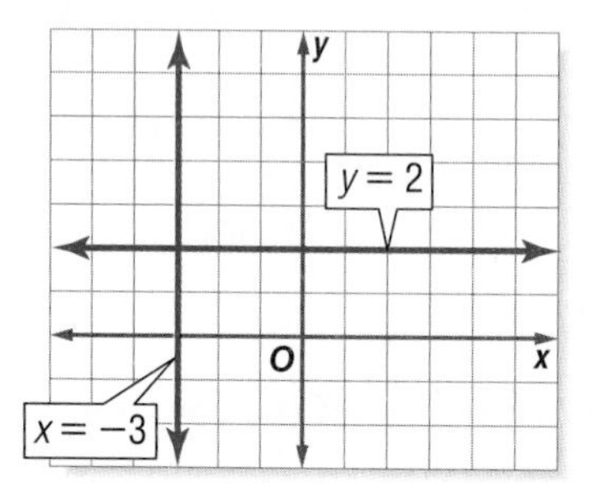

The equation of a vertical line is $x = a$, where a is the x-intercept of the line. In the graph above, the vertical line has the equation $x = -3$. This equation is not in slope-intercept form because y is not isolated on one side of the equation. A vertical line has no slope; therefore, the equation of a vertical line cannot be written in slope-intercept form.

Check It Out

Give the slope and y-intercept of each line. Graph the line.

27. $y = -3$
28. $x = 4$
29. $y = 1$
30. $x = -2$

Writing the Equation of a Line

If you know the slope and the y-intercept of a line, you can write the equation of the line. If a line has a slope of 3 and a y-intercept of −2, substitute 3 for m and −2 for b into the slope-intercept form. The equation of the line is $y = 3x - 2$.

EXAMPLE Writing the Equation of a Line

Write the equation of the line in slope-intercept form.

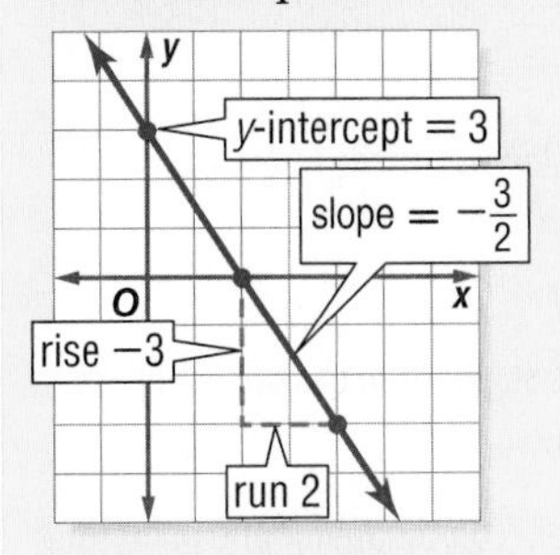

- Identify the y-intercept (b).
- Find the slope ($m = \frac{\text{rise}}{\text{run}}$).
- Substitute the y-intercept and slope into the slope-intercept form. ($y = mx + b$)

$y = -\frac{3}{2}x + 3$

So, the equation of the line in slope-intercept form is $y = -\frac{3}{2}x + 3$.

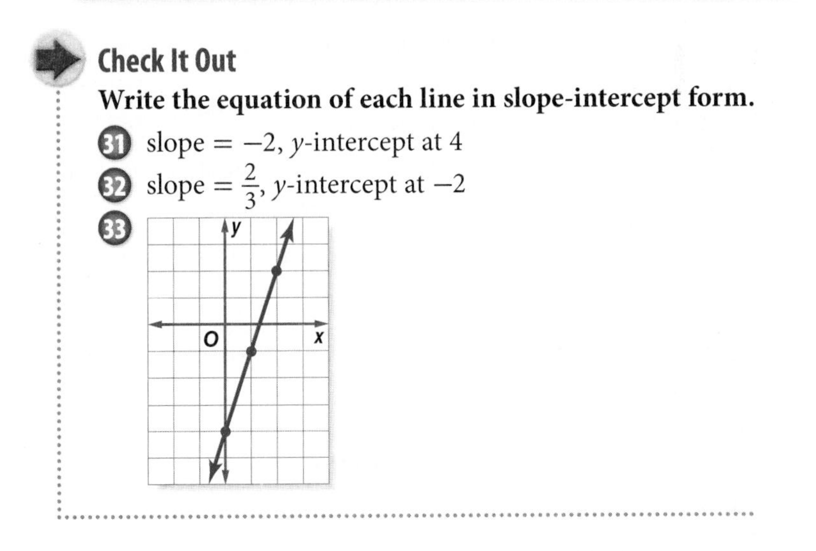

Check It Out

Write the equation of each line in slope-intercept form.

31. slope $= -2$, y-intercept at 4
32. slope $= \frac{2}{3}$, y-intercept at -2
33.

Writing the Equation of a Line from Two Points

If you know two points on a line, you can write its equation. First find the slope. Then find the y-intercept.

EXAMPLE Writing the Equation of a Line from Two Points

Write the equation of the line that contains the points (6, −1) and (−2, 3).

$\frac{3-(-1)}{-2-6} = \frac{4}{-8} = -\frac{1}{2}$ • Calculate the slope using the formula $m = \frac{\text{rise}}{\text{run}}$.

slope $= -\frac{1}{2}$

$y = -\frac{1}{2}x + b$ • Substitute the slope for m in slope-intercept form. ($y = mx + b$)

$-1 = -\frac{1}{2}(6) + b$ • Substitute the y-coordinate for one point for y and the x-coordinate of the *same* point for x.

$-1 = -\frac{6}{2} + b$ • Simplify.

$-1 = -3 + b$

$-1 + 3 = -3 + b + 3$ • Add or subtract to isolate b.

$2 = b$ • Combine like terms.

$y = -\frac{1}{2}x + 2$ • Substitute the values you found for m and b into the slope-intercept form.

So, the equation of the line through points (6, −1) and (−2, 3) is $y = -\frac{1}{2}x + 2$.

Check It Out

Write the equation of the line with the given points.

34. (1, −1) and (5, 3)
35. (−2, 9) and (3, −1)
36. (8, 3) and (−4, −6)
37. (−1, 2) and (4, 2)

6-8 Exercises

Determine the slope of each line.

1. slope of a

2. slope of b

3. slope of c

4. contains $(-3, 1)$ and $(5, -3)$

5. contains $(0, -5)$ and $(2, 6)$

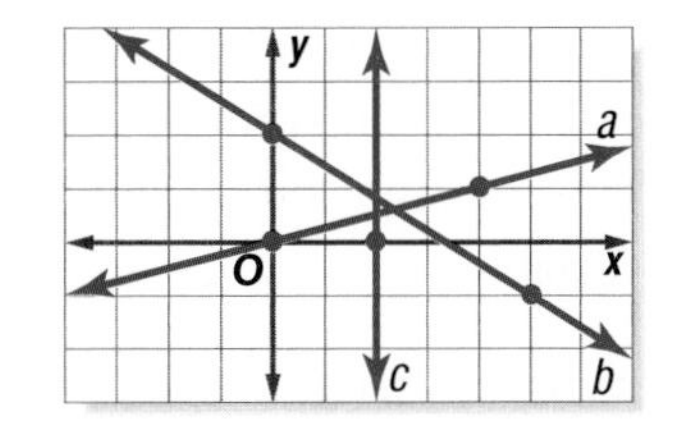

Graph each line.

6. $x = 5$

7. $y = -2$

8. slope $= -\frac{1}{3}$; y-intercept is 2

9. slope $= 4$; y-intercept is -3

Determine the slope and y-intercept from each linear equation.

10. $y = -3x - 2$

11. $y = -\frac{3}{4}x + 3$

12. $y = x + 2$

13. $y = 6$

14. $x = -2$

Write each equation in slope-intercept form. Graph the line.

15. $2x + y = 4$

16. $x - y = 1$

Write the equation of each line.

17. slope $= 3$; y-intercept is -7

18. slope $= -\frac{1}{3}$; y-intercept is 2

19. line a in graph above

Write the equation of the line that contains the given points.

20. $(-4, -5)$ and $(6, 0)$

21. $(-4, 2)$ and $(-3, 1)$

22. $(-6, 4)$ and $(3, -2)$

6·9 Direct Variation

When the ratio of two variable quantities is constant, their relationship is called a **direct variation**. The constant ratio is called the **constant of variation**. In a direct variation equation, the constant rate of change, or slope, is assigned a special variable, k.

A direct variation is a relationship in which the ratio of y *to* x is a constant, k. We say y varies directly with x.

$k = \frac{y}{x}$ or $y = kx$, where $k \neq 0$.

Consider the graph of gas mileage below.

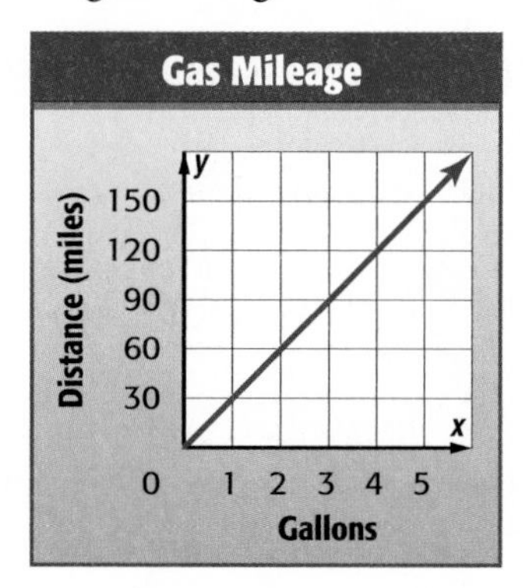

Since the graph of the data forms a line, the rate of change is constant. Use the graph to find the constant ratio.

$$k = \frac{miles\ (y)}{gallon\ (x)} \rightarrow \frac{60}{2}\ or\ \frac{30}{1}, \frac{90}{3}\ or\ \frac{30}{1}, \frac{120}{4}\ or\ \frac{30}{1}, \frac{150}{5}\ or\ \frac{30}{1}$$

Therefore, the slope $(k) = \frac{30}{1}$.

In this example, the ratio of miles traveled to gallons of gas used remains constant. The car travels 30 miles for every gallon of gas.

EXAMPLE Determining Direct Variation

Determine whether the linear function is a direct variation. If so, state the constant of variation.

Hours, x	2	4	6	8
Earnings, y	16	32	48	64

Compare the ratio of y to x.

$$k = \frac{earnings}{hours} \rightarrow \frac{16}{2} \text{ or } \frac{8}{1}, \frac{32}{4} \text{ or } \frac{8}{1}, \frac{48}{6} \text{ or } \frac{8}{1}, \frac{64}{8} \text{ or } \frac{8}{1}$$

Because the ratios are the same, the function is a direct variation. So, the constant of variation, k, is $\frac{8}{1}$.

Check It Out

Solve.

1. The Shelby Super Car (SSC) can travel 13.77 kilometers in 2 minutes and 41.31 kilometers in 6 minutes. If the distance varies directly with time, how many kilometers per hour can the SSC travel?
2. At a farm in Georgia, you can pick 4 peaches for $1.75. How much would it cost to pick 9 peaches?
3. Determine whether the linear function is a direct variation. If so, state the constant of variation.

Minutes, x	20	40	60	80
Profit, y	35	55	75	95

6•9 Exercises

Determine whether each linear function is a direct variation. If so, state the constant of variation.

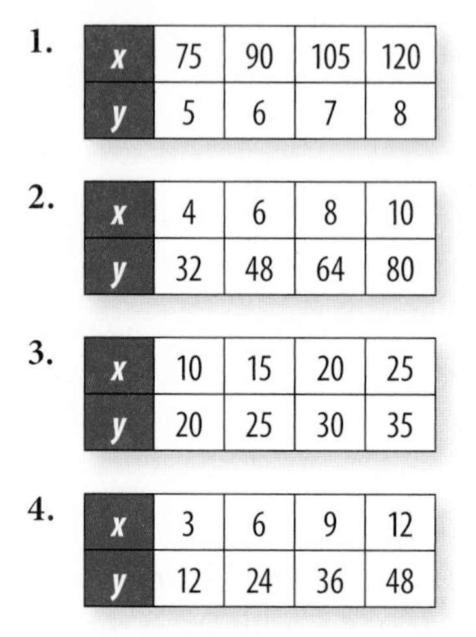

1.

x	75	90	105	120
y	5	6	7	8

2.

x	4	6	8	10
y	32	48	64	80

3.

x	10	15	20	25
y	20	25	30	35

4.

x	3	6	9	12
y	12	24	36	48

If y varies directly with x, write an equation for the direct variation. Then find each value.

5. If $y = 45$ when $x = 15$, find y when $x = 30$.

6. Find y when $x = 20$ if $y = 4$ when $x = 40$.

7. A cupcake recipe requires $2\frac{1}{4}$ cups of flour to make 24 cupcakes. How much flour is required to make 36 cupcakes?

6•10 Systems of Equations

Solving a System of Equations with One Solution

A **system of equations** is a set of two or more equations with the same variables. The equations $y = x + 2$ and $y = -1x + 2$ each have two different unknowns, x and y. The solution of a system of equations is an ordered pair that satisfies both equations. That ordered pair represents the point of intersection of the graphs of the equations.

EXAMPLE Solving Systems of Equations with One Solution

Solve the system $y = x + 2$ and $y = -1x + 2$ by graphing.

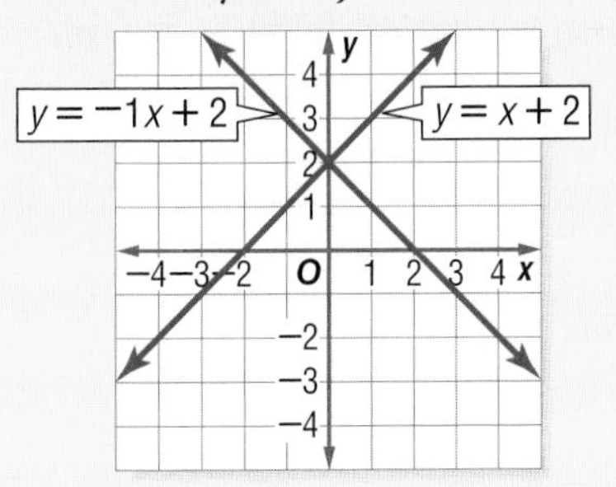

- Graph each equation on the same coordinate plane.

The lines appear to intersect at (0, 2).

$y = x + 2$	$y = -1x + 2$
$2 = 0 + 2$	$2 = -1(0) + 2$
$2 = 2$	$2 = 2$

- Check both equations by replacing x with 0 and y with 2.

So, the solution of the system is (0, 2).

Check It Out

Solve each system of equations by graphing.

1. $y = x + 3$
   $y = \frac{1}{2}x + 1$

2. $3x + y = 4$
   $y = x - 4$

Solving a System of Equations with No Solution

You may find a linear system of equations where the graphs of the equations are parallel. Because the equations have no points in common (no point of intersection), the system has no solution.

EXAMPLE **Evaluating Systems of Equations with No Solution**

Solve the system $y = x + 2$ and $y = x + 4$ by graphing.

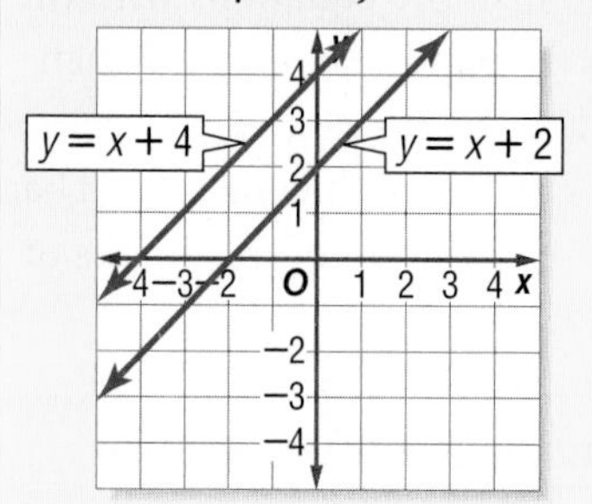

- Graph each equation on the same coordinate plane.

The lines are parallel. So, there is no solution for this system of equations.

Check It Out

Solve each system of equations by graphing.

3. $y = 2x - 1$
   $y - 2x = -4$
4. $y = 2x - 2$
   $y = -x + 1$
5. $2y = 4x + 2$
   $y = 2x + 4$

Solving a System of Equations with an Infinitely Many Solutions

When the equations of a linear system have the same slope and the same y-intercept, they graph the same line. Because the lines intersect at every point, there is an infinite number of solutions.

EXAMPLE Solving Systems of Equations with Many Solutions

Solve the system $y = x - 2$ and $y + 4 = x + 2$ by graphing.

$y + 4 - 4 = x + 2 - 4$
- Write the equation in slope-intercept form by subtracting 4 from each side.

$y = x - 2$
- Simplify.

Both equations are the same.

- Graph the equation.

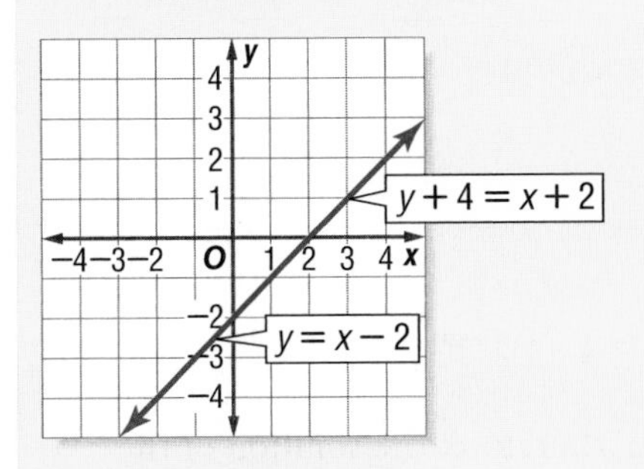

So, the solution of the system is all ordered pairs of the points on line $y = x - 2$.

Check It Out

Solve each system of equations by graphing.

6. $y = 3x - 2$
   $y + 2 = 3x$

7. $y = 4x + 6$
   $y = 2(2x + 3)$

Writing and Solving Systems of Equations

We can use what we know about systems of equations to solve problems that involve two or more different functions. One method of solving a system of equations is to graph the equations on the same coordinate grid and find their point of intersection.

EXAMPLE Writing and Solving Systems of Equations

Two movie channels sell service at different rates. Company A charges $2 per month plus $2 per hour of viewing. Company B charges $7 per month plus $1 per hour of viewing. Write a system of equations that represents the cost of each service for a given amount of time.

Company A: $y = 2x + 2$
Company B: $y = x + 7$

- Write an equation for each company. Let x = number of hours and y = total cost.
- Graph the equations on the same coordinate plane.

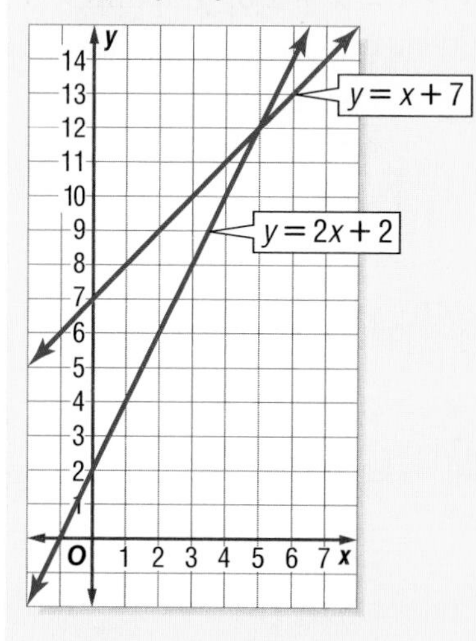

The lines intersect at (5, 12). Therefore, the solution of the system is (5, 12). This solution indicates that the cost for 5 hours of viewing is $12 for each company.

Check It Out

Solve.

8. A group of adults and children went to a museum. There were 9 people in the group. The number of children was three more than the number of adults. Write a system of equations that represents the number of adults and children. Solve by graphing.

6•10 SYSTEMS OF EQUATIONS

6•10 Exercises

Solve each system of equations by graphing.

1. $6y = 4x - 12$
$2x + 2y = 6$

2. $3y = -2x + 1$
$4x + 6y = 2$

3. $x - 2y = -1$
$-3y = 4x - 7$

4. $3x + 2y = 6$
$2y = -3x + 12$

5. $2y = x + 2$
$-x + 2y = -2$

6. $4x - 8y = -8$
$-2y = -x - 2$

Write and solve a system of equations that represents the situation.

7. Carol and Allison made quilts for a bazaar. Together they made 10 quilts. Carol made 2 more than Allison. Find the number of quilts each girl made.

8. Debbie and Kellie together have seventy-five cents in their pockets. Debbie has fifteen more cents than Kellie. Write a system of equations that represents the amount of money each has in their pocket. Solve by graphing.

Algebra

What have you learned?

You can use the problems and the list of words that follow to see what you learned in this chapter. You can find out more about a particular problem or word by referring to the topic number (*for example,* Lesson 6•2).

Problem Set

Write an equation for each sentence. (Lesson 6•1)

1. If seven is subtracted from the product of three and a number, the result is 5 more than the number.

Factor out the greatest common factor in each expression. (Lesson 6•2)

2. $4x + 28$
3. $9n - 6$

Simplify each expression. (Lesson 6•2)

4. $11a - b - 4a + 7b$
5. $8(2n - 1) - (2n + 5)$

6. Find the distance traveled by an in-line skater who skates at 12 miles per hour for $1\frac{1}{2}$ hours. Use the formula $d = rt$. (Lesson 6•3)

Solve each equation. Check your solution. (Lesson 6•4)

7. $\frac{y}{2} - 5 = 1$
8. $y - 10 = 7y + 8$

Use a proportion to solve the exercise. (Lesson 6•5)

9. In a class, the ratio of boys to girls is $\frac{3}{2}$. If there are 12 girls in the class, how many boys are there?

Solve each inequality. Graph the solution. (Lesson 6•6)

10. $x + 9 \leq 6$
11. $4x + 10 > 2$

Draw each point on the coordinate plane and tell in which quadrant or on which axis it lies. (Lesson 6•7)

12. $A(1, 5)$ **13.** $B(4, 0)$ **14.** $C(0, -2)$ **15.** $D(-2, 3)$

16. Write the equation of the line that contains the points (3, −2) and (3, 5). (Lesson 6•8)

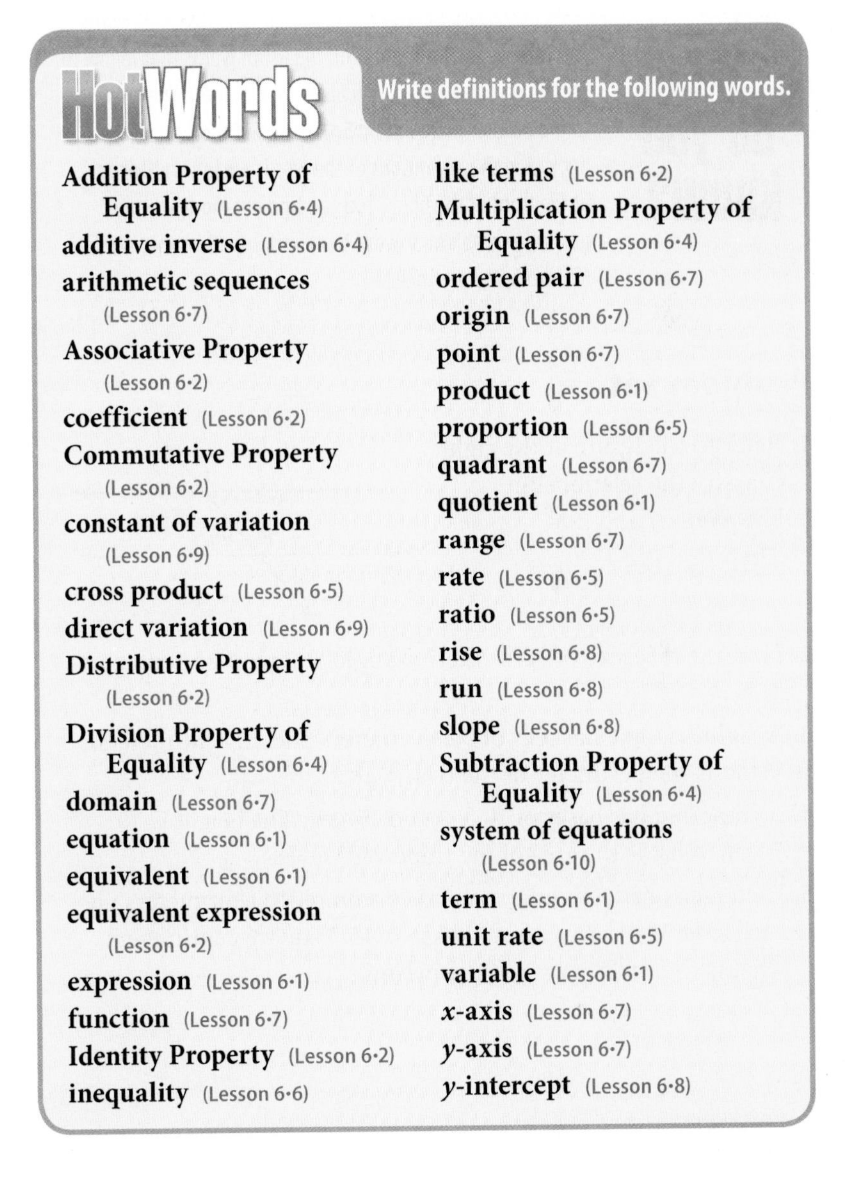

HotWords

Write definitions for the following words.

Addition Property of Equality (Lesson 6•4)
additive inverse (Lesson 6•4)
arithmetic sequences (Lesson 6•7)
Associative Property (Lesson 6•2)
coefficient (Lesson 6•2)
Commutative Property (Lesson 6•2)
constant of variation (Lesson 6•9)
cross product (Lesson 6•5)
direct variation (Lesson 6•9)
Distributive Property (Lesson 6•2)
Division Property of Equality (Lesson 6•4)
domain (Lesson 6•7)
equation (Lesson 6•1)
equivalent (Lesson 6•1)
equivalent expression (Lesson 6•2)
expression (Lesson 6•1)
function (Lesson 6•7)
Identity Property (Lesson 6•2)
inequality (Lesson 6•6)
like terms (Lesson 6•2)
Multiplication Property of Equality (Lesson 6•4)
ordered pair (Lesson 6•7)
origin (Lesson 6•7)
point (Lesson 6•7)
product (Lesson 6•1)
proportion (Lesson 6•5)
quadrant (Lesson 6•7)
quotient (Lesson 6•1)
range (Lesson 6•7)
rate (Lesson 6•5)
ratio (Lesson 6•5)
rise (Lesson 6•8)
run (Lesson 6•8)
slope (Lesson 6•8)
Subtraction Property of Equality (Lesson 6•4)
system of equations (Lesson 6•10)
term (Lesson 6•1)
unit rate (Lesson 6•5)
variable (Lesson 6•1)
***x*-axis** (Lesson 6•7)
***y*-axis** (Lesson 6•7)
***y*-intercept** (Lesson 6•8)

HotTopic 7

Geometry

What do you know?

You can use the problems and the list of words that follow to see what you already know about this chapter. The answers to the problems are in **HotSolutions** at the back of the book, and the definitions of the words are in **HotWords** at the front of the book. You can find out more about a particular problem or word by referring to the topic number (*for example,* Lesson 7·2).

Problem Set

1. Refer to the figure at the right. Classify the relationship between $\angle J$ and $\angle K$. Then find $m\angle J$. (Lesson 7·1)

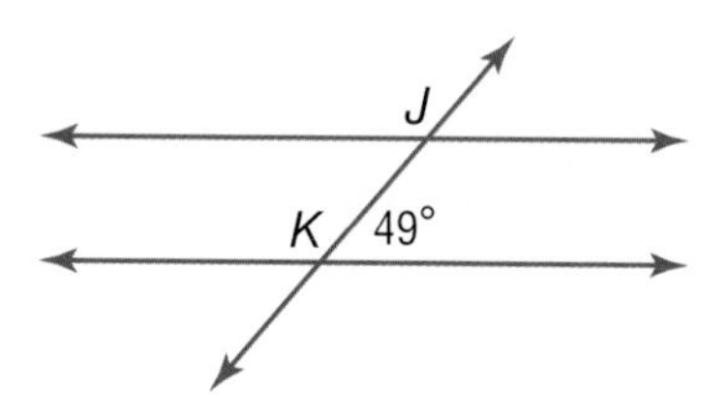

2. Find the measure of each interior angle of a regular pentagon. (Lesson 7·2)
3. A right triangle has legs of 5 centimeters and 12 centimeters. What is the perimeter of the triangle? (Lesson 7·4)
4. A trapezoid has bases of 10 feet and 16 feet. The height is 5 feet. What is the area of the trapezoid? (Lesson 7·5)
5. Each face of a triangular prism is a square, 10 centimeters on each side. If the area of each base is 43.3 square centimeters, what is the surface area of the prism? (Lesson 7·6)

6. Find the volume of a cylinder that has a 5-inch diameter and is 6 inches high. Round to the nearest cubic inch. (Lesson 7•7)
7. What is the circumference and area of a circle with radius 25 feet? Round to the nearest foot or square foot. (Lesson 7•8)
8. Graph the ordered pairs (−1, −1) and (2, 1). Then find the distance between the points. (Lesson 7•9)

HotWords

alternate exterior angles (Lesson 7•1)
alternate interior angles (Lesson 7•1)
arc (Lesson 7•8)
base (Lesson 7•2)
circumference (Lesson 7•8)
complementary angles (Lesson 7•1)
congruent (Lesson 7•1)
corresponding angles (Lesson 7•1)
cube (Lesson 7•2)
diagonal (Lesson 7•2)
diameter (Lesson 7•8)
face (Lesson 7•2)
hypotenuse (Lesson 7•9)
line of symmetry (Lesson 7•3)
net (Lesson 7•6)
parallelogram (Lesson 7•2)
pi (Lesson 7•7)
polygon (Lesson 7•2)
polyhedron (Lesson 7•2)
prism (Lesson 7•2)
pyramid (Lesson 7•2)
Pythagorean Theorem (Lesson 7•9)
Pythagorean triple (Lesson 7•9)
quadrilateral (Lesson 7•2)
radius (Lesson 7•8)
rectangular prism (Lesson 7•2)
reflection (Lesson 7•3)
regular polygon (Lesson 7•2)
rhombus (Lesson 7•2)
rotation (Lesson 7•3)
segment (Lesson 7•2)
supplementary angles (Lesson 7•1)
surface area (Lesson 7•6)
tetrahedron (Lesson 7•2)
transformation (Lesson 7•3)
translation (Lesson 7•3)
transversal (Lesson 7•1)
trapezoid (Lesson 7•2)
triangular prism (Lesson 7•6)
vertical angles (Lesson 7•1)
volume (Lesson 7•7)

7•1 Classifying Angles and Triangles

Classifying Angles

You can classify angles by their measures.

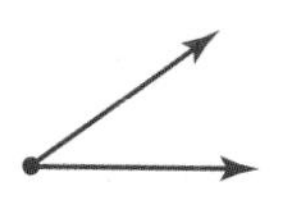

Acute angle
measures less than 90°

Right angle
measures 90°

Obtuse angle
measures greater than 90° and less than 180°

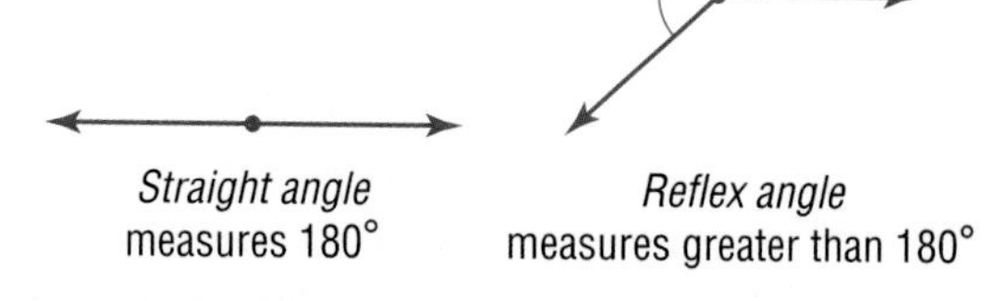

Straight angle
measures 180°

Reflex angle
measures greater than 180°

Angles that share a side are called *adjacent angles.* You can add measures if the angles are adjacent.

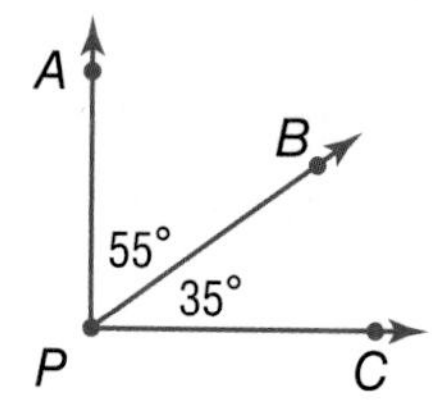

$m\angle APB = 55°$
$m\angle BPC = 35°$
$m\angle APC = 55° + 35° = 90°$
$\angle APC$ is a right angle.

Check It Out

Classify each angle.

1. 230°
2. 180°
3. 40°
4. 135°

Special Pairs of Angles

You can also classify angles by their relationship to each other. **Vertical angles** are opposite angles formed by the intersection of two lines.

$\angle 1$ and $\angle 3$ are vertical angles.
$\angle 2$ and $\angle 4$ are vertical angles.

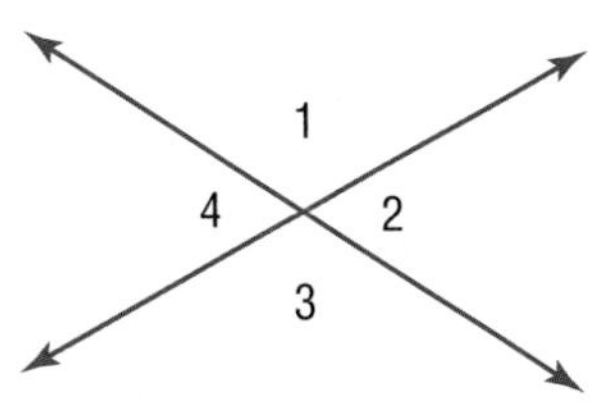

When two angles have the same angle measure, they are **congruent**. Since $m\angle 1 + m\angle 2 = 180°$ and $m\angle 1 + m\angle 4 = 180°$, then $m\angle 2 = m\angle 4$. Vertical angles are congruent.

Two angles are **complementary angles** if the sum of their measures is 90°.

$\angle EFG$ and $\angle GFH$ are complementary angles.

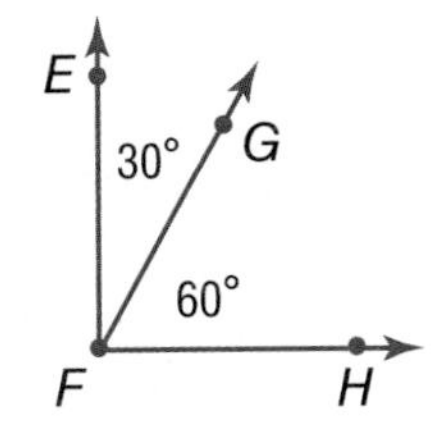

Two angles are **supplementary angles** if the sum of their measures is 180°.

$\angle MNO$ and $\angle ONQ$ are supplementary angles.

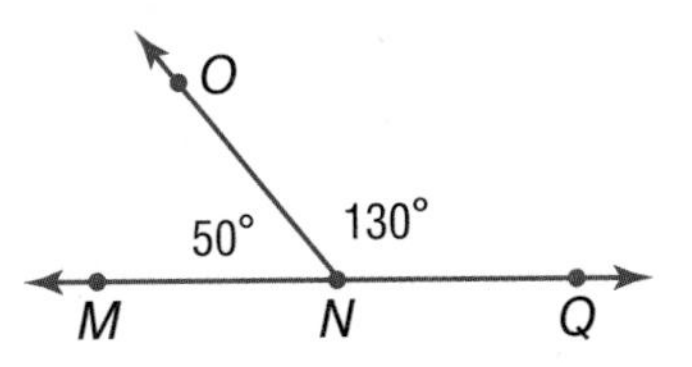

EXAMPLE Finding Missing Angle Measures

$\angle RSU$ is a right angle.

Find the value of x in the figure.

$m\angle RST + m\angle TSU = 90°$	• Write an equation.
$55 + x = 90$	• Substitute the known angle measure.
$55 - 55 + x = 90 - 55$	• Subtract 55 from both sides.
	• Simplify.

So, $x = 35°$.

Check It Out

Find the value of x in each figure.

5

6

Line and Angle Relationships

Lines that lie in the same plane that never intersect are called parallel lines. In the figure, lines b and c are parallel ($b \parallel c$). Two lines that intersect at right angles are called perpendicular lines. In the figure, lines a and b are perpendicular ($a \perp b$) and lines a and c are perpendicular ($a \perp c$).

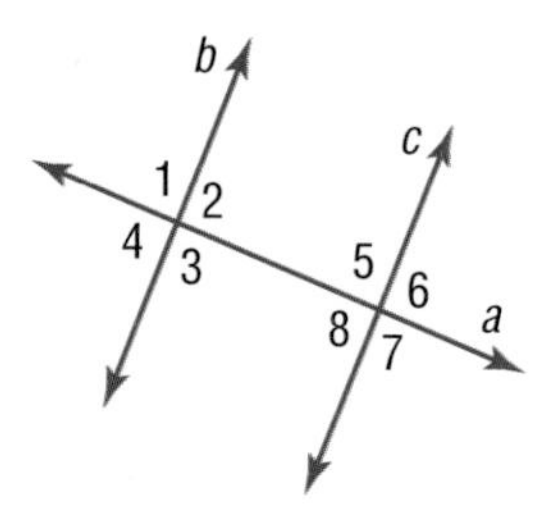

A **transversal** is a line that intersects two or more other lines. In the figure below, line *a* is a transversal of lines *b* and *c*. Eight angles are formed by a transversal: four interior angles and four exterior angles.

$\angle 2$, $\angle 3$, $\angle 5$, and $\angle 8$ are interior angles.
$\angle 1$, $\angle 4$, $\angle 6$, and $\angle 7$ are exterior angles.

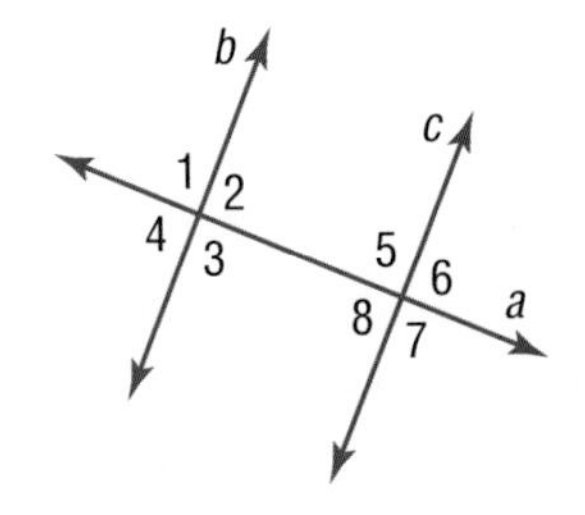

Alternate interior angles are interior angles that lie on opposite sides of the transversal.

In the figure below, $\angle 3$ and $\angle 5$ are alternate interior angles, and $\angle 4$ and $\angle 6$ are alternate interior angles.

When a transversal intersects parallel lines, alternate interior angles are congruent.

Alternate exterior angles are exterior angles that lie on different lines on opposite sides of the transversal. In the figure at the right, $\angle 1$ and $\angle 7$ are alternate exterior angles, and $\angle 2$ and $\angle 8$ are alternate exterior angles. When a transversal intersects parallel lines, alternate exterior angles are congruent.

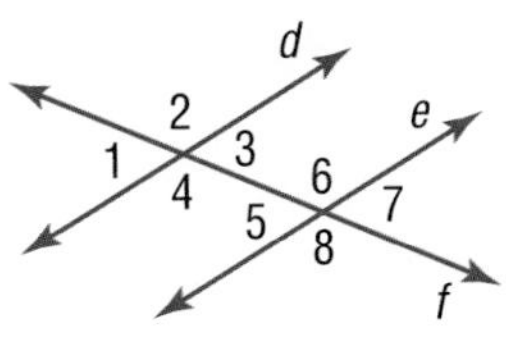

Corresponding angles are angles that lie in the same position in relation to the transversal on the parallel lines.

There are four pairs of corresponding angles in the figure above: $\angle 1$ and $\angle 5$, $\angle 2$ and $\angle 6$, $\angle 3$ and $\angle 7$, and $\angle 4$ and $\angle 8$. When a transversal intersects parallel lines, corresponding angles are congruent.

Check It Out

In the figure at the right, the two parallel lines are cut by the transversal n. $\ell \parallel m$

7. Classify the relationship between $\angle 4$ and $\angle 8$.
8. Classify the relationship between $\angle 3$ and $\angle 5$.
9. Name a pair of alternate exterior angles.
10. Name a pair of congruent angles.

Triangles

Triangles are *polygons* (p. 328) that have three sides, three vertices, and three angles.

You name a triangle by naming the three vertices, in any order. $\triangle ABC$ is read "triangle *ABC*."

Classifying Triangles

Like angles, triangles are classified by their angle measures. They can also be classified by the number of **congruent** sides, which are sides of equal length.

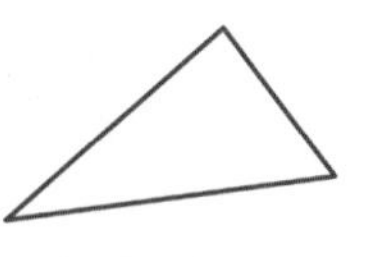

Acute triangle
three acute angles

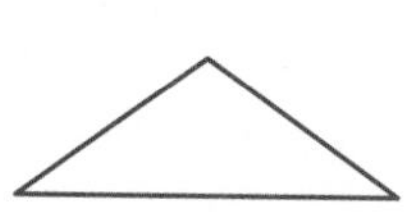

Obtuse triangle
one obtuse angle

Right triangle
one right angle

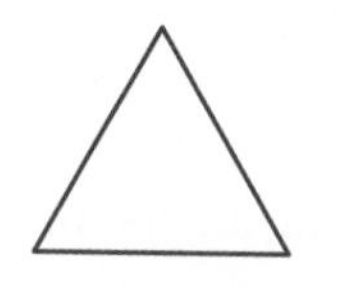

Equilateral triangle
three congruent sides;
three congruent angles

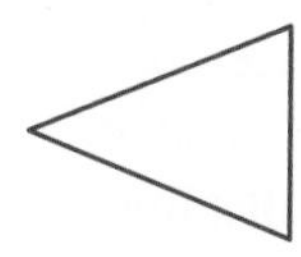

Isosceles triangle
at least two congruent sides;
at least two congruent angles

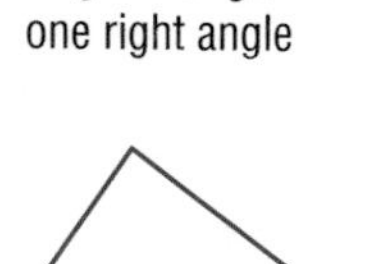

Scalene triangle
no congruent sides

The sum of the measures of the three angles in a triangle is always 180°.

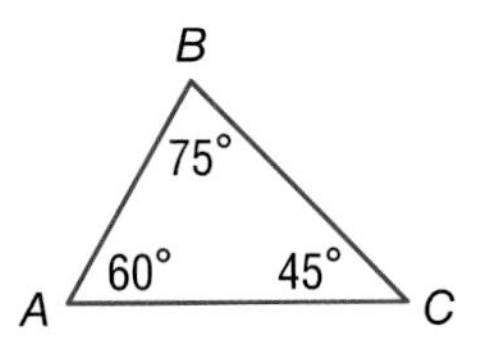

In $\triangle ABC$, $m\angle A = 60°$, $m\angle B = 75°$, and $m\angle C = 45°$.

$60° + 75° + 45° = 180°$

So, the sum of the angles of $\triangle ABC$ is 180°.

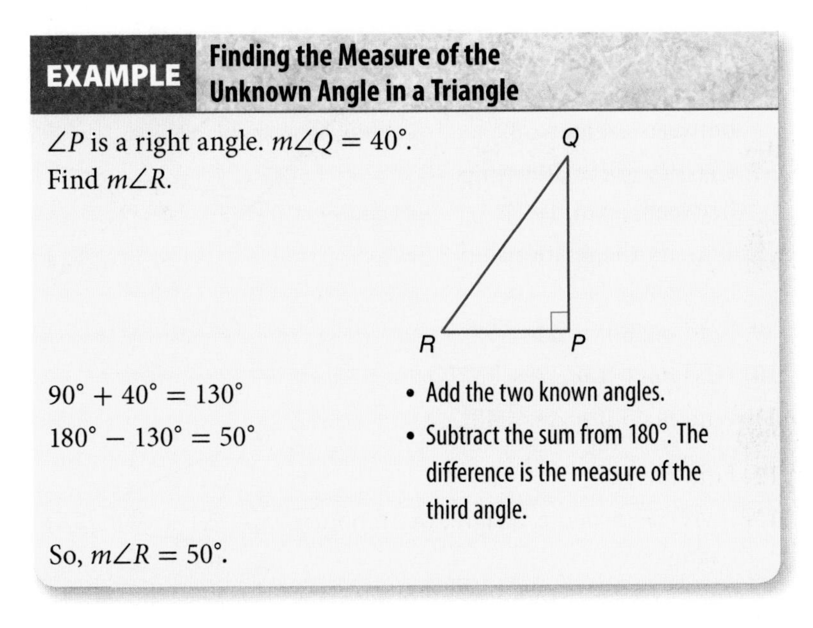

EXAMPLE Finding the Measure of the Unknown Angle in a Triangle

$\angle P$ is a right angle. $m\angle Q = 40°$. Find $m\angle R$.

$90° + 40° = 130°$ • Add the two known angles.

$180° - 130° = 50°$ • Subtract the sum from 180°. The difference is the measure of the third angle.

So, $m\angle R = 50°$.

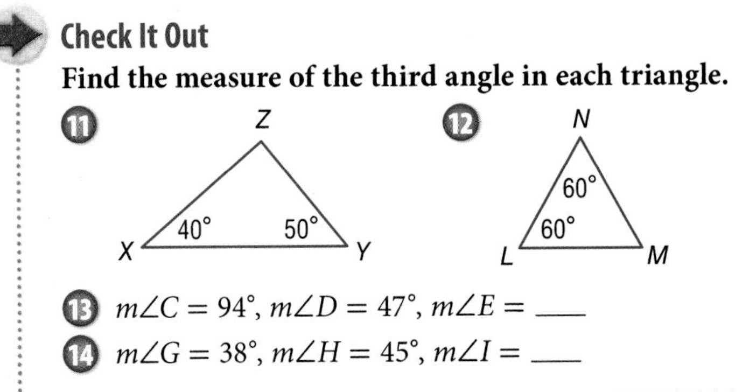

Check It Out

Find the measure of the third angle in each triangle.

11

12

13 $m\angle C = 94°$, $m\angle D = 47°$, $m\angle E =$ ____

14 $m\angle G = 38°$, $m\angle H = 45°$, $m\angle I =$ ____

7•1 Exercises

Use the figure at the right to answer Exercises 1–3.

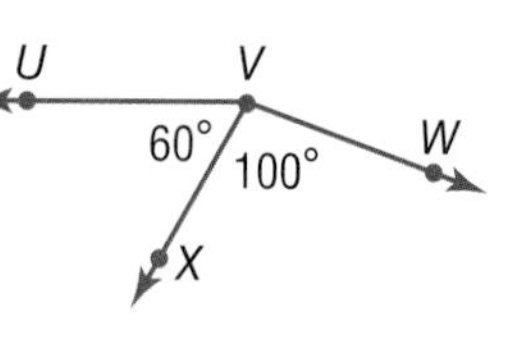

1. Name an acute angle.
2. Name two obtuse angles.
3. What is the measure of $\angle UVW$?

For Exercises 4–11, use the figure below. In the figure, $j \parallel k$ and v is a transversal. Justify your answer.

4. Classify the relationship between $\angle 1$ and $\angle 3$.
5. Classify the relationship between $\angle 5$ and $\angle 8$.
6. Classify the relationship between $\angle 4$ and $\angle 8$.
7. Classify the relationship between $\angle 3$ and $\angle 7$.
8. Classify the relationship between $\angle 2$ and $\angle 6$.

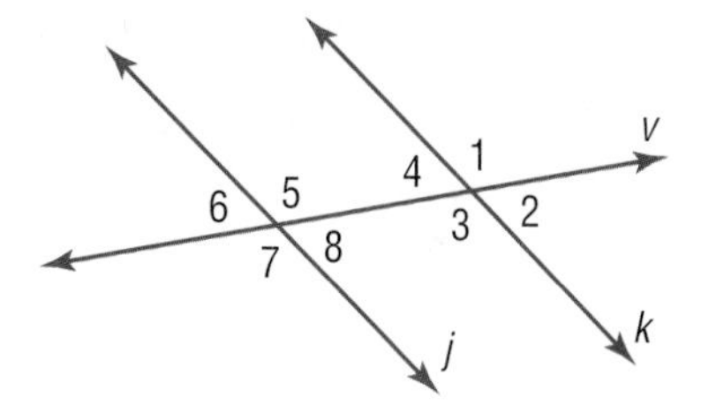

9. Find $m\angle 8$ if $m\angle 4 = 55°$.
10. Find $m\angle 1$ if $m\angle 7 = 137°$.
11. Find $m\angle 5$ if $m\angle 4 = 48°$.

12. Find $m\angle D$.
13. Is $\triangle DEF$ an acute, a right, or an obtuse triangle?

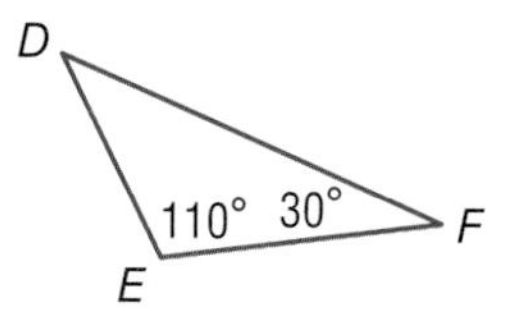

14. Find $m\angle T$.
15. Is $\triangle RST$ an acute, a right, or an obtuse triangle?

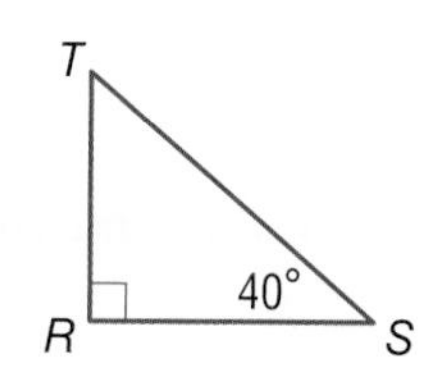

CLASSIFYING ANGLES AND TRIANGLES 7•1

7•2 Naming and Classifying Polygons and Polyhedrons

Quadrilaterals

Four-sided figures are called **quadrilaterals**. Some quadrilaterals have specific names based on the relationship between their sides and/or angles.

To name a quadrilateral, list the four vertices, either clockwise or counterclockwise. One name for the figure at the right is quadrilateral *ISHF.*

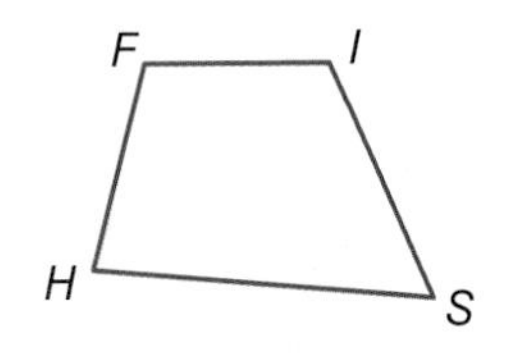

Angles of a Quadrilateral

The sum of the angles of a quadrilateral is 360°. If you know the measures of three angles in a quadrilateral, you can find the measure of the fourth angle.

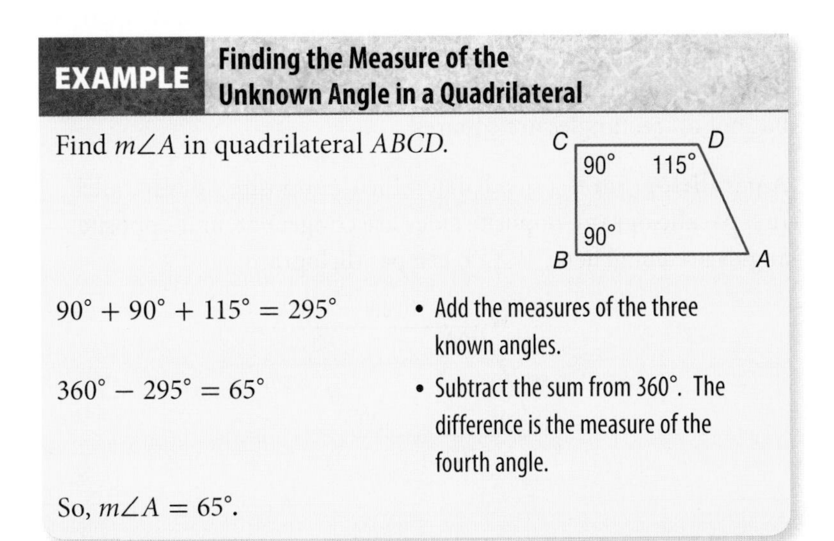

EXAMPLE **Finding the Measure of the Unknown Angle in a Quadrilateral**

Find $m\angle A$ in quadrilateral *ABCD.*

$90° + 90° + 115° = 295°$
- Add the measures of the three known angles.

$360° - 295° = 65°$
- Subtract the sum from 360°. The difference is the measure of the fourth angle.

So, $m\angle A = 65°$.

Check It Out

Use the figure below to answer Exercises 1–3.

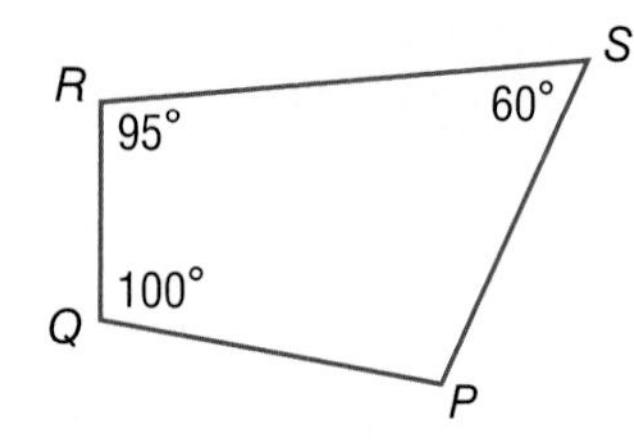

1. Name the quadrilateral in two ways.
2. What is the sum of the angles of a quadrilateral?
3. Find $m\angle P$.

Types of Quadrilaterals

A rectangle is a quadrilateral with four right angles. *ABCD* is a rectangle. Its length is 10 centimeters and its width is 6 centimeters.

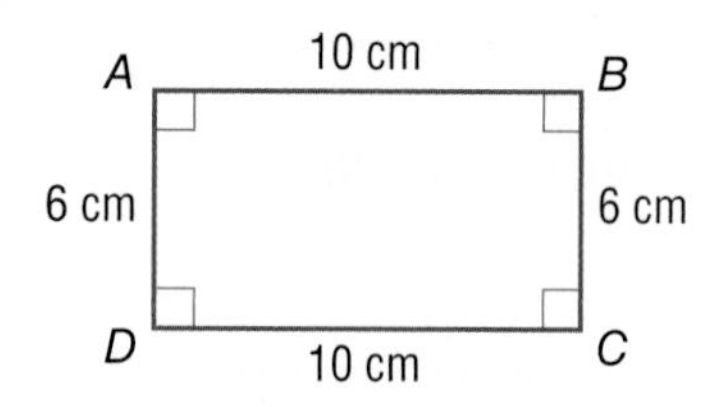

Opposite sides of a rectangle are congruent. If all four sides of the rectangle are congruent, the rectangle is a square. A square is a **regular polygon** because all of the sides are congruent and all of the interior angles are congruent. *All* squares are rectangles, but not all rectangles are squares.

A **parallelogram** is a quadrilateral with opposite sides parallel. In a parallelogram, opposite sides are congruent, and opposite angles are congruent. *WXYZ* is a parallelogram.

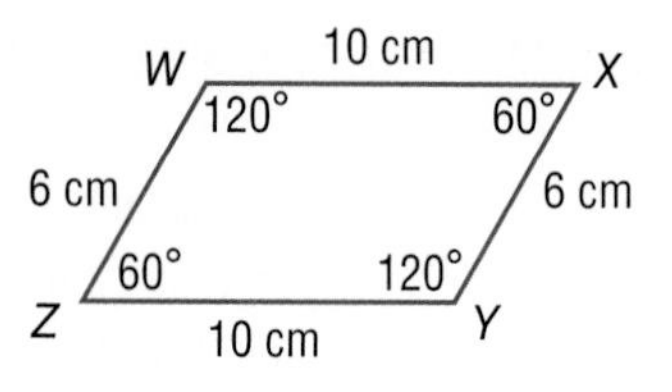

Not all parallelograms are rectangles, but all rectangles are parallelograms. Therefore, all squares are also parallelograms. If all four sides of a parallelogram are congruent, the parallelogram is a **rhombus**. *EFGH* is a rhombus.

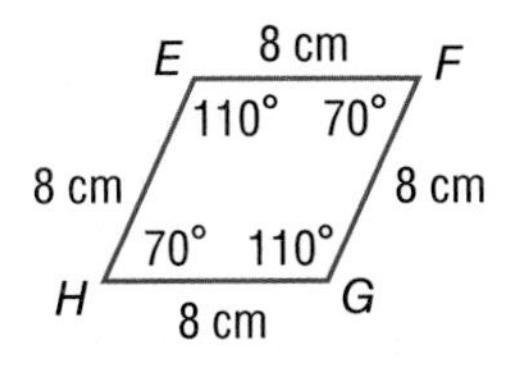

Every square is a rhombus, but not every rhombus is a square because a square must also have congruent angles.

In a **trapezoid**, two sides are parallel and two are not. A trapezoid is a quadrilateral, but it is not a parallelogram. *ACKJ* is a trapezoid.

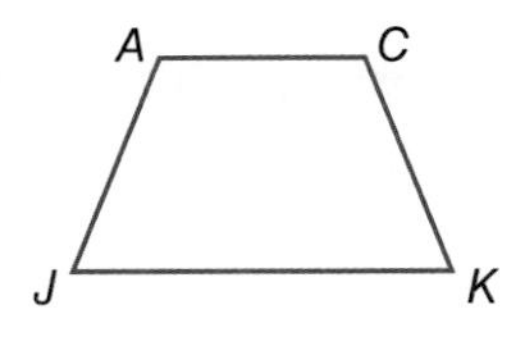

Check It Out

Complete Exercises 4–7.

4. Is quadrilateral *RSTU* a rectangle? a parallelogram? a square? a rhombus? a trapezoid?

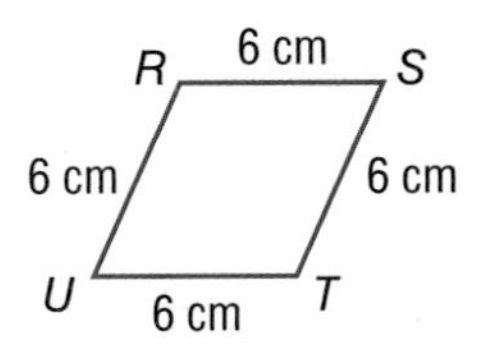

5. Is a square a rhombus? Why or why not?
6. Is a rectangle always a square? Why or why not?
7. Is a parallelogram always a rectangle? Why or why not?

Polygons

A **polygon** is a closed figure that has three or more sides. Each side is a line **segment**, and the sides meet only at the endpoints, or vertices.

This figure is a polygon. These figures are not.

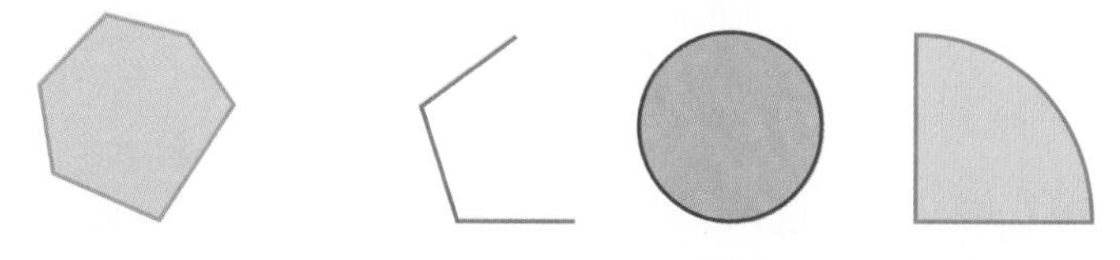

A rectangle, a square, a parallelogram, a rhombus, a trapezoid, and a triangle are all examples of polygons.

A regular polygon is a polygon with congruent sides and angles.

A polygon always has an equal number of sides, angles, and vertices.

For example, a polygon with three sides has three angles and three vertices. A polygon with eight sides has eight angles and eight vertices, and so on.

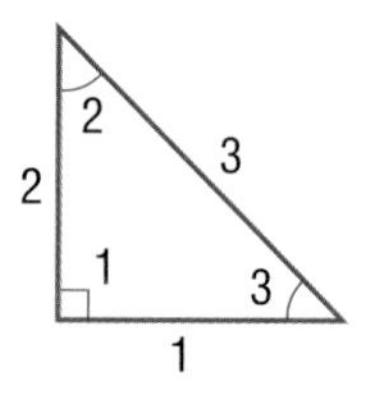

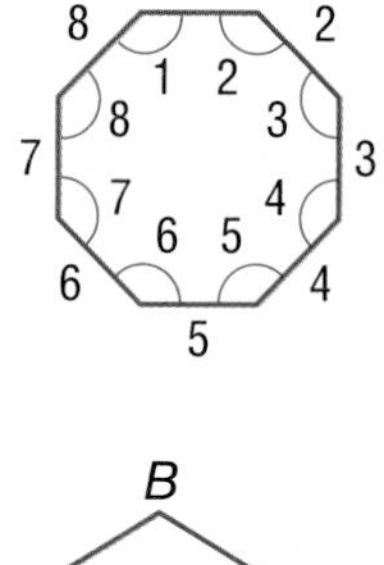

A line segment connecting two vertices of a polygon is either a side or a **diagonal**. $\overline{AE}$ is a side of polygon *ABCDE*. $\overline{AD}$ is a diagonal.

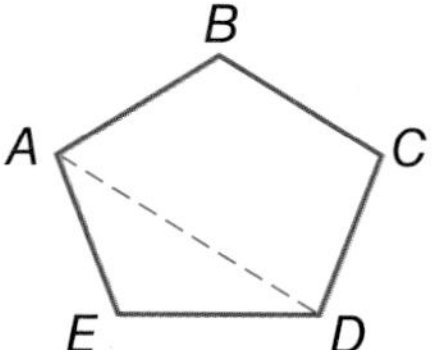

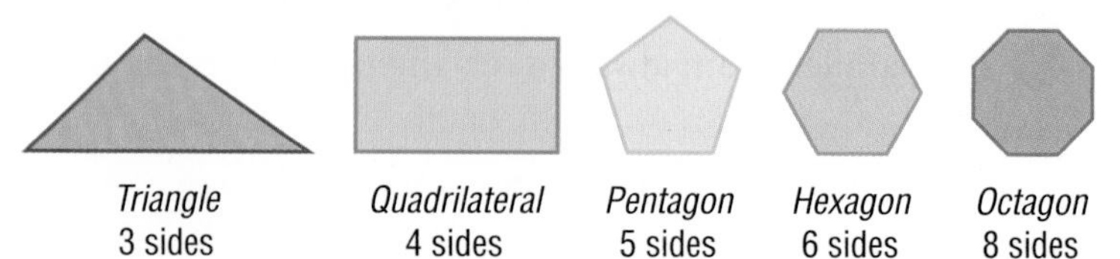

A seven-sided polygon is called a heptagon, a nine-sided polygon is called a nonagon, and a ten-sided polygon is called a decagon.

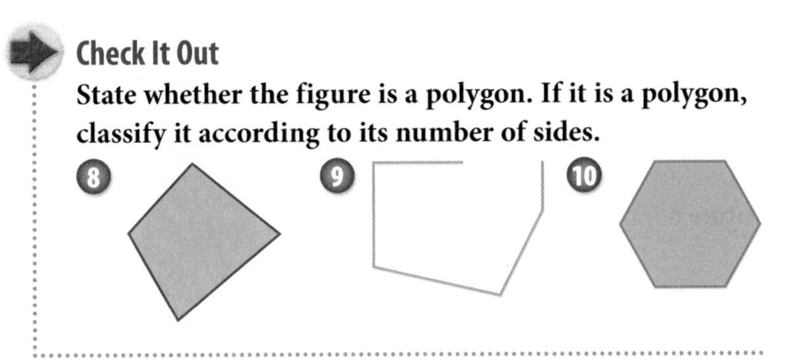

Angles of a Polygon

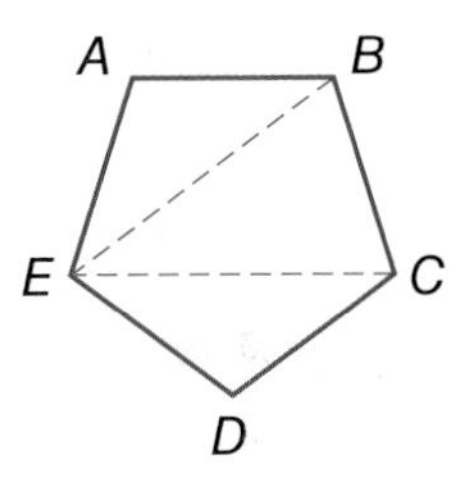

You know that the sum of the angles of a triangle is 180°. To find the sum of the interior angles of *any* polygon, add another 180° for each additional side to the measure of the first three angles. Look at pentagon *ABCDE*.

Diagonals $\overline{EB}$ and $\overline{EC}$ show that the sum of the angles of a pentagon is equal to the sum of the angles in three triangles.

$$3 \cdot 180° = 540°$$

So, the sum of the interior angles of a pentagon is 540°.

You can use the formula $(n - 2) \cdot 180°$ to find the sum of the interior angles of a polygon. Let n equal the number of sides of a polygon. The solution is equal to the sum of the measures of all the angles of the polygon.

EXAMPLE Finding the Sum of the Angles of a Polygon

Find the sum of the interior angles of an octagon.

$(n - 2) \cdot 180°$	• Use the formula.
$= (8 - 2) \cdot 180°$	• Substitute the number of sides.
$= 6 \cdot 180°$	• Simplify, using the order of operations.
$= 1{,}080°$	

So, the sum of the angles of an octagon is 1,080°.

You can use what you know about finding the sum of the angles of a polygon to find the measure of each angle of a regular polygon.

Begin by finding the sum of all the angles, using the formula $(n - 2) \cdot 180°$. For example, a hexagon has 6 sides, and so substitute 6 for n.

$$(6 - 2) \cdot 180° = 4 \cdot 180° = 720°$$

Then divide the sum of the angles by the total number of angles. Because a hexagon has 6 angles, divide by 6.

$$720° \div 6 = 120°$$

Therefore, each angle of a regular hexagon measures 120°.

Check It Out

Use the formula $(n - 2) \cdot 180°$.

11. Find the sum of the angles of a decagon.
12. Find the measure of each angle in a regular pentagon.

Polyhedrons

Some solid shapes are curved. These shapes are not polyhedrons.

Sphere

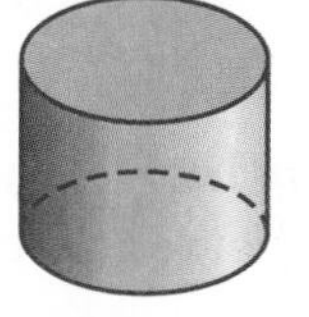

Cylinder

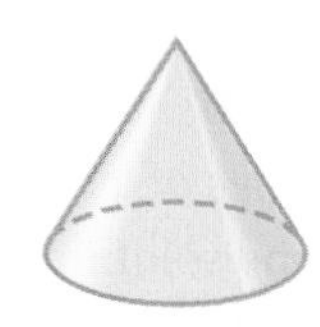

Cone

Some solid shapes have flat surfaces. Each of the figures below is a *polyhedron.*

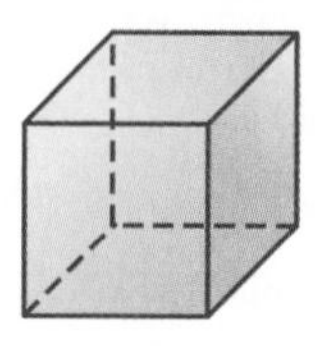

Cube

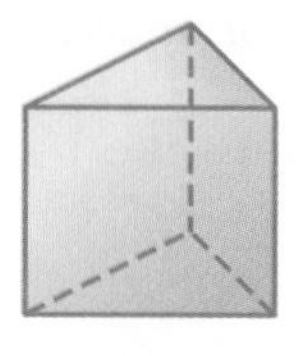

Prism

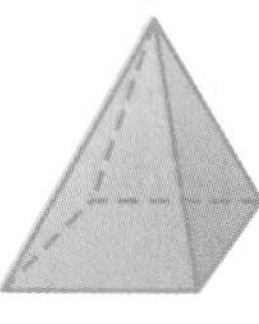

Pyramid

A **polyhedron** is a solid with flat surfaces that are polygons. Triangles, quadrilaterals, and pentagons make up the **faces** of the common polyhedrons below.

A **prism** has two bases, or "end" faces. The **bases** of a prism are polygons that are congruent and parallel to each other. The other faces are parallelograms. The bases of the prisms shown below are shaded. When all six faces of a **rectangular prism** are square, the figure is called a **cube**.

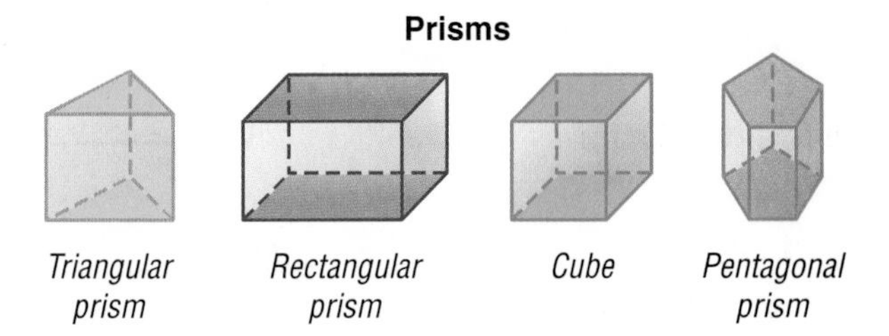

A **pyramid** is a structure that has one polygonal base. It has triangular faces that meet at a point called the *apex*. The base of each pyramid shown below is shaded. A triangular pyramid is a **tetrahedron**. A tetrahedron has four faces. Each face is triangular.

Pyramids

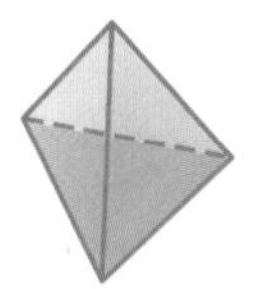
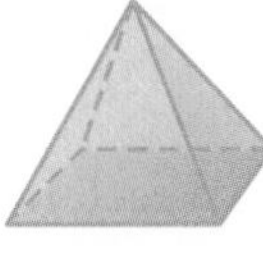

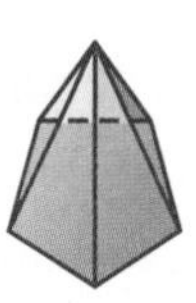

Triangular pyramid (tetrahedron) | *Rectangular pyramid* | *Square pyramid* | *Pentagonal pyramid*

Check It Out

Identify each polyhedron.

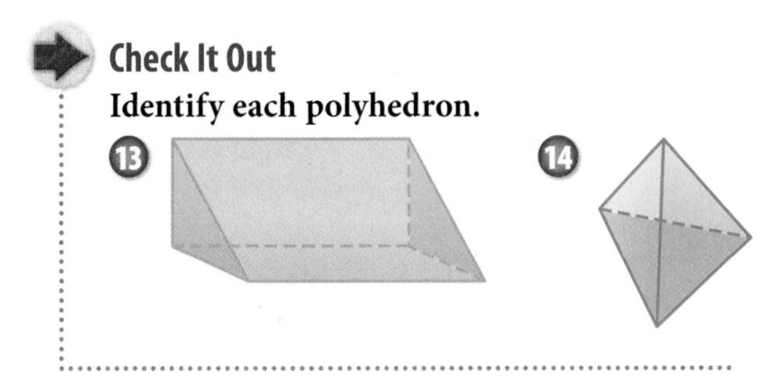

7•2 Exercises

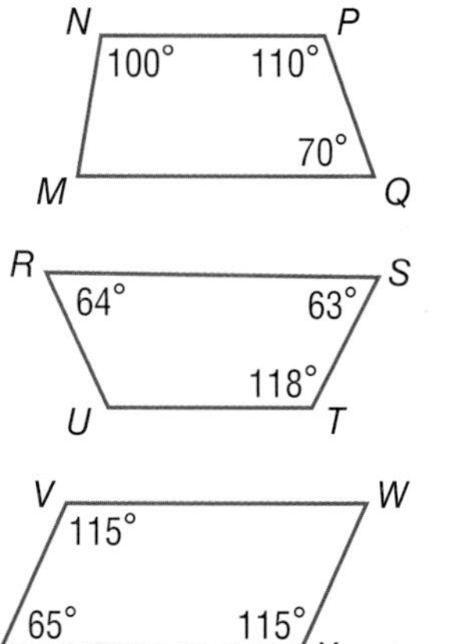

1. Give two other names for quadrilateral $MNPQ$.
2. Find $m\angle M$.
3. Give two other names for quadrilateral $RSTU$.
4. Find $m\angle U$.
5. Give two other names for quadrilateral $VWXY$.
6. Find $m\angle W$.

Tell whether each statement below is *true* or *false*.

7. A square is a parallelogram.
8. Every rectangle is a parallelogram.
9. Not all rectangles are squares.
10. Some trapezoids are parallelograms.
11. Every square is a rhombus.
12. All rhombuses are quadrilaterals.
13. A quadrilateral cannot be both a rectangle and a rhombus.

Identify each polygon.

14. 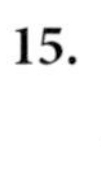

15.

16. 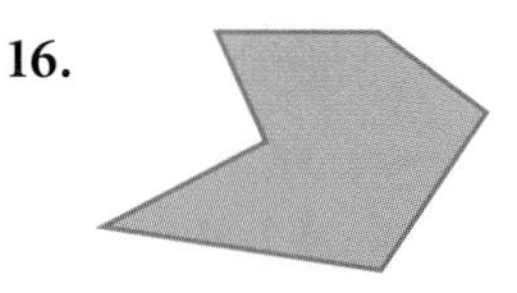

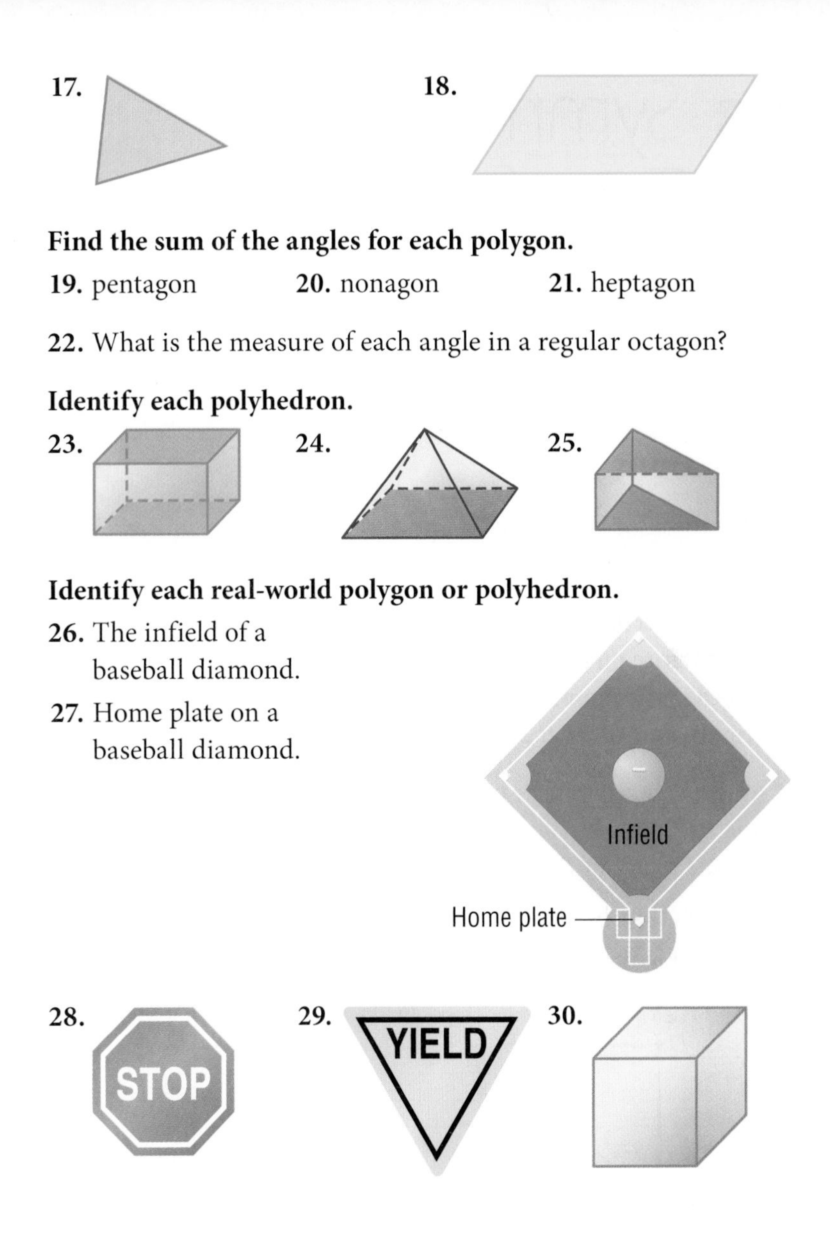

17.

18.

Find the sum of the angles for each polygon.

19. pentagon **20.** nonagon **21.** heptagon

22. What is the measure of each angle in a regular octagon?

Identify each polyhedron.

23. **24.** **25.**

Identify each real-world polygon or polyhedron.

26. The infield of a baseball diamond.

27. Home plate on a baseball diamond.

28. **29.** **30.**

7•3 Symmetry and Transformations

Whenever you move a shape that is in a plane, you are performing a **transformation**. There are three basic types of transformations: reflections, rotations, and translations.

Reflections

A **reflection** (or flip) is the mirror image, or reverse image, of a point, a line, or a shape.

A *line of reflection* is a line such that the figure on one side of the line is the reflection image of the figure on the other side.

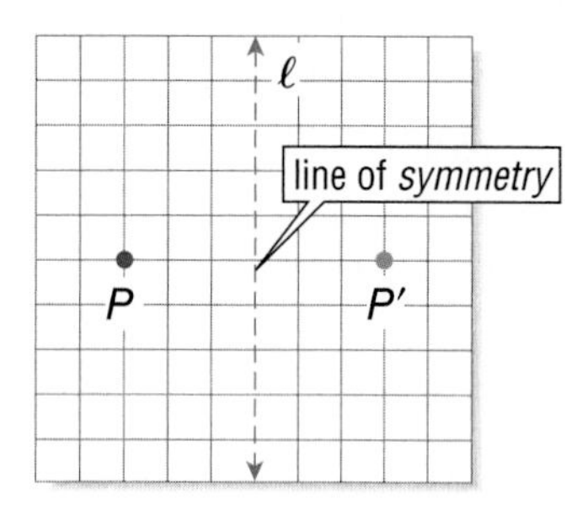

P' reflects point P on the other side of line ℓ. P' is read "P-prime." P' is called the *image* of P.

Any point, line, or polygon can be reflected. Quadrilateral $DEFG$ is reflected across line m. The image of $DEFG$ is $D'E'F'G'$.

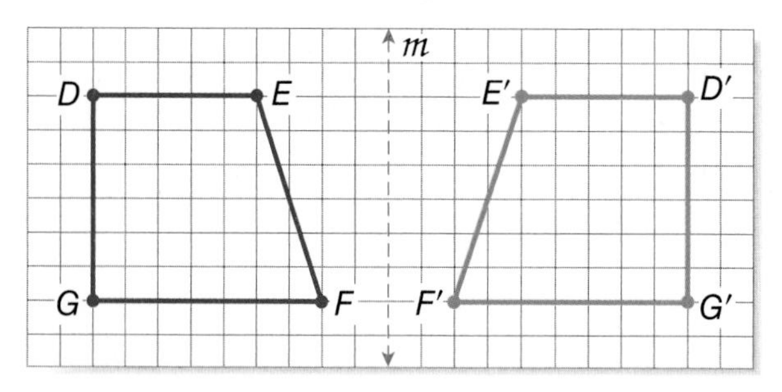

To find an image of a shape, measure the horizontal distance of each point to the line of symmetry. The image of each point will be the same horizontal distance from the line of symmetry on the opposite side.

SYMMETRY AND TRANSFORMATIONS
7•3

Check It Out

Copy each shape on grid paper. Then draw and label the reflection.

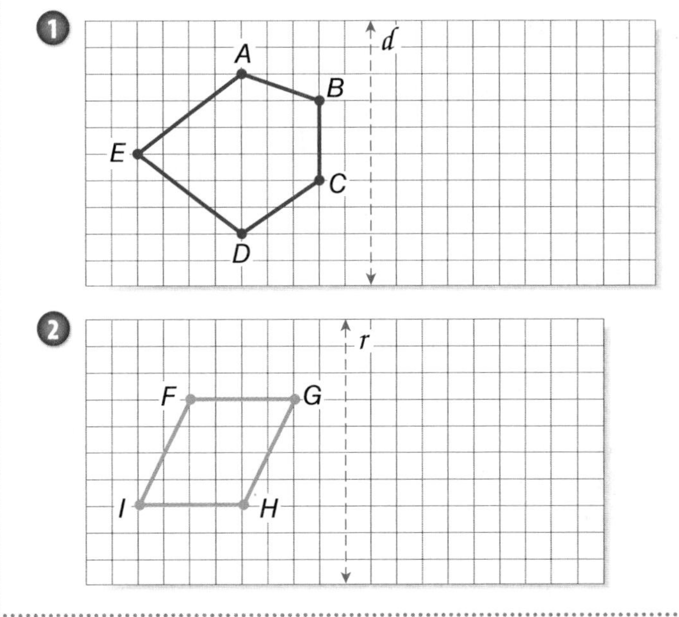

Reflection Symmetry

A **line of symmetry** is a line along which a figure can be folded so that the two resulting halves match. Each of these figures has line symmetry.

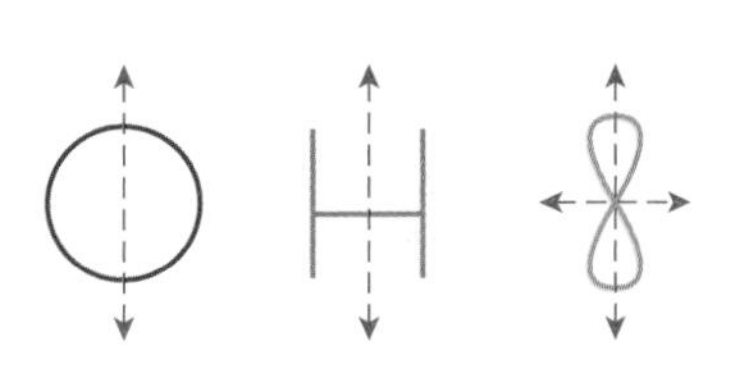

Some figures have more than one line of symmetry.

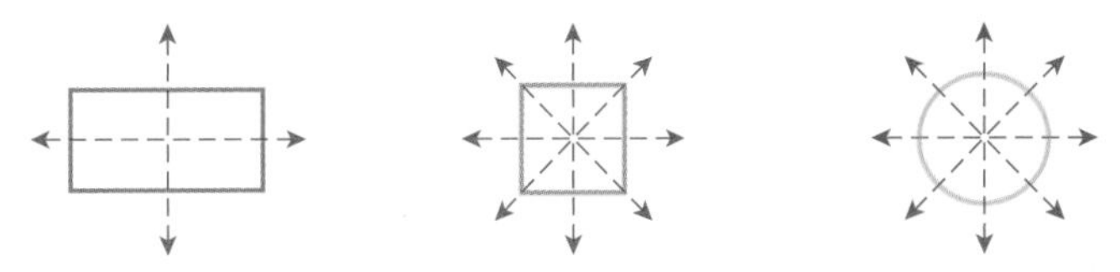

A rectangle has two lines of symmetry. A square has four lines of symmetry. Any line through the center of a circle is a line of symmetry. So, a circle has an infinite number of lines of symmetry.

Check It Out

Tell whether each figure has reflection symmetry. If your answer is *yes*, tell how many lines of symmetry can be drawn through the figure.

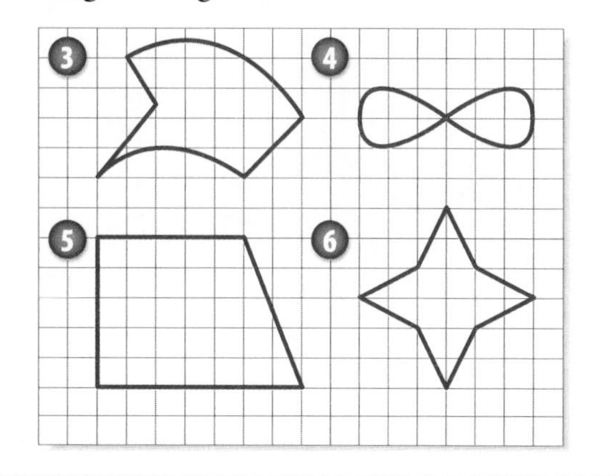

Rotations

A **rotation** (or turn) is a transformation that turns a line or a shape around a fixed point called the *center of rotation*.

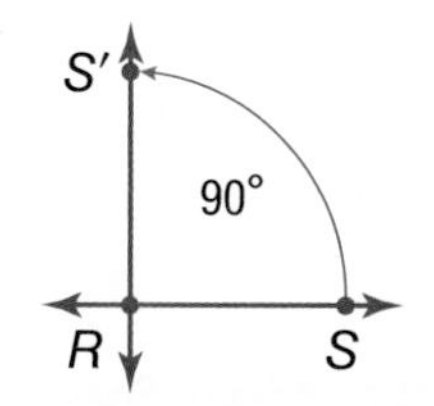

$\overrightarrow{RS}$ is rotated 90° around point *R*.

If you rotate a figure 360°, its position is unchanged.

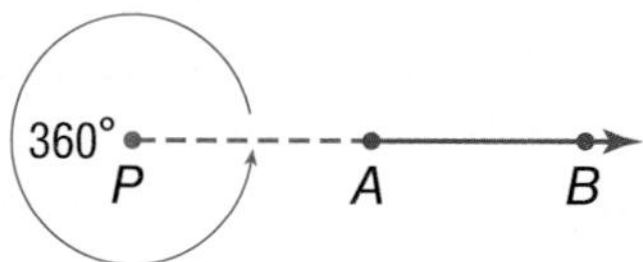

Check It Out

Find the degrees of rotation.

7. How many degrees has $\overrightarrow{PQ}$ been rotated?

Q′ P′ P Q

8. How many degrees has ΔTSR been rotated?

T S R R′ T′

Translations

A **translation** (or slide) is another kind of transformation. When you move a figure to a new position without turning it, you are performing a translation.

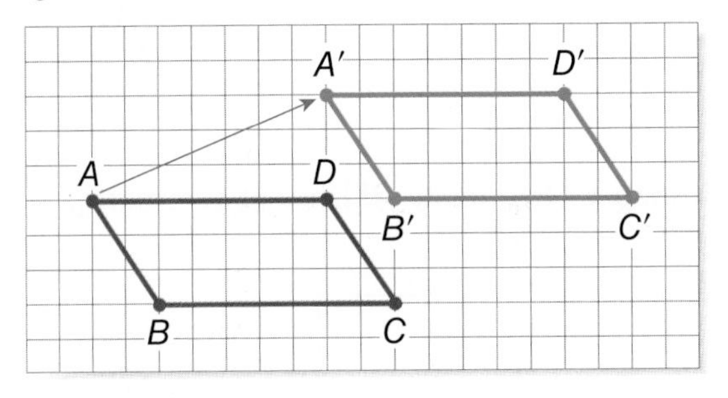

Rectangle *ABCD* moves right and up. *A′B′C′D′* is the image of *ABCD* under a translation. *A′* is 7 units to the right and 3 units up from *A*. All other points on the rectangle moved the same way.

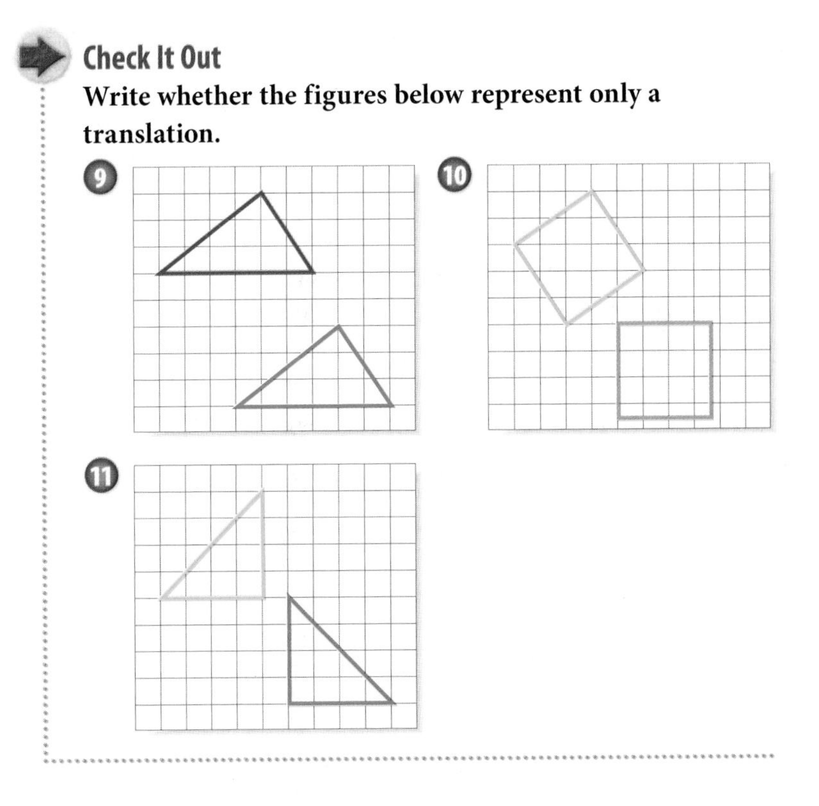

Check It Out

Write whether the figures below represent only a translation.

9

10

11

7•3 Exercises

1. Copy the shape on grid paper. Then translate and label the image 4 units down and 3 units left.

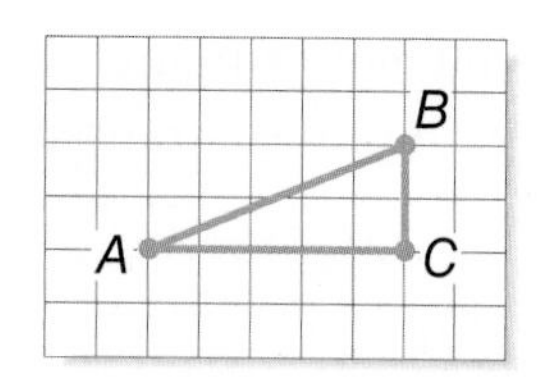

Copy each shape. Then draw all lines of symmetry.

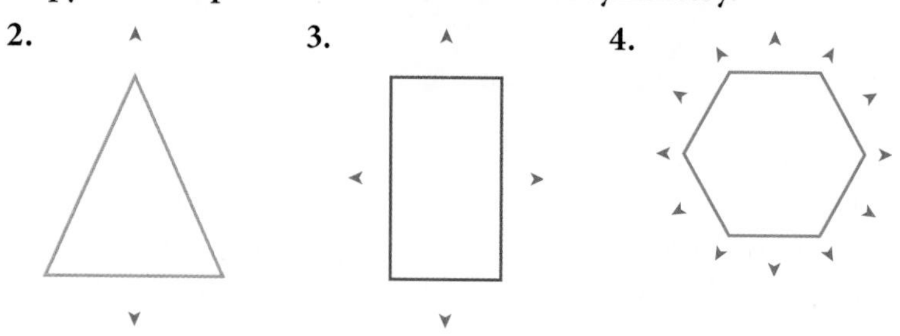

Which type of transformation is illustrated?

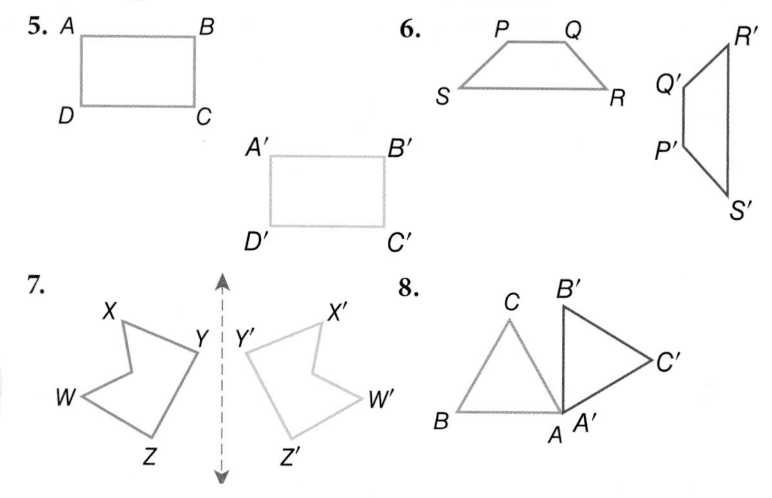

7•4 Perimeter

Perimeter of a Polygon

The perimeter of a polygon is the sum of the lengths of the sides. To find the perimeter of a regular polygon, you can multiply the length of one side by the total number of sides.

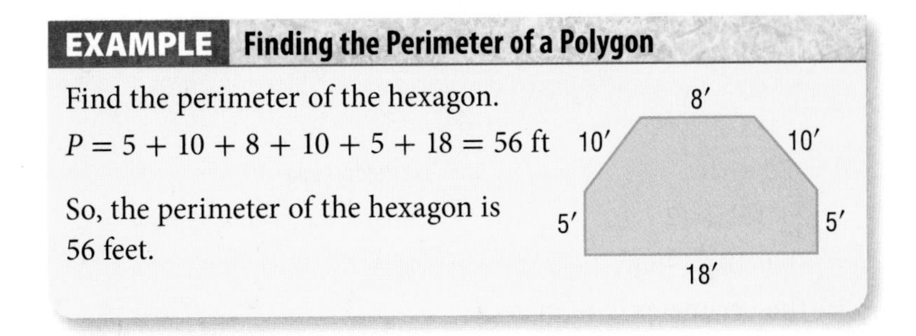

EXAMPLE **Finding the Perimeter of a Polygon**

Find the perimeter of the hexagon.

$P = 5 + 10 + 8 + 10 + 5 + 18 = 56$ ft

So, the perimeter of the hexagon is 56 feet.

Regular Polygon Perimeters

If you know the perimeter of a regular polygon, you can find the length of each side.

Assume a regular octagon has a perimeter of 36 centimeters.

Let x = length of a side. $36 \text{ cm} = 8x$

$4.5 \text{ cm} = x$

Each side is 4.5 centimeters long.

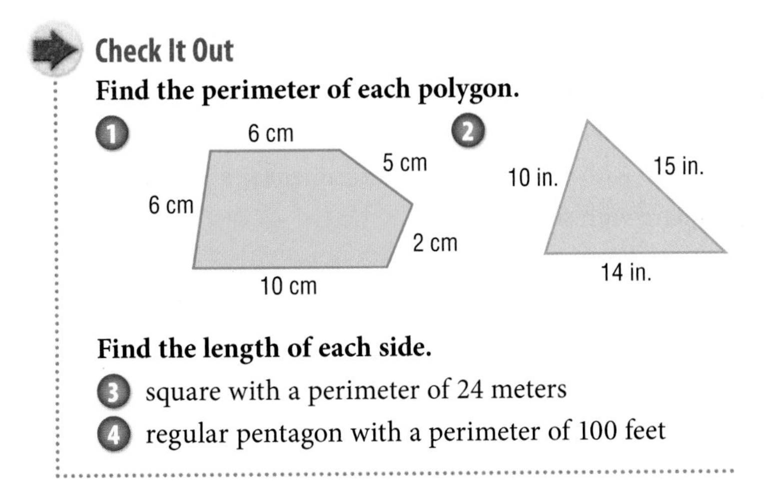

Check It Out

Find the perimeter of each polygon.

1

2

Find the length of each side.

3. square with a perimeter of 24 meters
4. regular pentagon with a perimeter of 100 feet

Perimeter of a Rectangle

Opposite sides of a rectangle are congruent. So to find the perimeter of a rectangle, you need to know only its length and width.

EXAMPLE **Finding the Perimeter of a Rectangle**

Find the perimeter of a rectangle with length 15 meters and width 9 meters.

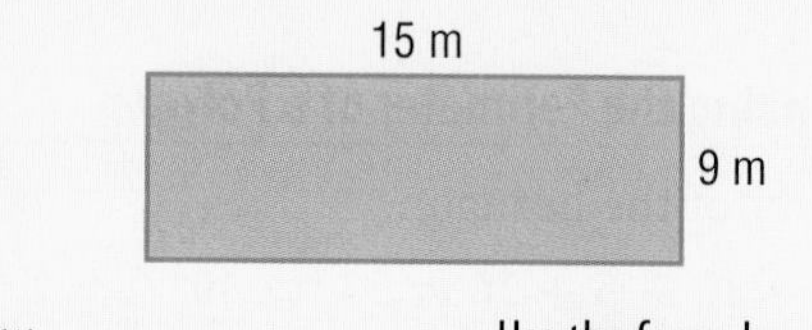

$P = 2\ell + 2w$ • Use the formula.

$= (2 \cdot 15) + (2 \cdot 9)$ • Substitute the length and width.

$= 30 + 18 = 48$ m • Simplify.

So, the perimeter is 48 meters.

A square is a rectangle whose length and width are congruent. Let $s =$ the length of one side. The formula for finding the perimeter of a square is $P = 4 \cdot s$ or $P = 4s$.

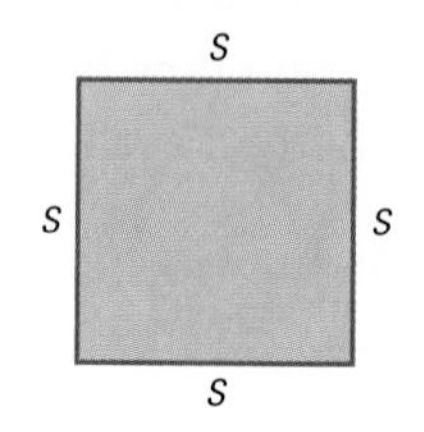

Check It Out

Find the perimeter.

5. rectangle with length 16 centimeters and width 14 centimeters
6. square with sides that are 12 centimeters
7. square with sides that are 1.3 meters

Perimeter of a Right Triangle

If you know lengths of two sides of a right triangle, you can find the length of the third side by using the Pythagorean Theorem.

For a review of the *Pythagorean Theorem*, see page 366.

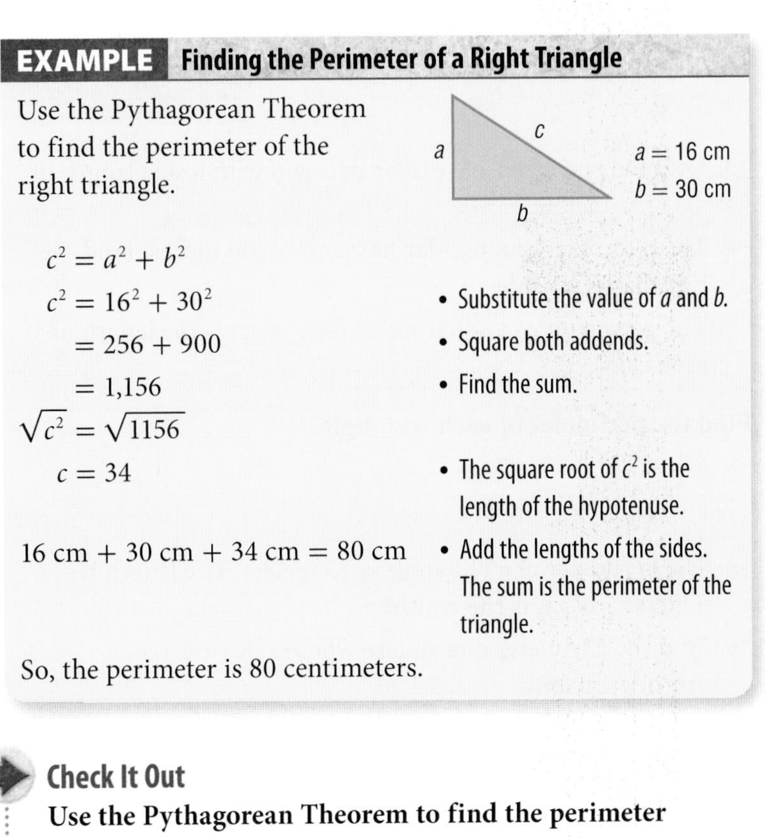

EXAMPLE **Finding the Perimeter of a Right Triangle**

Use the Pythagorean Theorem to find the perimeter of the right triangle.

$c^2 = a^2 + b^2$

$c^2 = 16^2 + 30^2$ • Substitute the value of *a* and *b*.

$= 256 + 900$ • Square both addends.

$= 1{,}156$ • Find the sum.

$\sqrt{c^2} = \sqrt{1156}$

$c = 34$ • The square root of c^2 is the length of the hypotenuse.

16 cm + 30 cm + 34 cm = 80 cm • Add the lengths of the sides. The sum is the perimeter of the triangle.

So, the perimeter is 80 centimeters.

Check It Out

Use the Pythagorean Theorem to find the perimeter of each triangle.

8 **9**

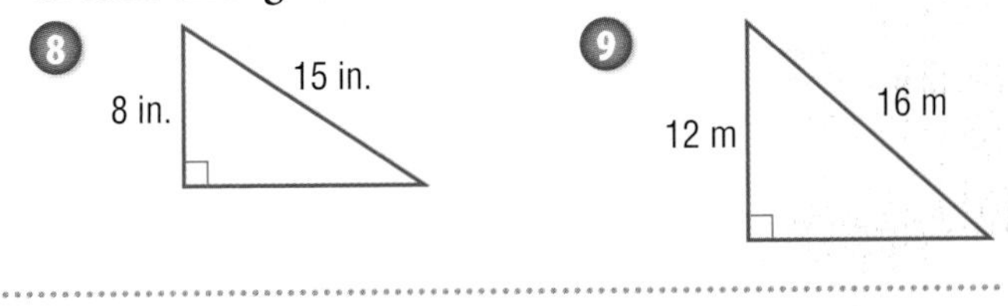

7•4 Exercises

Find the perimeter of each polygon.

1. **2.**

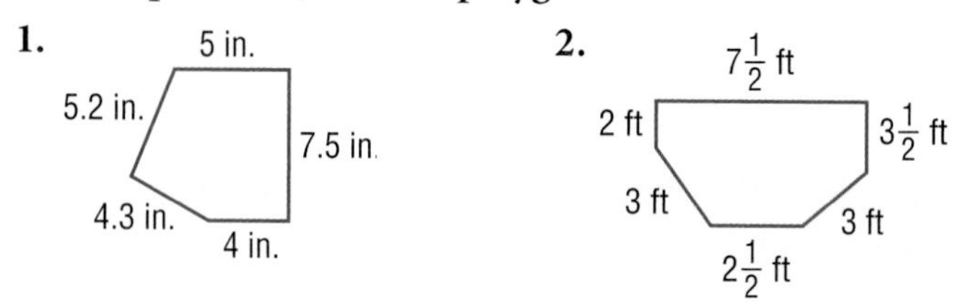

3. Find the perimeter of regular decagon with a side length of 4.8 centimeters.

4. The perimeter of a regular hexagon is 200 inches. Find the length of each side.

5. The perimeter of a square is 16 feet. What is the length of each side?

Find the perimeter of each rectangle.

6. $\ell = 6.1$ m, $w = 4.3$ m

7. $\ell = 2$ cm, $w = 1.5$ cm

8. The perimeter of a rectangle is 15 meters. The length is 6 meters. What is the width?

9. Find the perimeter of a square whose sides are 1.5 centimeters long.

10. Use the Pythagorean Theorem to find the perimeter of the triangle at the right.

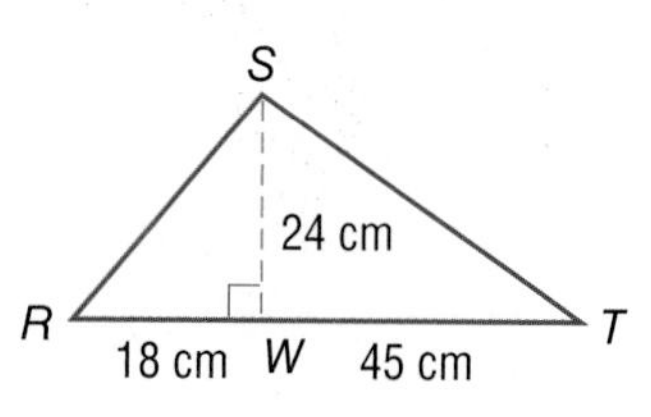

11. Two sides of a triangle are 9 inches and 7 inches. If it is an isosceles triangle, what are the two possible perimeters?

12. The perimeter of an equilateral triangle is 27 centimeters. What is the measure of each side?

13. If one side of a regular pentagon measures 18 inches, what is the perimeter?

14. If a side of a regular nonagon measures 8 centimeters, what is the perimeter?

15. A carpenter needs to install trim around the edges of the ceiling in the room shown in the diagram at the right. The trim is sold in 8 foot lengths. How many pieces must the carpenter buy?

For Exercises 16 and 17, use the race course below.

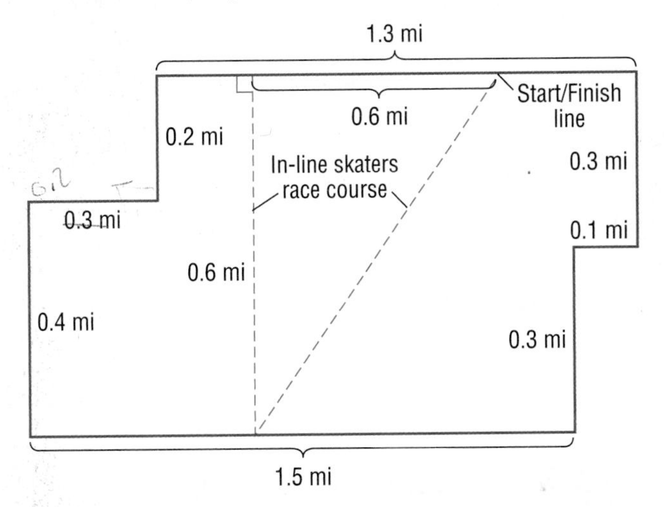

16. How long is the race course?

17. If they changed the course to go around the edge of the lot, how long would it be?

18. Cleve wants to mark off a square with sides that are 60 feet long for a baseball diamond. In addition, he wants to mark off two batters' boxes, each with a length of 5 feet and a width of 3 feet. The chalkbag will make 375 feet of chalk line. How many feet does Cleve need to chalk? Does he have enough chalk to complete the job?

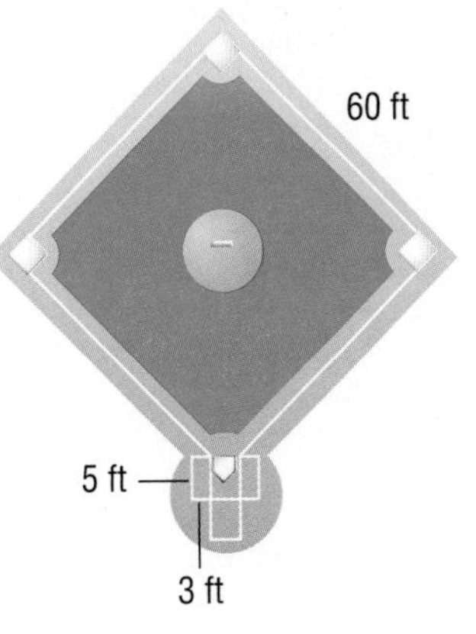

7•5 Area

What Is Area?

Area measures the size of a surface. Instead of measuring with units of length, such as inches, centimeters, feet, and kilometers, you measure area in square units, such as square inches (in^2) and square centimeters (cm^2).

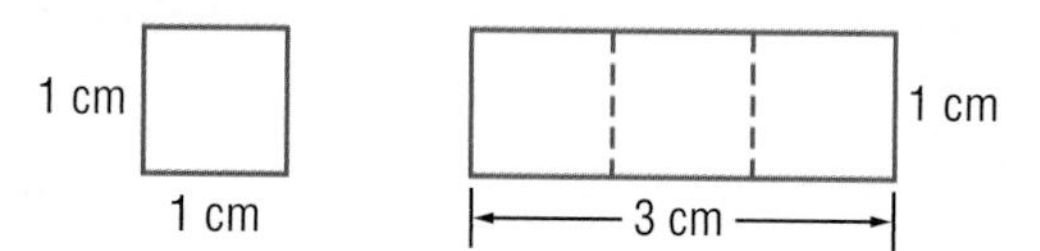

This square has an area of one square centimeter. It takes exactly three of these squares to cover the rectangle. The area of the rectangle is three square centimeters, or 3 cm^2.

Area of a Rectangle

The formula for finding the area of a rectangle is
$A = \ell \cdot w$, or $A = \ell w$.

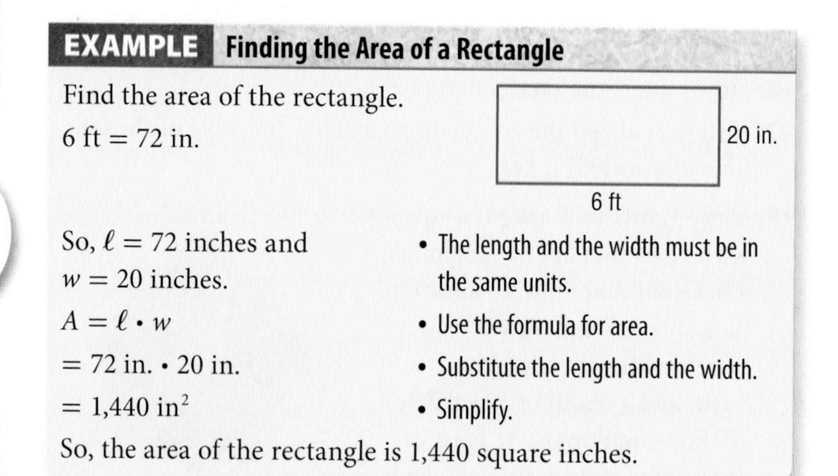

EXAMPLE **Finding the Area of a Rectangle**

Find the area of the rectangle.

6 ft = 72 in.

So, $\ell = 72$ inches and $w = 20$ inches. • The length and the width must be in the same units.

$A = \ell \cdot w$ • Use the formula for area.

$= 72$ in. • 20 in. • Substitute the length and the width.

$= 1{,}440$ in^2 • Simplify.

So, the area of the rectangle is 1,440 square inches.

For a square whose sides measure s units, you can use the formula $A = s \cdot s$, or $A = s^2$.

Check It Out

Use the formula $A = \ell w$.

1. Find the area of a rectangle if $\ell = 40$ inches and $w = 2$ feet.
2. Find the area of a square whose sides are 6 centimeters.

Area of a Parallelogram

To find the area of a parallelogram, you multiply the base by the height.

Area = base • height

$A = b \cdot h$

or $A = bh$

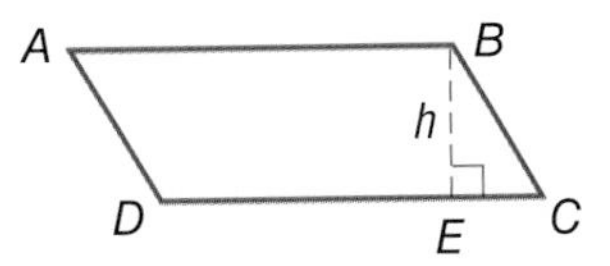

The height of a parallelogram is always perpendicular to the base. In parallelogram *ABCD,* the height, *h,* is equal to *BE,* not *BC.* The base, *b,* is equal to *DC.*

EXAMPLE Finding the Area of a Parallelogram

Find the area of a parallelogram with a base of 8 inches and a height of 5 inches.

$A = b \cdot h$	• Use the formula for area.
$= 8 \text{ in.} \cdot 5 \text{ in.}$	• Substitute the base and the height.
$= 40 \text{ in}^2$	• Simplify.

So, the area of the parallelogram is 40 square inches.

Check It Out

Use the formula $A = bh$.

3. Find the area of a parallelogram if $b = 9$ meters and $h = 6$ meters.
4. Find the length of the base of a parallelogram if the area is 32 square meters and the height is 4 meters.

Area of a Triangle

If you were to cut a parallelogram along a diagonal, you would have two triangles with equal bases, b, and equal height, h.

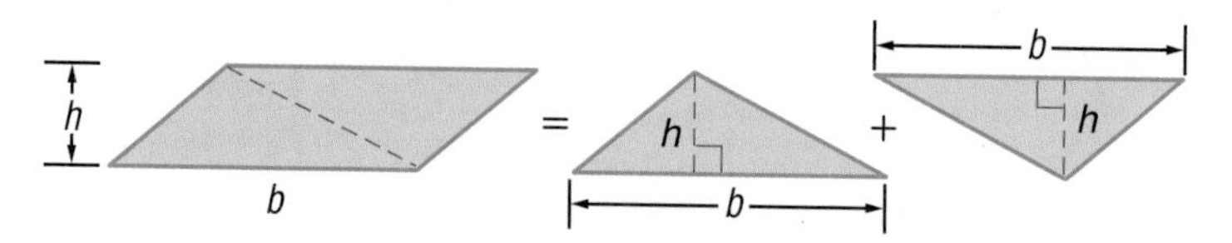

So, the formula for the area of a triangle is half the area of a parallelogram:
$A = \frac{1}{2} \cdot b \cdot h$, or $A = \frac{1}{2}bh$.

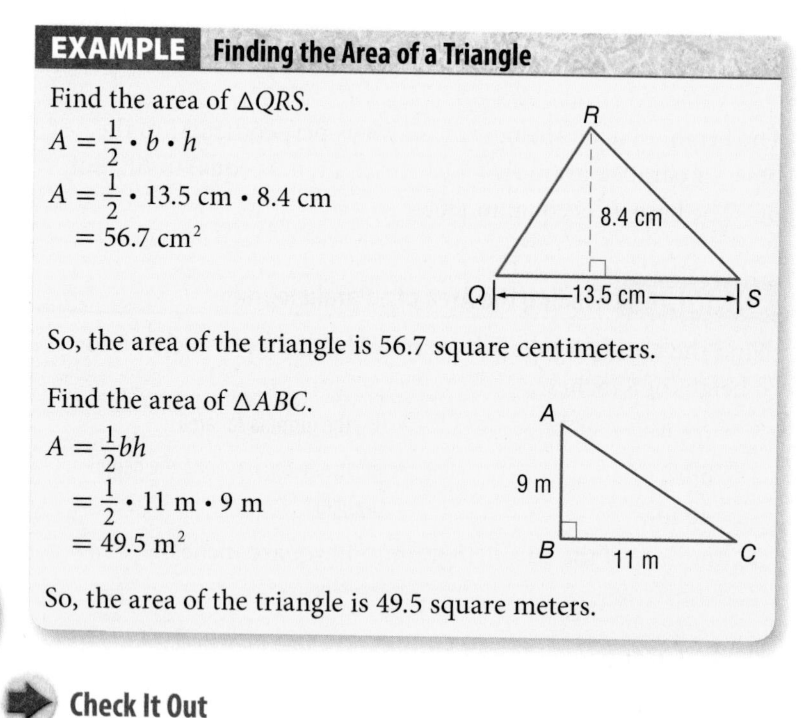

EXAMPLE **Finding the Area of a Triangle**

Find the area of $\triangle QRS$.

$A = \frac{1}{2} \cdot b \cdot h$

$A = \frac{1}{2} \cdot 13.5 \text{ cm} \cdot 8.4 \text{ cm}$
$= 56.7 \text{ cm}^2$

So, the area of the triangle is 56.7 square centimeters.

Find the area of $\triangle ABC$.

$A = \frac{1}{2}bh$
$= \frac{1}{2} \cdot 11 \text{ m} \cdot 9 \text{ m}$
$= 49.5 \text{ m}^2$

So, the area of the triangle is 49.5 square meters.

Check It Out

Use the formula $A = \frac{1}{2}bh$.

5. Find the area of a triangle where $b = 20$ inches and $h = 6$ inches.

6. Find the area of a right triangle whose sides are 24 centimeters, 45 centimeters, and 51 centimeters.

Area of a Trapezoid

A trapezoid has two bases, which are labeled b_1 and b_2. You read b_1 as "b sub-one." The area of a trapezoid is equal to the sum of the areas of two triangles with different base lengths.

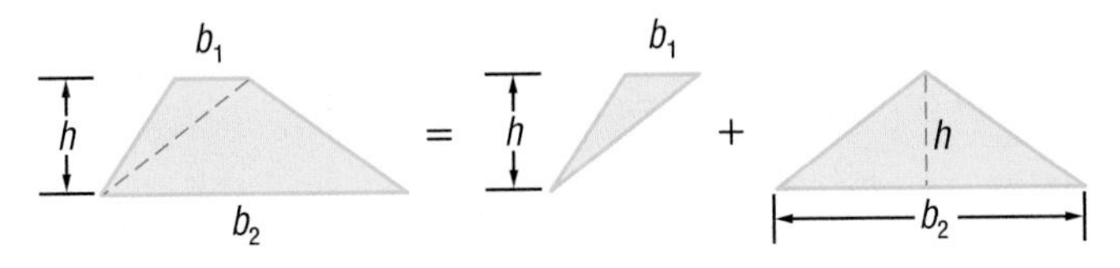

So, the formula for the area of a trapezoid is $A = \frac{1}{2}b_1h + \frac{1}{2}b_2h$ or, in simplified form, $A = \frac{1}{2}h(b_1 + b_2)$.

EXAMPLE Finding the Area of a Trapezoid

Find the area of trapezoid *EFGH*.

$$A = \frac{1}{2}h(b_1 + b_2)$$
$$= \frac{1}{2} \cdot 5(6 + 12)$$
$$= 2.5 \cdot 18$$
$$= 45 \text{ cm}^2$$

E $b_1 = 6$ cm F
$h = 5$ cm
H $b_2 = 12$ cm G

So, the area of the trapezoid is 45 square centimeters.

Since $\frac{1}{2}h(b_1 + b_2)$ is equal to $h \cdot \frac{b_1 + b_2}{2}$, you can also say $A =$ height $\cdot$ the average of the bases.

For a review of how to find an *average* or *mean*, see page 201.

Check It Out

Use the formula $A = \frac{1}{2}h(b_1 + b_2)$.

7. The height of a trapezoid is 3 feet. The bases are 2 feet and 6 feet. What is the area?

8. The height of a trapezoid is 4 feet. The bases are 8 feet and 7 feet. What is the area?

7•5 Exercises

Find the area of each rectangle given the length, ℓ, and the width, w.

1. $\ell = 3$ meters, $w = 2.5$ meters

2. $\ell = 200$ centimeters, $w = 1.5$ meters

Find the area of each parallelogram.

3.

4.

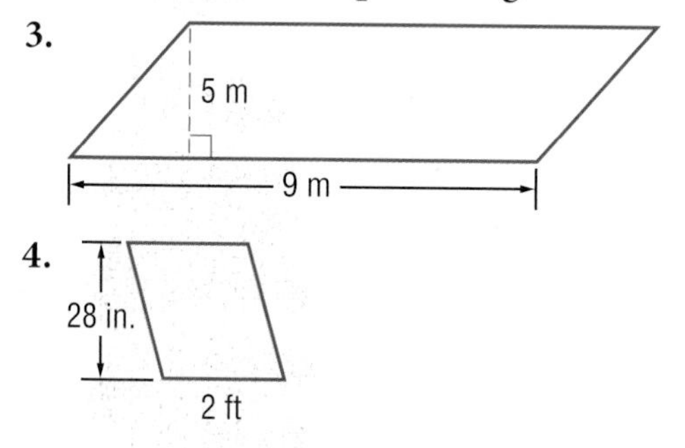

Find the area of each triangle given the base, b, and the height, h.

5. $b = 5$ inches, $h = 4$ inches

6. $b = 6.8$ centimeters, $h = 1.5$ centimeters

7. Find the area of a trapezoid with bases 7 inches and 9 inches and a height of 1 foot.

8. Mr. Lopez plans to give the plot of land shown below to his two daughters. How many square yards of land will each daughter receive if the land is divided evenly between them?

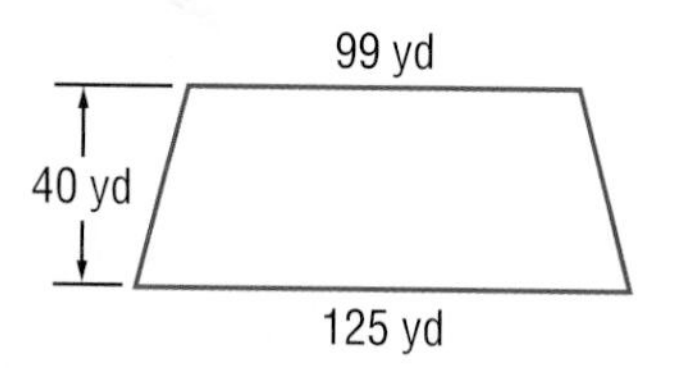

7·6 Surface Area

The **surface area** of a solid is the sum of area of the exterior surfaces. Like area, surface area is expressed in square units.

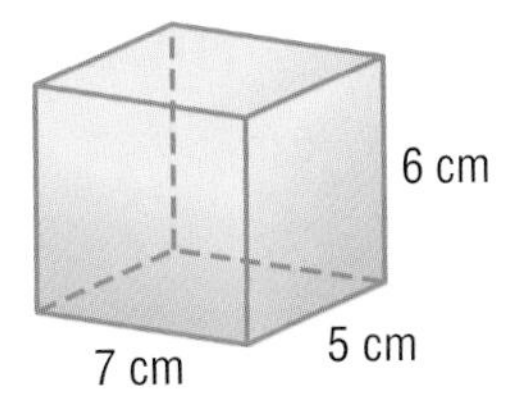

An unfolded 3-dimensional figure is a **net**. This is a net of the rectangular prism shown above. The sum of areas of each section of the net is equal to the surface area of the figure.

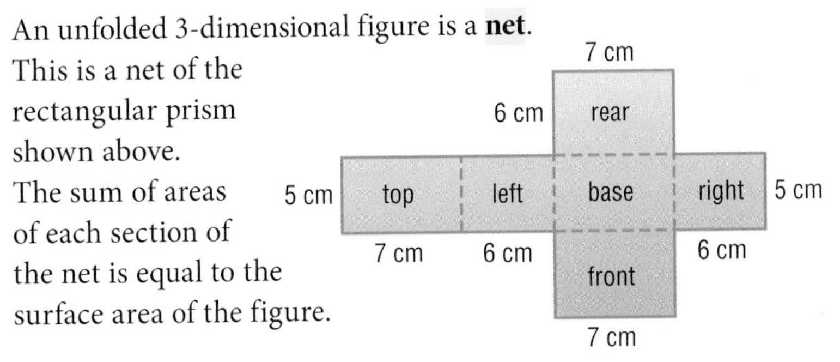

Surface Area of a Rectangular Prism

To find the surface area of a rectangular prism, find the sum of the areas of the six faces, or rectangles. *Remember:* Opposite faces are equal. For a review of *polyhedrons* and *prisms*, see page 331.

EXAMPLE **Finding the Surface Area of a Rectangular Prism**

Use the net to find the area of the rectangular prism above.

- Use the formula $A = \ell w$ to find the area of each face.
- Add the six areas.
- Express the answer in square units.

Area	=	top + base	+	left + right	+	front + rear
	=	$2 \cdot (7 \cdot 5)$	+	$2 \cdot (6 \cdot 5)$	+	$2 \cdot (7 \cdot 6)$
	=	$2 \cdot 35$	+	$2 \cdot 30$	+	$2 \cdot 42$
	=	70	+	60	+	84
	=	214 cm^2				

So, the surface area of the rectangular prism is 214 square centimeters.

Check It Out

Find the surface area of each shape.

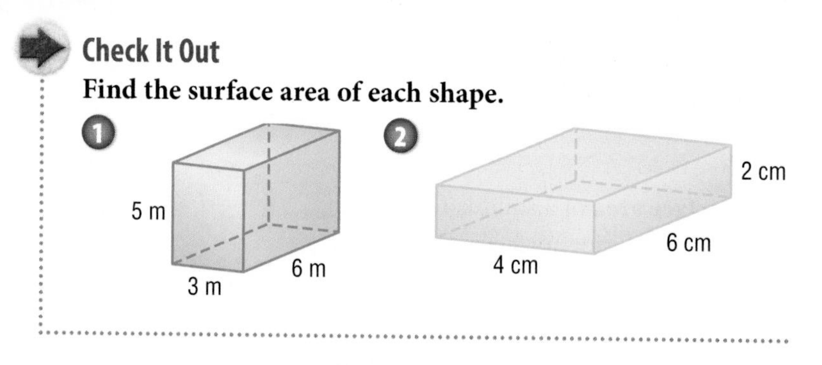

Surface Areas of Other Solids

Nets can be used to find the surface area of any polyhedron. Look at the **triangular prism** and its net.

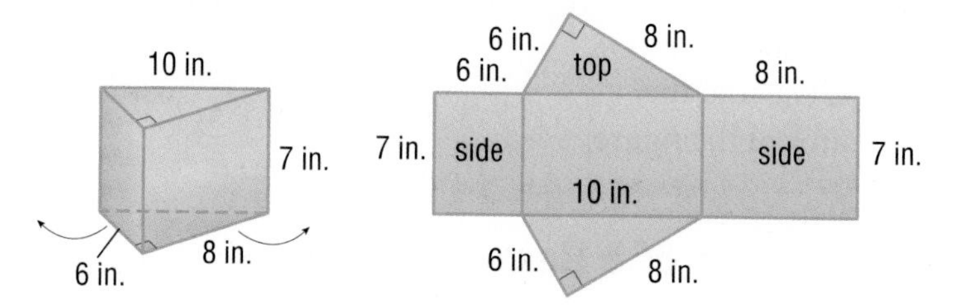

To find the surface area of this solid, use the area formulas for a rectangle ($A = \ell w$) and a triangle ($A = \frac{1}{2}bh$). Find the areas of the five faces and then find the sum of the areas.

Below are two pyramids and their nets. To find the surface area of these polyhedrons, you would again use the area formulas for a rectangle ($A = \ell w$) and a triangle ($A = \frac{1}{2}bh$).

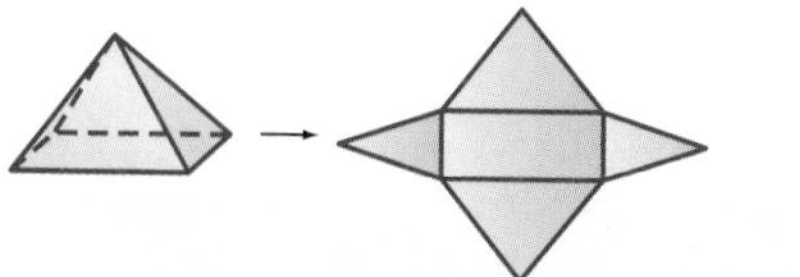

Rectangular pyramid

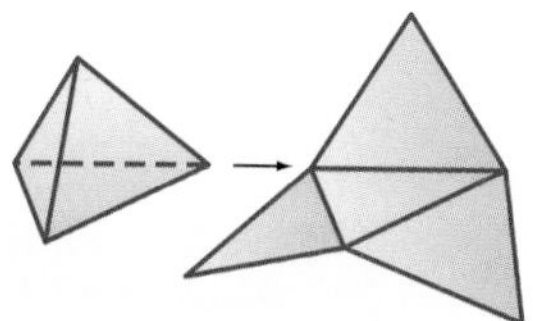

Tetrahedron (triangular pyramid)

The surface area of a cylinder is the sum of the areas of two circles and a rectangle. The height of the rectangle is equal to the height of the cylinder. The length of the rectangle is equal to the *circumference* (p. 360) of the cylinder.

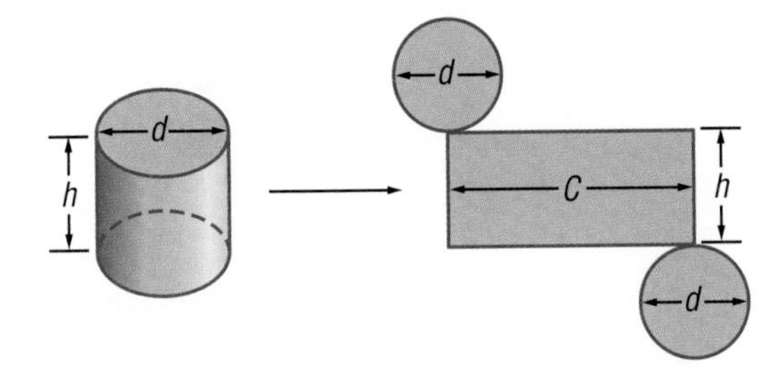

Find the surface area of a cylinder:

- Use the formula for the area of a circle, $A = \pi r^2$, to find the area of each base.
- Find the area of the rectangle using the formula $h \cdot (2\pi r)$.
- Add the area of the circles and the area of the rectangle. You can use the formula $S = 2\pi r^2 + 2\pi rh$.

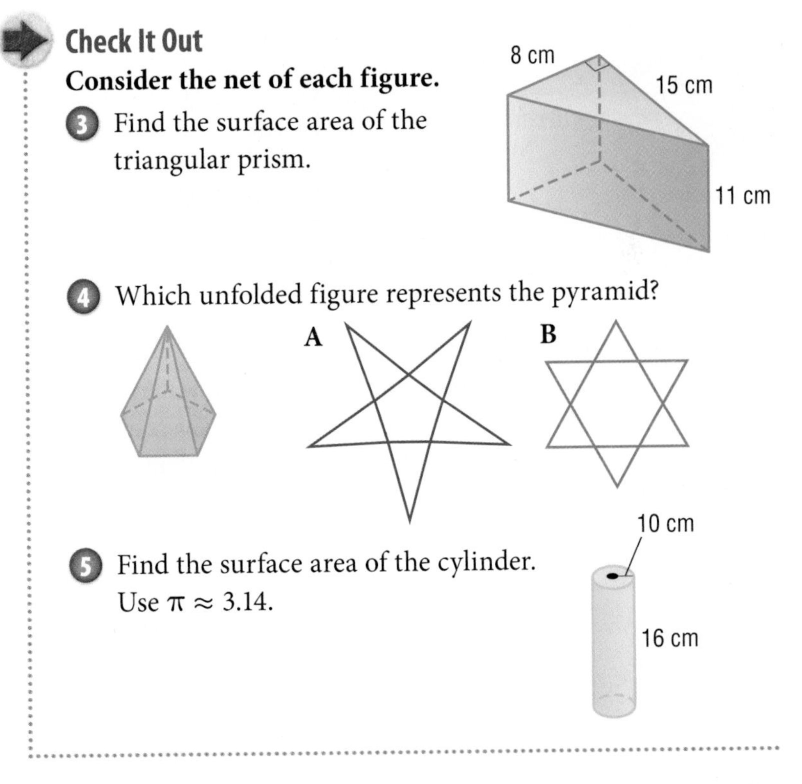

Check It Out

Consider the net of each figure.

3. Find the surface area of the triangular prism.

4. Which unfolded figure represents the pyramid?

5. Find the surface area of the cylinder. Use $\pi \approx 3.14$.

7•6 Exercises

Find the surface area of each shape. Round decimal answers to the nearest tenth.

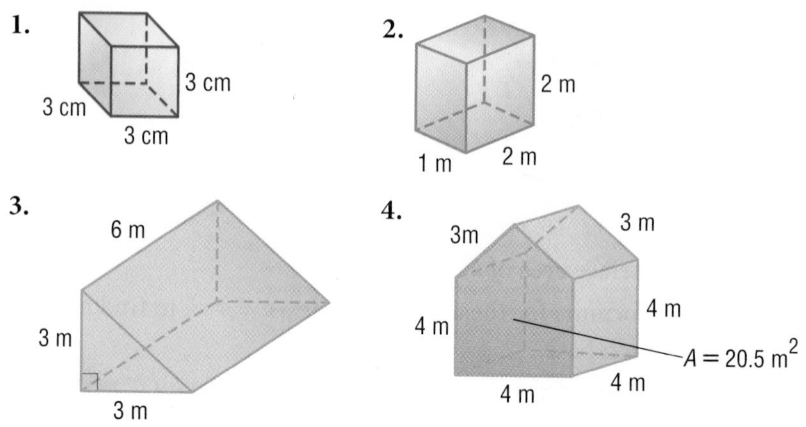

Find the surface area of each cylinder. Round to the nearest tenth.

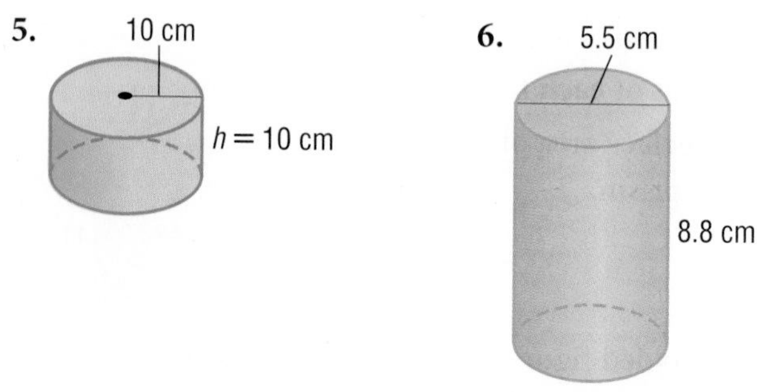

7. Rita and Derrick are building a 3 foot by 3 foot by 6 foot platform to use for skateboarding. They plan to waterproof all six sides of the platform, using sealant that covers about 50 square feet per quart. How many quarts of sealant will they need?

7•7 Volume

What Is Volume?

Volume is the amount of space an object occupies. One way to measure volume is to count the number of cubic units it would take to fill the space inside the object.

The volume of the small cube is 1 cubic inch.

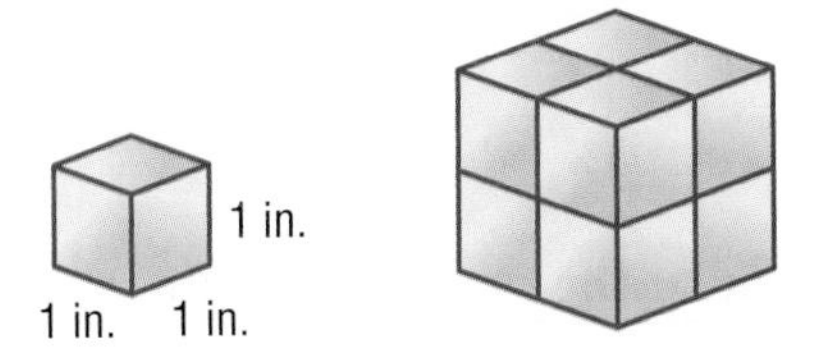

It takes 8 smaller cubes to fill the space inside the larger cube, so the volume of the larger cube is 8 cubic inches.

Volume is measured in *cubic* units. For example, 1 cubic inch is written as 1 in^3, and 1 cubic meter is written as 1 m^3.

For a review of *cubes,* see page 330.

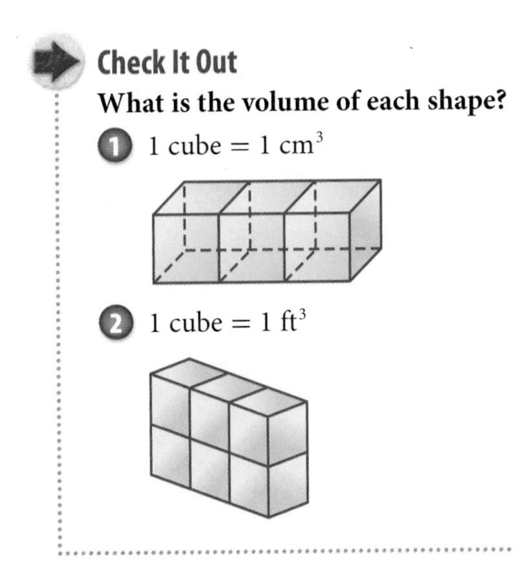

Volume of a Prism

To find the volume of a prism, multiply the *area* (pp. 344–347) of the *base, B,* by the *height, h.*

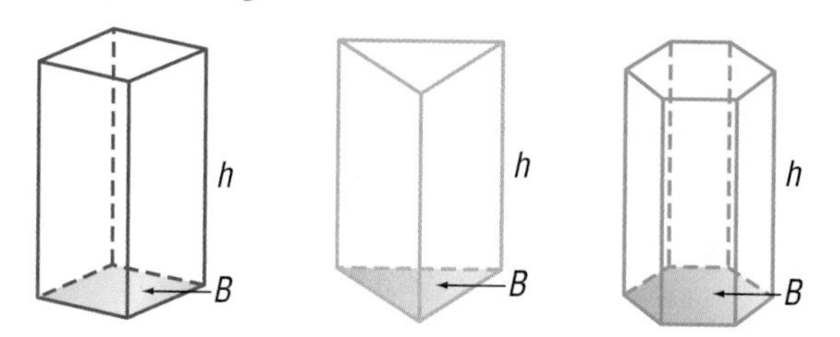

Volume = Bh
See *Formulas,* page 64.

EXAMPLE Finding the Volume of a Prism

Find the volume of the rectangular prism. The base is 14 inches long and 12 inches wide. The height is 17 inches.

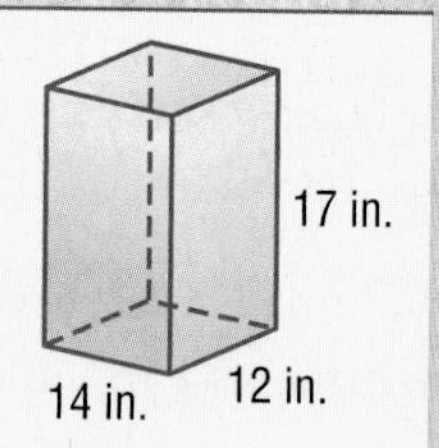

base $A = 14 \text{ in.} \cdot 12 \text{ in.}$ • Find the area of the base.
$= 168 \text{ in}^2$
$V = 168 \text{ in}^2 \cdot 17 \text{ in.}$ • Multiply the base and the height.
$= 2{,}856 \text{ in}^3$

So, the volume of the prism is 2,856 cubic inches.

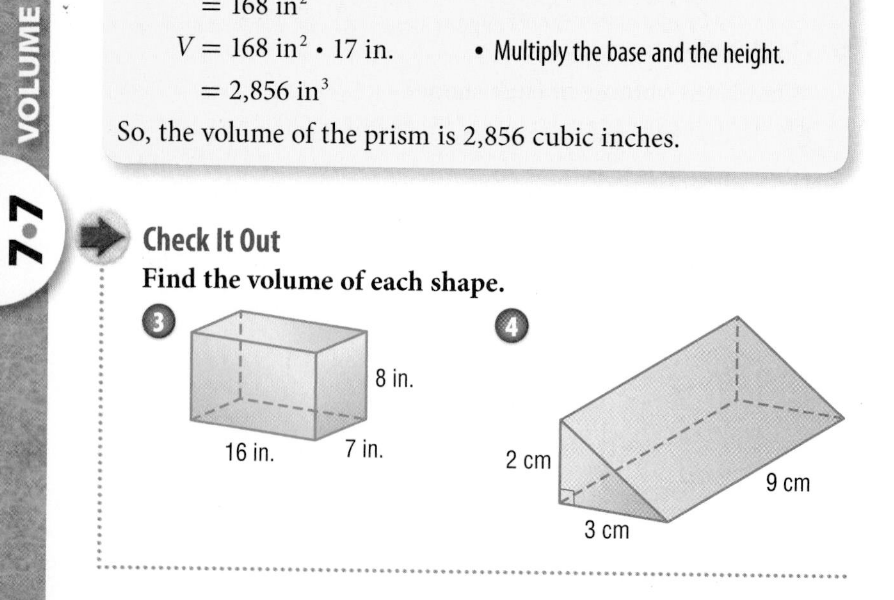

Check It Out

Find the volume of each shape.

3

4

7•7 VOLUME

Volume of a Cylinder

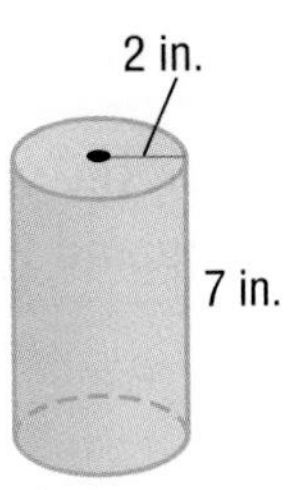

You can use the same formula to find the volume of a cylinder: $V = Bh$. *Remember:* The base of a cylinder is a circle.

The base has a radius of 2 inches. Estimating **pi** (π) at 3.14, you will find that the area of the base is about 12.56 square inches. Multiply the area of the base by the height to find the volume.

$V = 12.56 \text{ in}^2 \cdot 7 \text{ in.}$

$= 87.92 \text{ in}^3$

The volume of the cylinder is 87.92 cubic inches.

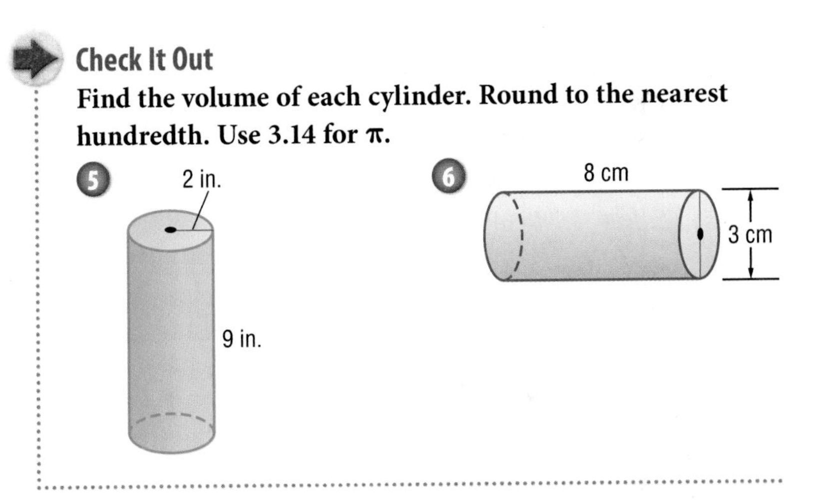

Volume of a Pyramid and a Cone

The formula for the volume of a pyramid or a cone is $V = \frac{1}{3}Bh$.

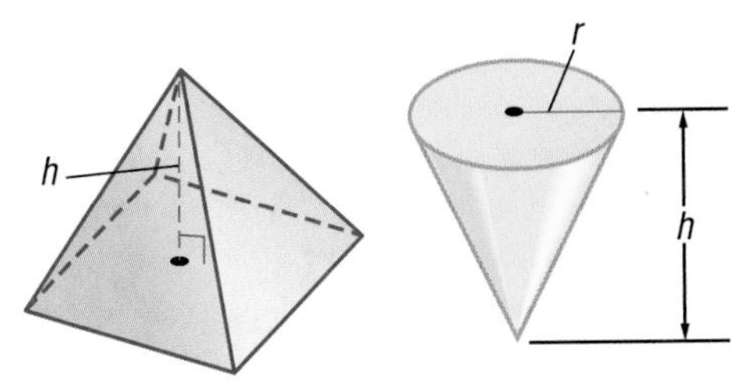

EXAMPLE Finding the Volume of a Pyramid

Find the volume of the pyramid. The base is 175 centimeters long and 90 centimeters wide. The height is 200 centimeters.

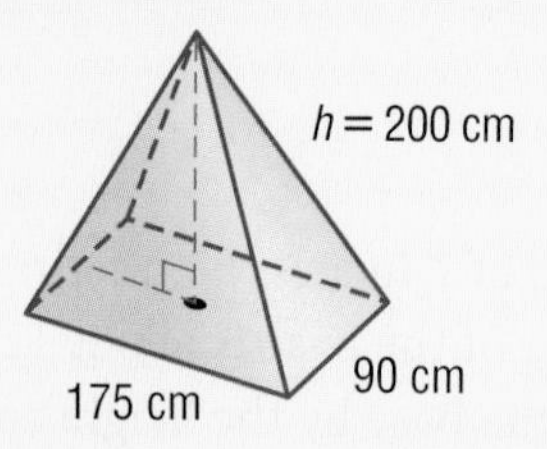

base $A = 175 \cdot 90$ • Find the area of the base.

$= 15{,}750 \text{ cm}^2$

$V = \frac{1}{3}(15{,}750 \cdot 200)$ • Multiply the base by the height and then by $\frac{1}{3}$.

$= 1{,}050{,}000$

So, the volume is 1,050,000 cubic centimeters.

To find the volume of a cone, you follow the same procedure as above. For example, a cone has a base with a radius of 3 centimeters and a height of 10 centimeters. What is the volume of the cone to the nearest tenth?

Square the radius and multiply by π to find the area of the base. Then multiply by the height and divide by 3 to find the volume. The volume of the cone is 94.2 cubic centimeters.

Using a calculator, you would press [π] [×] 9 [=] [28.27433] [×] 10 [÷] 3 [=] [94.24778].

For other volume *Formulas,* see page 64.

Check It Out

Find the volume of the shapes below, rounded to the nearest tenth.

7. 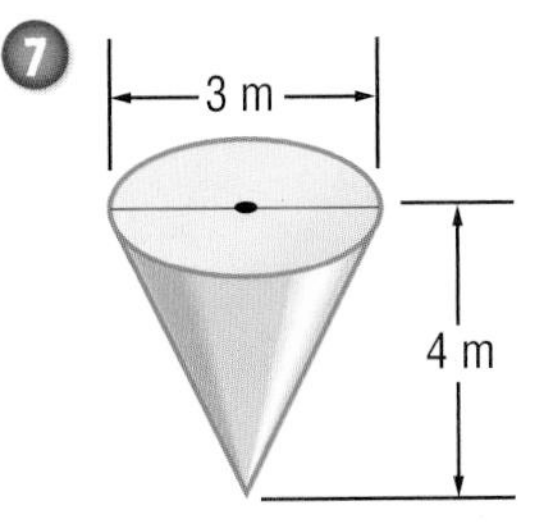

8. 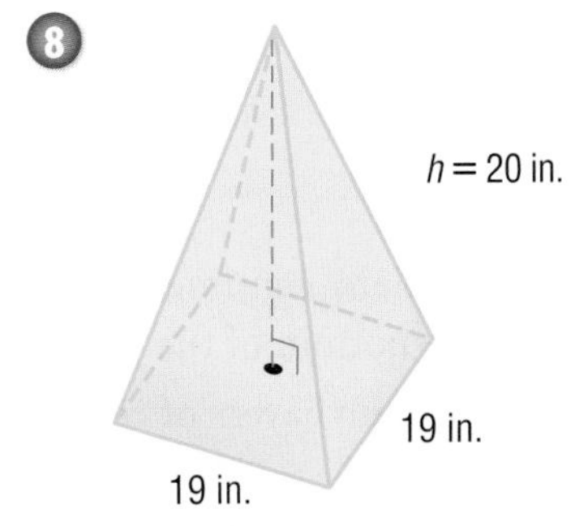

APPLICATION Good Night, T. Rex

Why did the dinosaurs disappear? New evidence from the ocean floor points to a giant asteroid that collided with Earth about 65 million years ago.

The asteroid, 6 to 12 mi in diameter, hit Earth somewhere in the Gulf of Mexico. It was traveling at a speed of thousands of miles per hour.

The collision sent billions of tons of debris into the atmosphere. The debris rained down on the planet, obscuring the Sun. Global temperatures plummeted. The fossil record shows that most of the species that were alive before the collision disappeared.

Assume the crater left by the asteroid had the shape of a hemisphere with a diameter of 165 mi. About how many cubic miles of debris would have been flung from the crater into the air? See page 64 for the formula for the volume of a sphere. See **HotSolutions** for the answer.

7•7 Exercises

Use the rectangular prism to answer Exercises 1–4.

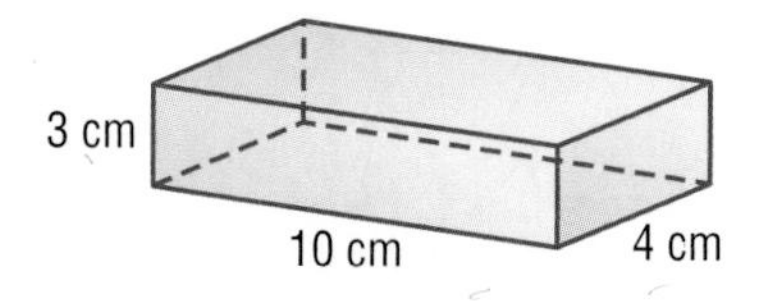

1. How many centimeter cubes would it take to make one layer in the bottom of the prism?
2. How many layers of centimeter cubes would you need to fill the prism?
3. How many centimeter cubes do you need to fill the prism?
4. Each cube has a volume of 1 cubic centimeter. What is the volume of the prism?

5. Find the volume of a rectangular prism with base 10 centimeters, width 10 centimeters, and height 8 centimeters.
6. The base of a cylinder has an area of 5 square centimeters Its height is 7 centimeters. What is its volume?
7. Find the volume of a cylinder 8.2 meters high when its base has a radius of 2.1 meters. Round your answer to the nearest tenth.
8. Find the volume of a pyramid with a height of 4 inches and a rectangular base that measures 6 inches by 3.5 inches.

9. Look at the cone and the rectangular pyramid below. Which has the greater volume and by how many cubic inches?

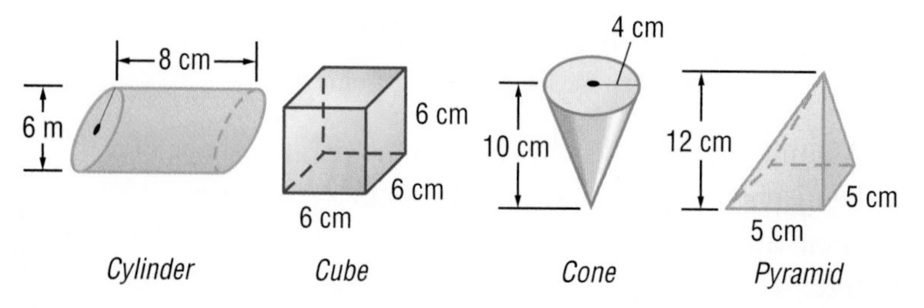

10. List the shapes above from the least volume to the greatest volume.

7•8 Circles

Parts of a Circle

Circles differ from other geometric shapes in several ways. For instance, all circles are the same shape; polygons vary in shape. Circles do not have any sides; polygons are named and classified by the number of sides they have. The *only* thing that makes one circle different from another is size.

A circle is a set of points equidistant from a given point that is the center of the circle. A circle is named by its center point.

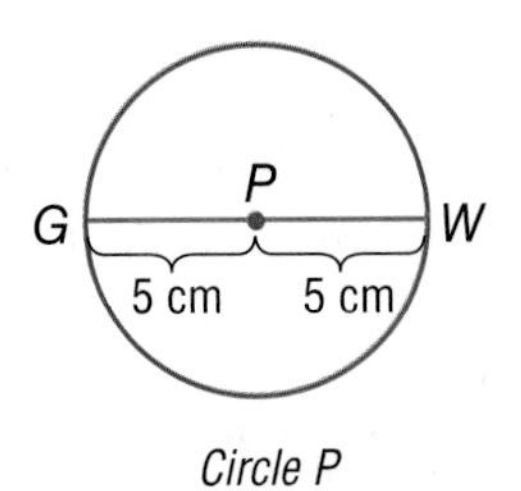

Circle P

A **radius** is any segment that has one endpoint at the center and the other endpoint on the circle. In circle *P*, $\overline{PW}$ is a *radius,* and so is $\overline{PG}$.

A **diameter** is any line segment that passes through the center of the circle and has both endpoints on the circle. $\overline{GW}$ is the diameter of circle *P*. Notice that the length of the diameter $\overline{GW}$ is equal to the sum of $\overline{PW}$ and $\overline{PG}$. The diameter, *d*, is twice the radius, *r*. So, the diameter of circle *P* is 2(5) or 10 centimeters.

Check It Out

Solve.

1. Find the radius of a circle with diameter 18 inches.
2. Find the radius of a circle with diameter 3 meters.
3. Find the radius of a circle in which $d = x$.
4. Find the diameter of a circle with radius 6 centimeters.
5. Find the diameter of a circle with radius 16 meters.
6. Find the diameter of a circle where $r = y$.

Circumference

The **circumference** of a circle is the distance around the circle. The ratio of every circle's circumference to its diameter is always the same. The circumference of every circle is about 3.14 times the diameter. The symbol π, which is read as *pi*, is used to represent the ratio $\frac{C}{d}$.

$\frac{C}{d} \approx 3.141592\ldots$

Circumference = pi • diameter, or $C = \pi d$

Look at the illustration below. The circumference of the circle is a little bit longer than the length of three diameters. This is true for any circle.

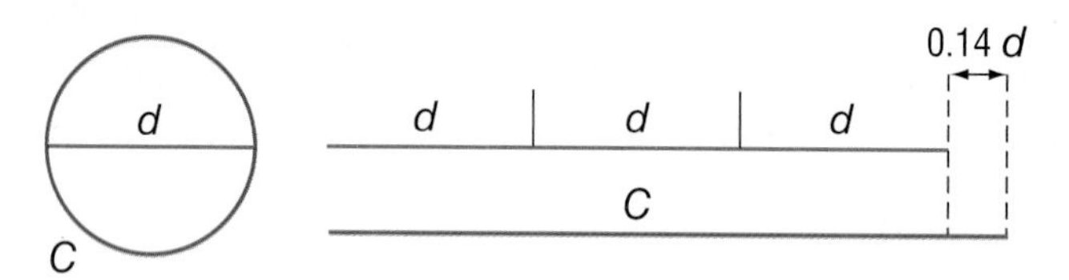

Since $d = 2r$, Circumference = 2 • pi • radius, or $C = 2\pi r$.

The π key on a calculator gives an approximation for π to several decimal places: $\pi \approx 3.141592\ldots$ For practical purposes, however, π is often rounded to 3.14 or left in terms of π.

EXAMPLE Finding the Circumference of a Circle

Find the circumference of a circle with radius 13 meters.

$C = 2\pi r$ • Use the formula for circumference.

$C = 13 \cdot 2 \cdot \pi$ • Substitute the radius.

$C = 26\pi$ • Simplify.

The exact circumference is 26π meters.

$C = 26 \cdot 3.14$ • Substitute 3.14 for π.

≈ 81.64 • Simplify.

So, to the nearest tenth, the circumference is 81.6 meters.

You can find the diameter if you know the circumference. Divide both sides by π.

$$C = \pi d$$
$$\frac{C}{\pi} = \frac{\pi d}{\pi}$$
$$\frac{C}{\pi} = d$$

Check It Out

Solve.

7. Find the circumference of a circle with a diameter of 5 inches. Give the answer in terms of π.
8. Find the circumference of a circle with a radius of 3.2 centimeters. Round to the nearest tenth.
9. Find the diameter of a circle with circumference 25 meters. Round to the nearest hundredth of a meter.
10. Using the π key on your calculator or $\pi \approx 3.141592$, find the radius of a circle with a circumference of 35 inches. Round your answer to the nearest half inch.

Central Angles

A central angle is an angle whose vertex is at the center of a circle. The sum of the central angles in any circle is 360°.

The part of a circle where a central angle intercepts the circle is called an **arc**. The measure of the arc, in degrees, is equal to the measure of the central angle.

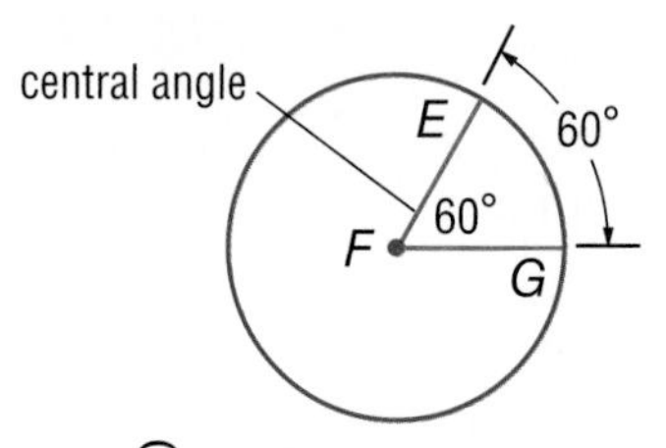

$\widehat{EG} = 60°$ and $m\angle EFG = 60°$.

Check It Out

Use circle *B* to answer Exercises 11–13.

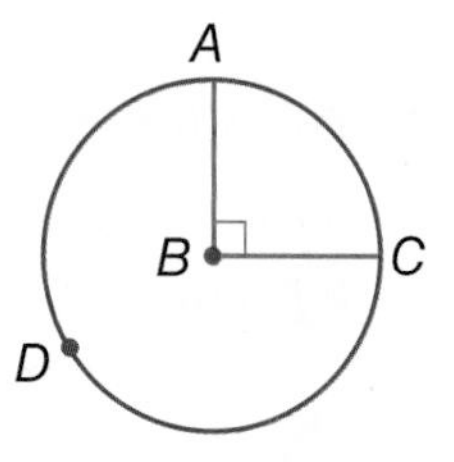

11. Name a central angle in circle *B*.
12. What is the measure of $\widehat{AC}$?
13. What is the measure of $\widehat{ADC}$?

Use circle *M* to answer Exercises 14 and 15.

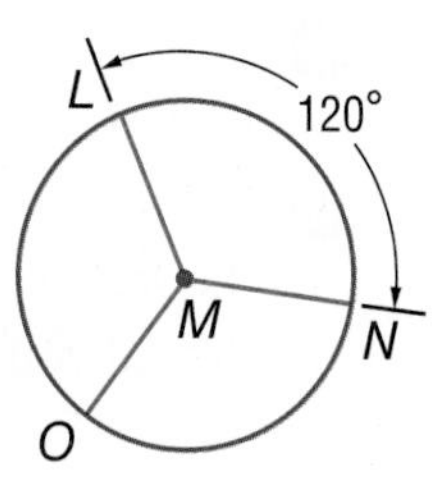

14. What is the measure of $\angle LMN$?

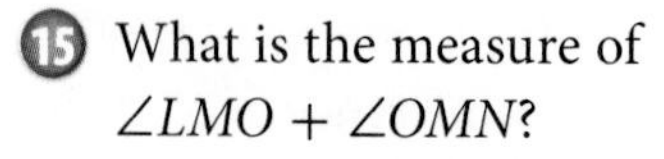

15. What is the measure of $\angle LMO + \angle OMN$?

Area of a Circle

To find the area of a circle, you use the formula $A = \pi r^2$. As with the area of polygons, the area of a circle is expressed in square units.

For a review of *area* and *square units,* see page 344.

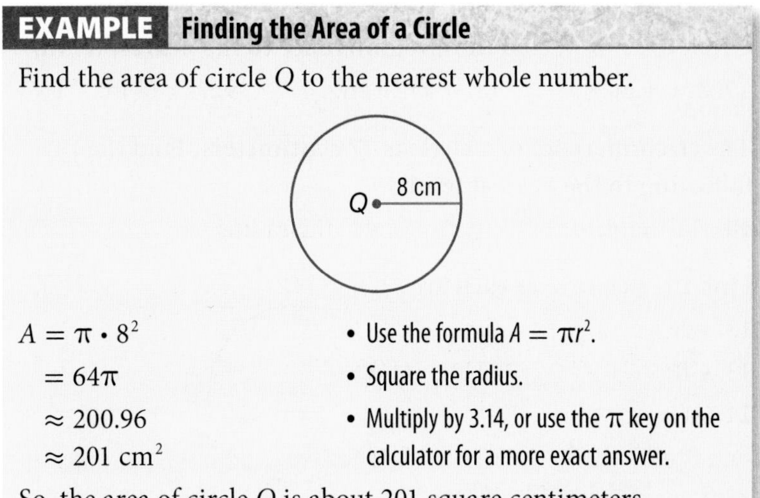

EXAMPLE **Finding the Area of a Circle**

Find the area of circle Q to the nearest whole number.

$A = \pi \cdot 8^2$	• Use the formula $A = \pi r^2$.
$= 64\pi$	• Square the radius.
≈ 200.96	• Multiply by 3.14, or use the π key on the calculator for a more exact answer.
$\approx 201 \text{ cm}^2$	

So, the area of circle Q is about 201 square centimeters.

If you are given the diameter instead of the radius, remember to divide the diameter by 2.

Check It Out

Find the area of each circle.

16. Find the area of a circle with radius 6.5 meters. Use 3.14 for π and round to the nearest tenth.
17. The diameter of a circle is 9 inches. Find the area. Give your answer in terms of π; then multiply and round to the nearest tenth.
18. Use your calculator to find the area of a circle with a diameter of 15 centimeters. Use 3.14 or the calculator key for π and round to the nearest square centimeter.

7•8 Exercises

Find the diameter of each circle with the given radius.

1. $r = 11$ ft **2.** $r = 7.2$ cm **3.** $r = x$

Find the radius with the given diameter.

4. $d = 7$ in. **5.** $d = 2.6$ m **6.** $d = y$

Given the *r* or *d*, find the circumference to the nearest tenth.

7. $d = 1$ m **8.** $d = 7.9$ cm **9.** $r = 18$ in.

The circumference of a circle is 47 centimeters. Find the following to the nearest tenth.

10. the diameter **11.** the radius

Find the measure of each arc.

12. $\overset{\frown}{AB}$

13. $\overset{\frown}{CB}$

14. $\overset{\frown}{AC}$

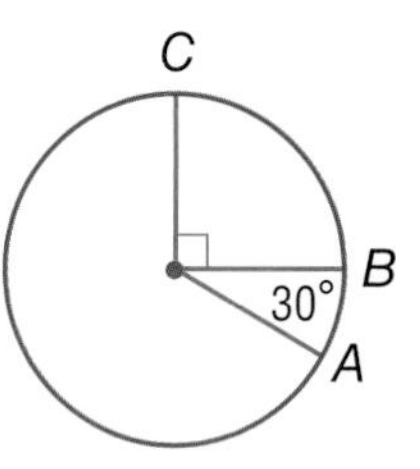

Find the area of each circle, given the radius *r* or the diameter *d*. Round to the nearest whole number.

15. $r = 2$ m **16.** $r = 35$ in.

17. $d = 50$ cm **18.** $d = 10$ ft

19. A dog is tied to a stake. The rope is 20 meters long, so the dog can roam up to 20 meters from the stake. Find the area within which the dog can roam. (If you use a calculator, round to the nearest whole number.)

20. Tony's Pizza Palace sells a large pizza with a diameter of 14 inches. Pizza Emporium sells a large pizza with a diameter of 15 inches for the same price. How much more pizza do you get for your money at Pizza Emporium?

7•9 Pythagorean Theorem

Right Triangles

The smaller illustration at the right shows a right triangle on a geoboard. You can count that leg a is 3 units long and leg b is 4 units long. The **hypotenuse**, side c, is always opposite the right angle.

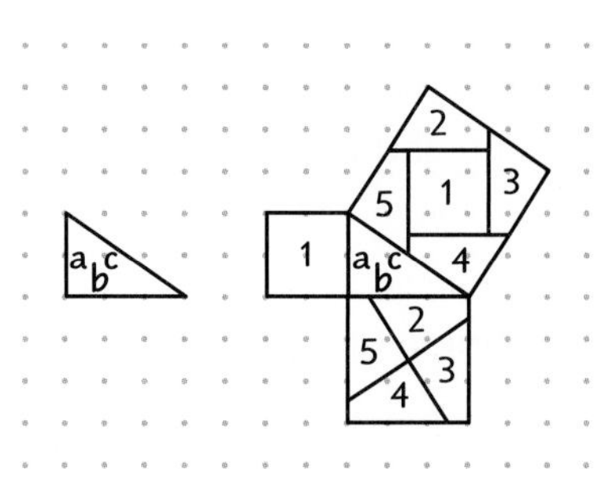

Now look at the larger illustration on the right. A square is drawn on each of the three sides of the triangle.

Using the formula for the area of a square ($A = s^2$), the area of the square on leg a is $A = 3^2 = 9$ square units. The area of the square on leg b is $A = 4^2 = 16$ square units. Both the squares on legs a and b combine to make the square on leg c. The area of the square on leg c must be equal to the sum of the areas of the squares on legs a and b. The area of the square on leg c is $A = 3^2 + 4^2 = 9 + 16 = 25$ square units.

Area of square a + Area of square b = Area of square c

This relationship holds true for all right triangles.

Check It Out

For Exercises 1 and 2, use the illustration below.

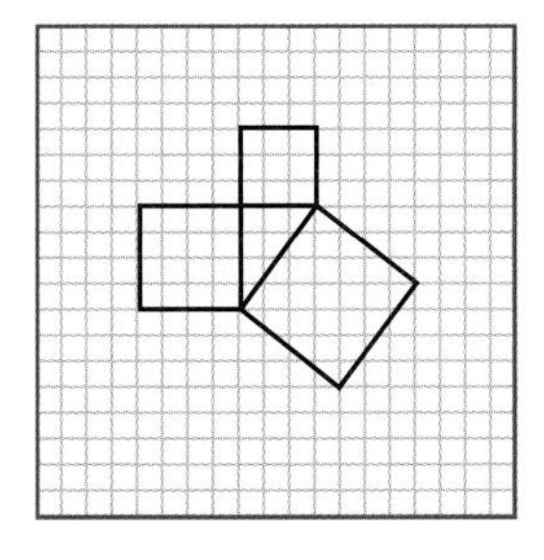

1. What is the area of each of the squares?
2. Does the sum of the areas of the two smaller squares equal the area of the largest square?

The Pythagorean Theorem

In every right triangle, there is a relationship between the area of the square of the hypotenuse (the side opposite the right angle) and the areas of the squares of the legs. A Greek mathematician named Pythagoras noticed the relationship about 2,500 years ago and drew a conclusion. That conclusion, known as the **Pythagorean Theorem**, can be stated as follows: In a right triangle, the square of the length of the hypotenuse is equal to the sum of the squares of the lengths of the legs.

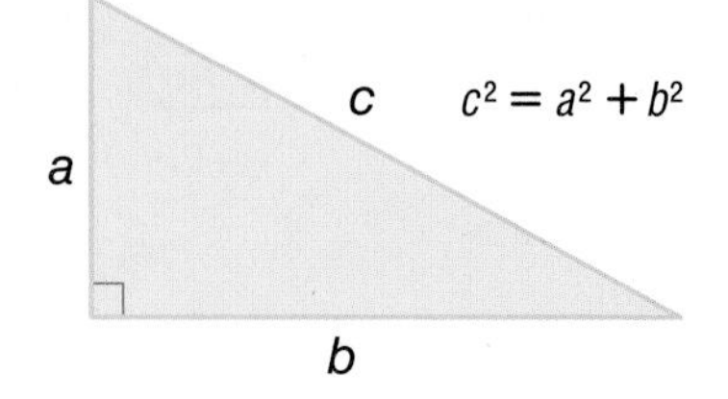

You can use the Pythagorean Theorem to find the length of the third side of a right triangle if you know the length of two sides.

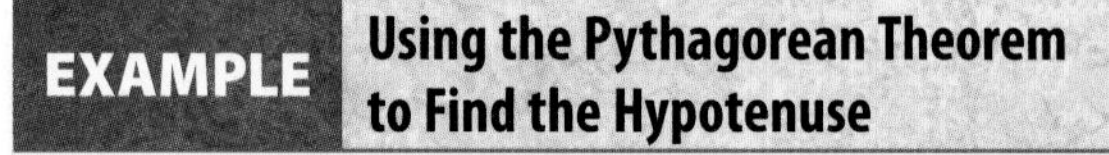

EXAMPLE Using the Pythagorean Theorem to Find the Hypotenuse

Use the Pythagorean Theorem to find the length of the hypotenuse, c, of ΔEFG.

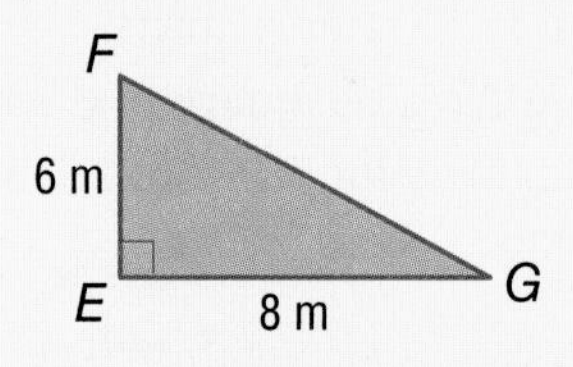

$c^2 = a^2 + b^2$

$c^2 = 6^2 + 8^2$ • Substitute the two known lengths for a and b.

$c^2 = 36 + 64$ • Square the two known lengths.

$c^2 = 100$ • Find the sum of the squares of the two legs.

$c = 10$ m • Take the square root of the sum.

So, the length of the hypotenuse is 10 meters.

EXAMPLE Using the Pythagorean Theorem to Find a Side Length

Use the Pythagorean Theorem, $c^2 = a^2 + b^2$, to find the length of leg b of a right triangle with a hypotenuse 14 inches and leg a 5 inches.

$14^2 = 5^2 + b^2$	• Substitute the two known lengths for a and c.
$196 = 25 + b^2$	• Square the known lengths.
$196 - 25 = (25 - 25) + b^2$	• Subtract to isolate the unknown.
$171 = b^2$	• Use your calculator to find the square root. Round to the nearest tenth.
$13.1 = b$	

So, the length of the unknown side is 13.1 inches.

Check It Out

Use the Pythagorean Theorem to find the missing length.

3. To the nearest whole number, find the length of the hypotenuse of a right triangle with legs 9 centimeters and 11 centimeters.

4. Find the length of $\overline{SR}$ to the nearest whole number.

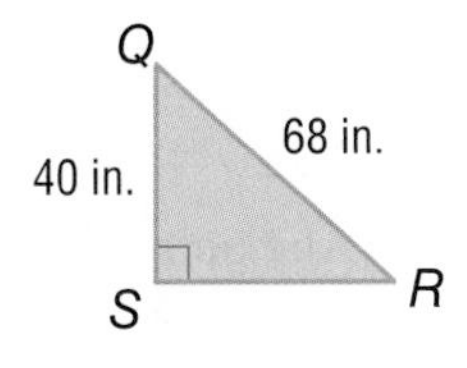

Pythagorean Triples

The numbers 3, 4, and 5 form a **Pythagorean triple** because $3^2 + 4^2 = 5^2$. Pythagorean triples are formed by whole numbers, so that $a^2 + b^2 = c^2$. There are many Pythagorean triples. Here are three:

5, 12, 13 8, 15, 17 7, 24, 25

If you multiply each number of a Pythagorean triple by the same number, you form another Pythagorean triple. 6, 8, 10 is a triple because it is 2(3), 2(4), 2(5).

Distance and the Pythagorean Theorem

To find the diagonal distance between two points on the coordinate plane, connect them with a line (the hypotenuse, *c*). Then draw the horizontal and vertical legs (*a* and *b*) to complete the right triangle. Use the Pythagorean Theorem to solve for *c*.

EXAMPLE Finding Distance on the Coordinate Grid

Find the distance *c* between the points (4, −2) and (2, −6).

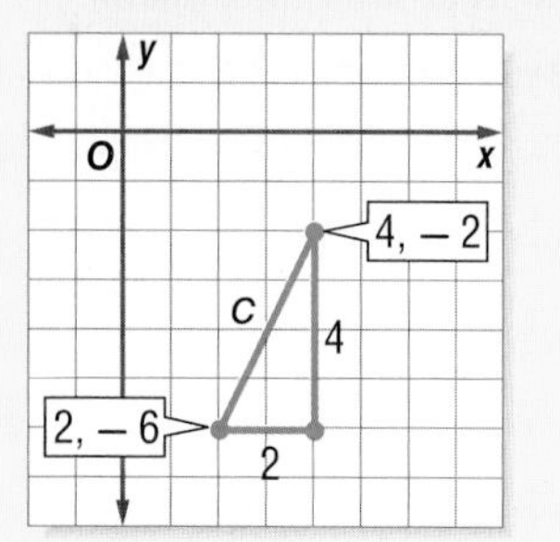

- Graph the ordered pairs (4, −2) and (2, −6).
- Connect the two points to create the hypotenuse (*c*).
- Draw the horizontal and vertical legs (sides *a* and *b*).

$c^2 = a^2 + b^2$

$c^2 = 2^2 + 4^2$

$c^2 = 4 + 16$

$c^2 = 20$

$c \approx 4.5$

- Using the Pythagorean Theorem, replace *a* with 2 and *b* with 4.
- Solve to find the distance *c* between the two points.

So, the points are about 4.5 units apart.

Check It Out

Find the distance between the points to the nearest tenth.

5. (1, 2), (3, 5)

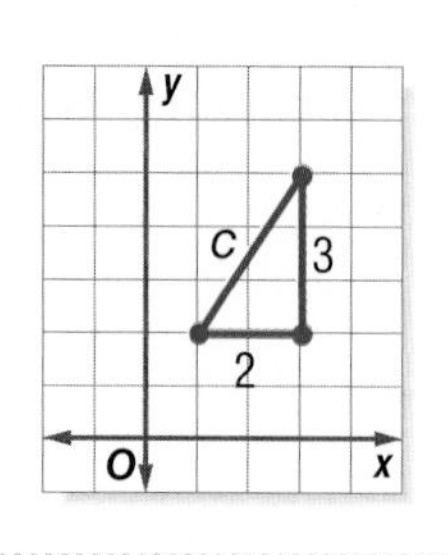

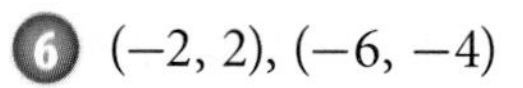

6. (−2, 2), (−6, −4)

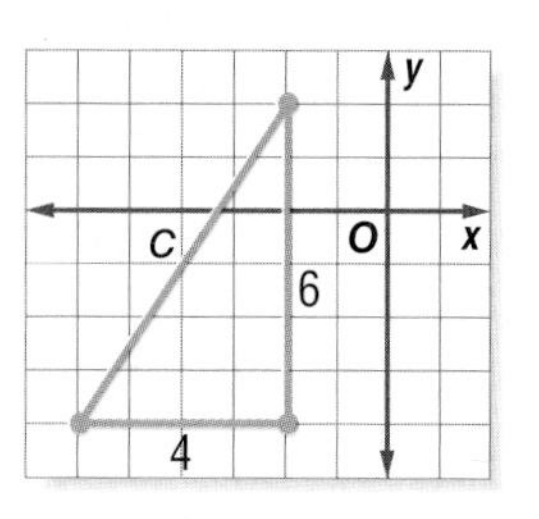

7-9 Exercises

Each side of the triangle has a square on it. The squares are labeled regions I, II, and III.

1. Find the areas of regions I, II, and III.
2. What relationship exists among the areas of regions I, II, and III?

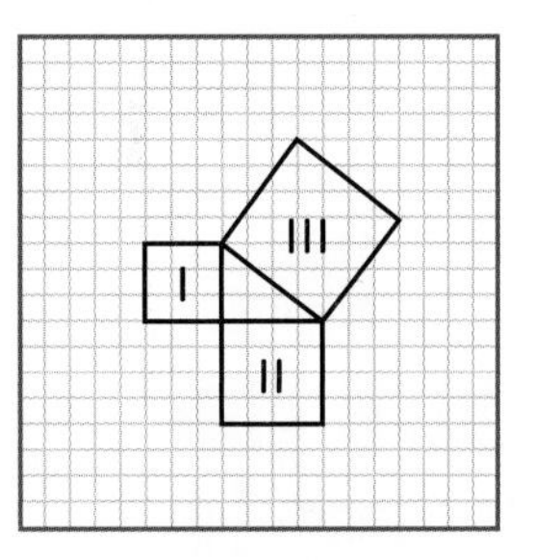

Find the missing length in each right triangle. Round to the nearest tenth.

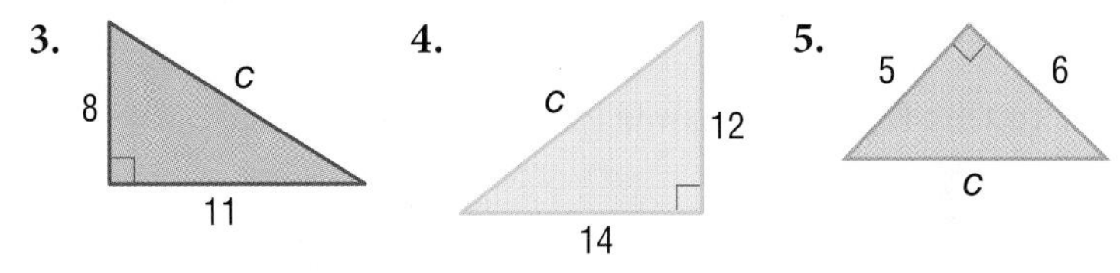

Are the numbers Pythagorean triples? Write *yes* or *no*.

6. 3, 4, 5
7. 4, 5, 6
8. 24, 45, 51

9. Find the length, to the nearest tenth, of the unknown leg of a right triangle with a hypotenuse of 16 inches and one leg measuring 9 inches.
10. To the nearest tenth, find the length of the hypotenuse of a right triangle with legs measuring 39 centimeters and 44 centimeters.

Find the distance between the points to the nearest tenth.

11. (−5, −7), (−1, 0)
12. (2, 1), (4, 8)

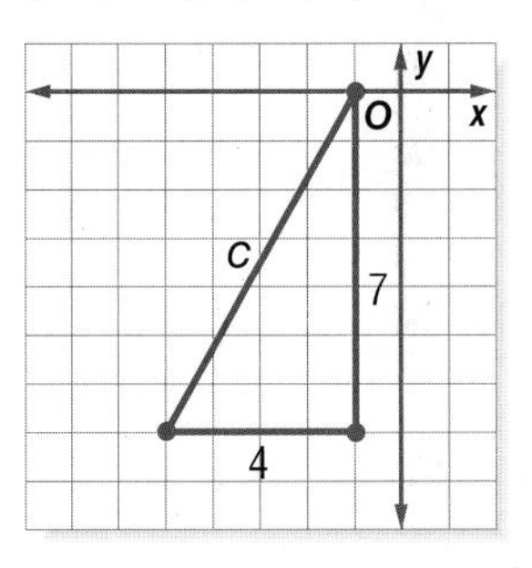

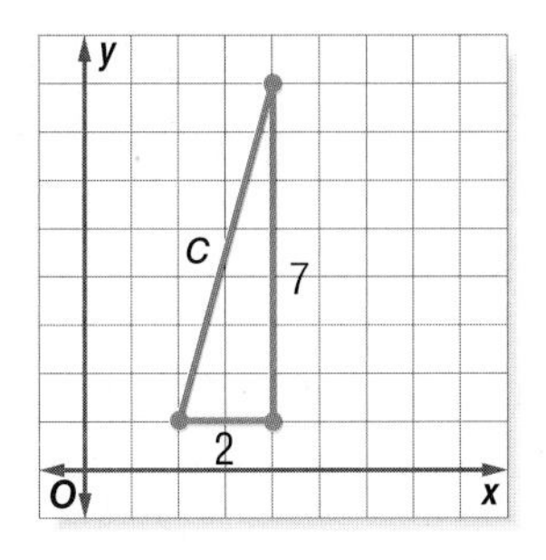

Geometry

What have you learned?

You can use the problems and the list of words that follow to see what you learned in this chapter. You can find out more about a particular problem or word by referring to the topic number (*for example,* Lesson 7·2).

Problem Set

Refer to this figure to answer Exercises 1–3. (Lesson 7·1)

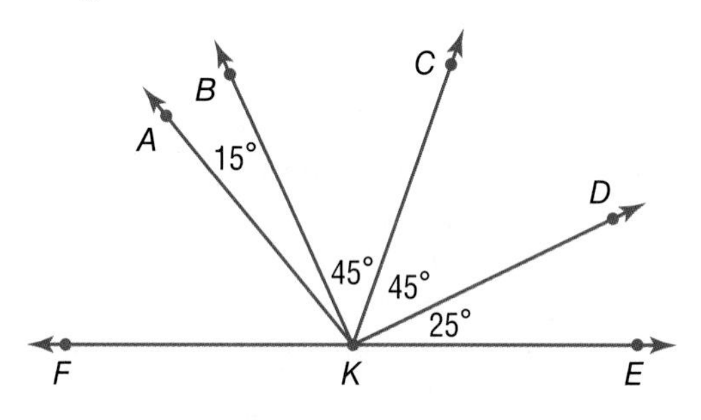

1. Name the angle adjacent to $\angle BKE$.
2. Name the right angle in this figure.
3. $\angle FKE$ is a straight angle. What is $m\angle FKA$?
4. Find the measure of each angle of a regular hexagon. (Lesson 7·2)
5. Find the perimeter of a right triangle with legs 3 centimeters and 4 centimeters. (Lesson 7·4)
6. Find the area of a right triangle with legs 20 meters and 5 meters. (Lesson 7·5)
7. A trapezoid has bases 12 feet and 20 feet and height 6 feet. What is the area? (Lesson 7·5)
8. Find the surface area of a cylinder if $h = 10$ feet and $C = 4.5$ feet. Round to the nearest square foot. (Lesson 7·6)

9. Find the volume of a cylinder that has an 8-inch diameter and is 6 inches high. Round to the nearest cubic inch. (Lesson 7•7)
10. A triangle has sides 15 centimeters, 16 centimeters, and 23 centimeters. Use the Pythagorean Theorem to determine whether this triangle is a right triangle. (Lesson 7•9)
11. Find the length of the unknown leg of a right triangle that has hypotenuse 18 meters and one leg that is 10 meters. (Lesson 7•9)

HotWords

Write definitions for the following words.

alternate exterior angles (Lesson 7•1)
alternate interior angles (Lesson 7•1)
arc (Lesson 7•8)
base (Lesson 7•2)
circumference (Lesson 7•8)
complementary angles (Lesson 7•1)
congruent (Lesson 7•1)
corresponding angles (Lesson 7•1)
cube (Lesson 7•2)
diagonal (Lesson 7•2)
diameter (Lesson 7•8)
face (Lesson 7•2)
hypotenuse (Lesson 7•9)
line of symmetry (Lesson 7•3)
net (Lesson 7•6)
parallelogram (Lesson 7•2)
pi (Lesson 7•7)
polygon (Lesson 7•2)
polyhedron (Lesson 7•2)
prism (Lesson 7•2)
pyramid (Lesson 7•2)
Pythagorean Theorem (Lesson 7•9)
Pythagorean triple (Lesson 7•9)
quadrilateral (Lesson 7•2)
radius (Lesson 7•8)
rectangular prism (Lesson 7•2)
reflection (Lesson 7•3)
regular polygon (Lesson 7•2)
rhombus (Lesson 7•2)
rotation (Lesson 7•3)
segment (Lesson 7•2)
supplementary angles (Lesson 7•1)
surface area (Lesson 7•6)
tetrahedron (Lesson 7•2)
transformation (Lesson 7•3)
translation (Lesson 7•3)
transversal (Lesson 7•1)
trapezoid (Lesson 7•2)
triangular prism (Lesson 7•6)
vertical angles (Lesson 7•1)
volume (Lesson 7•7)

HotTopic 8

Measurement

What do you know?

You can use the problems and the list of words that follow to see what you already know about this chapter. The answers to the problems are in **HotSolutions** at the back of the book, and the definitions of the words are in **HotWords** at the front of the book. You can find out more about a particular problem or word by referring to the topic number (*for example,* Lesson 8•2).

Problem Set

Give the meaning for each metric system prefix. (Lesson 8•1)

1. centi- **2.** kilo- **3.** milli-

Complete each of the following conversions. Round your answers to the nearest hundredth. (Lesson 8•2)

4. 800 mm = ___ m **5.** 5,500 m = ___ km

6. 3 mi = ___ ft **7.** 468 in. = ___ yd

Exercises 8–13 refer to the rectangle. Round to the nearest whole unit.

36 in.

18 in.

8. What is the perimeter in inches? (Lesson 8•2)

9. What is the perimeter in yards? (Lesson 8•2)

10. What is the perimeter in centimeters? (Lesson 8•2)

11. What is the approximate perimeter in meters? (Lesson 8•2)

12. What is the area in square inches? (Lesson 8•3)

13. What is the area in square centimeters? (Lesson 8•3)

Convert the following area and volume measurements as indicated. (Lesson 8•3)

14. $5\ m^2 = \_\_\_\_\ cm^2$

15. $10\ yd^2 = \_\_\_\_\ ft^2$

16. $3\ ft^3 = \_\_\_\_\ in^3$

17. $4\ cm^3 = \_\_\_\_\ mm^3$

18. You pour 6 pints of water into a gallon jar. What fraction of the jar is filled? (Lesson 8•3)

19. A perfume bottle holds $\frac{1}{2}$ fluid ounce. How many bottles would you need to fill 1 cup? (Lesson 8•3)

20. A can of juice holds 385 milliliters. About how many cans will it take to fill a 5 liter container? (Lesson 8•3)

21. You are allowed to take a 20 kilogram suitcase on a small airplane in Africa. About how many pounds will your suitcase weigh? (Lesson 8•4)

22. Your cookie recipe calls for 4 ounces of butter to make one batch. If you want to make 12 batches of cookies for the bake sale, how many pounds of butter do you buy? (Lesson 8•4)

A picture 5 inches high and 8 inches wide was enlarged to make a poster. The width of the poster is 1.5 feet.

23. What is the ratio of the width of the poster to the width of the original photo? (Lesson 8•5)

24. What is the scale factor? (Lesson 8•5)

HotWords

area (Lesson 8•3)
customary system (Lesson 8•1)
metric system (Lesson 8•1)
scale factor (Lesson 8•5)
similar figures (Lesson 8•5)
volume (Lesson 8•3)

8·1 Systems of Measurement

The most common system of measurement in the world is the **metric system**. The **customary system** of measurement is used in the United States. It may be useful to make conversions from one unit of measurement to another within each system, as well as to convert units between the two systems.

The metric system of measuring is based on powers of ten, such as 10, 100, and 1,000. To convert within the metric system, multiply and divide by powers of ten.

Prefixes in the metric system have consistent meaning.

Prefix	Meaning	Example
milli-	one thousandth	1 *milli*liter is 0.001 liter.
centi-	one hundredth	1 *centi*meter is 0.01 meter.
kilo-	one thousand	1 *kilo*gram is 1,000 grams.

The customary system of measurement is not based on powers of ten. It is based on numbers such as 12 and 16, which have many factors. Whereas the metric system uses decimals, you will frequently encounter fractions in the customary system.

Unfortunately, there are no convenient prefixes as in the metric system, so you will have to memorize the basic equivalent units: 16 ounces = 1 pound, 36 inches = 1 yard, 4 quarts = 1 gallon, and so on.

Check It Out

Use the words *metric* or *customary* to answer Exercises 1–3.

1. Which system is based on multiples of 10?
2. Which system uses fractions?
3. Which system is the most common system of measurement in the world?

8•1 Exercises

Give the meaning of each prefix.

1. centi-

2. kilo-

3. milli-

Write the system of measurement used for the following.

4. inches

5. meters

6. quarts

7. yards

8. liters

9. pounds

10. gallons

11. grams

12. tons

13. ounces

14. What system of measurement is based on powers of ten?

Write the unit you would use for the following.

15. dispenser of dental floss

16. football field

17. soft drink bottle

18. toothpaste

19. flour

20. lumber

21. gasoline

22. storage capacity in computer

8•2 Length and Distance

Metric and Customary Units

Both systems of measurement can be used to measure length and distance. The commonly used metric measures for length and distance are millimeter, centimeter, meter, and kilometer. The customary system uses inch, foot, yard, and mile.

Metric Equivalents						
1 km	=	1,000 m	=	100,000 cm	=	1,000,000 mm
0.001 km	=	1 m	=	100 cm	=	1,000 mm
		0.01 m	=	1 cm	=	10 mm
		0.001 m	=	0.1 cm	=	1 mm

Customary Equivalents						
1 mi	=	1,760 yd	=	5,280 ft	=	63,360 in.
$\frac{1}{1,760}$ mi	=	1 yd	=	3 ft	=	36 in.
		$\frac{1}{3}$ yd	=	1 ft	=	12 in.
		$\frac{1}{36}$ yd	=	$\frac{1}{12}$ ft	=	1 in.

EXAMPLE **Changing Units Within a System**

How many inches are in $\frac{1}{3}$ mile?

units you have

$1 \text{ mi} = 63{,}360 \text{ in.}$

conversion factor for new units

- Find the units you have equal to 1 on the equivalents chart.

$\frac{1}{3} \cdot 63{,}360 = 21{,}120$

- Find the conversion factor.
- Multiply to get new units.

So, there are 21,120 inches in $\frac{1}{3}$ mile.

Check It Out

Complete the conversions.

1. 8 m = ____ cm
2. 3,500 m = ____ km
3. 48 in. = ____ ft
4. 2 mi = ____ ft

Conversions Between Systems

You may need to convert between the metric system and the customary system. You can use this conversion table to help.

Conversion Table					
1 inch	=	25.4 millimeters	1 millimeter	=	0.0394 inch
1 inch	=	2.54 centimeters	1 centimeter	=	0.3937 inch
1 foot	=	0.3048 meter	1 meter	=	3.2808 feet
1 yard	=	0.914 meter	1 meter	=	1.0936 yards
1 mile	=	1.609 kilometers	1 kilometer	=	0.621 mile

To calculate a conversion, find the conversion factor in the table above.

Your friend in Costa Rica says that he can jump 119 centimeters. Should you be impressed? 1 centimeter = 0.39737 inch. So, $119 \cdot 0.3937 \approx 46.9$ inches. How far can you jump?

Most of the time, you need only to estimate the conversion from one system to the other to get an idea of the size of your item. Round numbers in the conversion table to simplify your thinking.

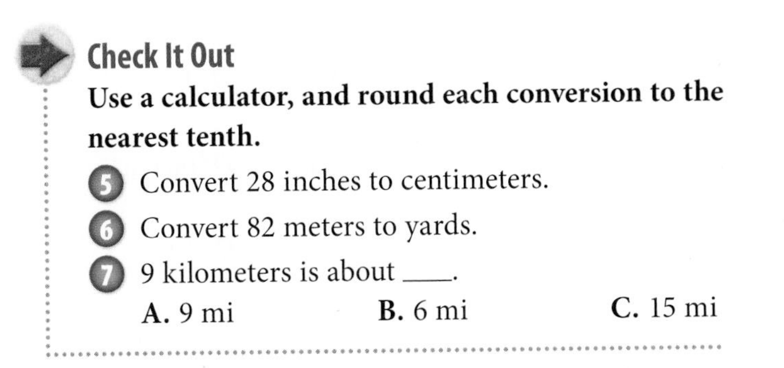

Check It Out

Use a calculator, and round each conversion to the nearest tenth.

5. Convert 28 inches to centimeters.
6. Convert 82 meters to yards.
7. 9 kilometers is about ____.

   A. 9 mi **B.** 6 mi **C.** 15 mi

8•2 Exercises

Complete the conversions.

1. 10 cm = ____ mm
2. 200 mm = ____ m
3. 3,000 mm = ____ cm
4. 2.4 km = ____ m
5. 11 yd = ____ in.
6. 7 mi = ____ ft
7. 400 in. = ____ ft
8. 3,024 in. = ____ yd
9. 0.5 yd = ____ ft
10. 520 yd = ____ mi

Use a calculator, and round each conversion to the nearest tenth.

11. Convert 6 inches to centimeters.
12. Convert 2 feet to centimeters.
13. Convert 200 millimeters to inches.

Choose the nearest conversion estimate for the following lengths.

14. 5 mm
    A. 5 in. **B.** 2 in. **C.** 5 yd **D.** $\frac{1}{5}$ in.
15. 1 ft
    A. 30 cm **B.** 1 m **C.** 50 cm **D.** 35 mm
16. 25 in.
    A. 25 cm **B.** 1 m **C.** 0.5 m **D.** 63.5 cm
17. 300 m
    A. $\frac{1}{2}$ mi **B.** 300 yd **C.** 600 ft **D.** 100 yd
18. 100 km
    A. 200 mi **B.** 1,000 yd **C.** 60 mi **D.** 600 mi
19. 36 in.
    A. 1 cm **B.** 1 mm **C.** 1 km **D.** 1 m
20. 6 ft
    A. 6 m **B.** 200 cm **C.** 600 cm **D.** 60 cm
21. 1 cm
    A. $\frac{1}{2}$ in. **B.** 1 in. **C.** 2 in. **D.** 1 ft
22. 2 mi
    A. 300 m **B.** 2,000 m **C.** 2 km **D.** 3 km

8•3 Area, Volume, and Capacity

Area

Area is the measure of the interior region of a 2-dimensional figure. Area is expressed in square units.

Area can be measured in metric units or customary units. You might need to convert within a measurement system. Below is a conversion table that provides the most common conversions.

Metric			Customary		
100 mm^2	=	1 cm^2	144 in^2	=	1 ft^2
10,000 cm^2	=	1 m^2	9 ft^2	=	1 yd^2
			4,840 yd^2	=	1 acre
			640 acres	=	1 mi^2

To convert to a new unit, find the conversion factor in the table above.

EXAMPLE Changing Area Units

The area of the United States is approximately 3,500,000 square miles. How many acres does the United States cover?

640 acres = 1 mi^2
- Find the units you have equal to 1 in the conversion table.
- Find the conversion factor.

3,500,000 • 640
- Multiply to get the new units.

So, there are 2,240,000,000 acres covered by the United States.

Check It Out

Solve.

1. How many square millimeters are equal to 16 square centimeters?
2. How many square inches are equal to 2 square feet?
3. How many square yards are equal to 3 acres?

Volume

Volume is the space occupied by a solid and is expressed in cubic units. Here are the basic relationships among units of volume.

Metric	Customary
1,000 $mm^3 = 1\ cm^3$	1,728 $in^3 = 1\ ft^3$
1,000,000 $cm^3 = 1\ m^3$	27 $ft^3 = 1\ yd^3$

EXAMPLE Converting Volume Within a System of Measurement

Express the volume of the carton in cubic meters.

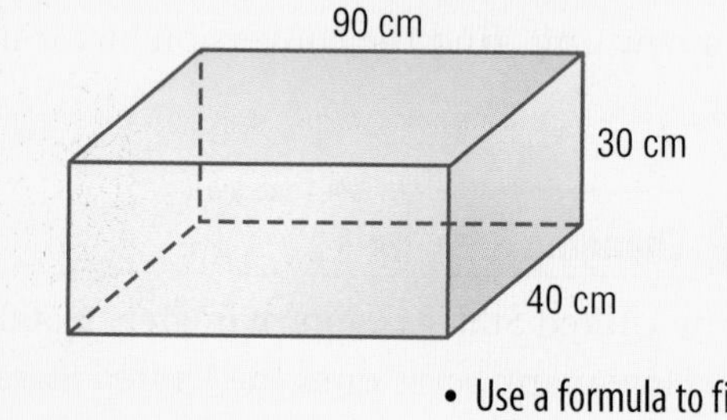

$V = \ell wh$
$= 90 \cdot 40 \cdot 30$
$= 108{,}000\ cm^3$

- Use a formula to find the *volume,* using the units of the dimensions.

$1{,}000{,}000\ cm^3 = 1\ m^3$

- Find the conversion factor.

$108{,}000 \div 1{,}000{,}000 = 0.108\ m^3$

- Divide to convert to larger units.
- Include the unit of measurement in your answer.

So, the volume of the carton is $0.108\ m^3$.

Check It Out

Solve.

4. Find the volume of a box that measures 9 feet by 6 feet by 6 feet. Convert to cubic yards.
5. Find the volume of a cube that measures 8 centimeters on a side. Convert to cubic millimeters.
6. How many cubic inches are equal to 15 cubic feet?
7. How many cubic centimeters are equal to 250 cubic millimeters?

Capacity

Capacity is closely related to volume, but there is a difference. The capacity of a container is a measure of the amount of liquid it can hold. A block of wood has volume but no capacity to hold liquid.

Conversion Table	
1 liter (L) = 1,000 milliliters (mL) 1 liter = 1.057 quarts (qt)	8 fluid ounces (fl oz) = 1 cup (c) 2 cups = 1 pint (pt) 2 pints = 1 quart (qt) 4 quarts = 1 gallon (gal)

Note the use of fluid ounce (*fl oz*) in the table. This is to distinguish it from ounce (*oz*) which is a unit of weight (16 oz = 1 lb). Fluid ounce is a unit of capacity (16 fl oz = 1 pint). There is a connection between ounce and fluid ounce. A pint of water weighs about a pound, so a fluid ounce of water weighs an ounce. For water, as well as for most other liquids used in cooking, *fluid ounce* and *ounce* are equivalent, and the "fl" is sometimes omitted (for example, "8 oz = 1 cup"). To be correct, use *ounce* for weight only and *fluid ounce* for capacity. For liquids that weigh considerably more or less than water, the difference is significant.

EXAMPLE Changing Capacity Units

Gasoline is priced at $0.92 per liter. What is the price per gallon?

1 gal = 4 qt	• Find the units you have equal to 1 in the conversion table.
1 qt = 1.057 L	• Find the conversion factor.
4 • 1.057 = 4.228 L	• Multiply to get the number of liters in a gallon.
$0.92 • 4.228	• Multiply to get the price per gallon.

So, the price of gasoline is $3.89 per gallon.

Check It Out

Solve.

8. If liters of cola are on sale for $0.99 each and you can buy a can of juice concentrate that makes 1 gallon for $3.49, which is the better buy?

APPLICATION In the Soup

One morning on a California freeway, a big-rig truck tipped over on its side. The truck was carrying 43,000 cans of cream of mushroom soup.

At 24 cans per carton, how many cartons of soup was the truck carrying? If each carton had a width of 11 inches, a length of 16 inches, and a height of 5 inches, what was the approximate carrying capacity of the truck in cubic feet? See **HotSolutions** for the answer.

8•3 Exercises

Tell whether the unit is a measure of distance, area, or volume.

1. cm

2. in^3

3. acre

4. mm^2

Calculate the volume of the cartons in each measurement unit below.

5. ft

6. in.

7. cm

8. m

9. mm

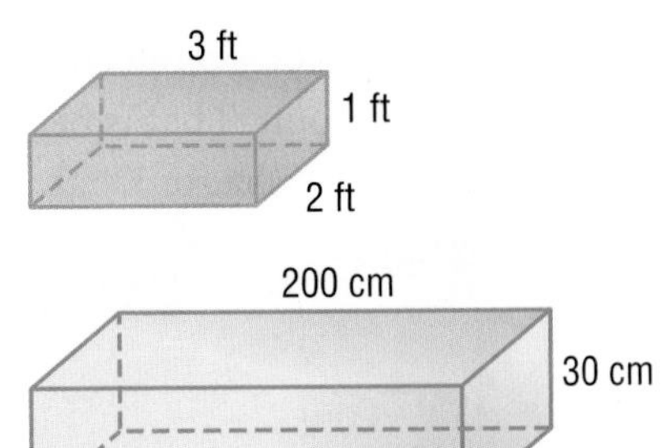

Convert to new units.

10. 1 gal = ____ c

11. 2 qt = ____ fl oz

12. 160 fl oz = ____ qt

13. 4 gal = ____ qt

14. 3 pt = ____ gal

15. 4 fl oz = ____ pt

16. 8 L = ____ mL

17. 24,500 mL = ____ L

18. 10 mL = ____ L

19. Krutika has a fish tank that holds 15 liters of water. One liter of water has evaporated. She has a 200-milliliter measuring cup. How many times will she have to fill the cup in order to refill the tank?

20. Estimate, to the nearest dime, the price per liter of gasoline selling for $3.68 per gallon.

8•4 Mass and Weight

Mass and weight are different. Mass is the amount of matter in an object. Weight is determined by the mass of an object and the effect of gravity on that object. On Earth, mass and weight are equal, but on the Moon, mass and weight are quite different. Your mass would be the same on the Moon as it is here on Earth. But, if you weigh 100 pounds on Earth, you weigh about $16\frac{2}{3}$ pounds on the Moon. That is because the gravitational pull of the Moon is only $\frac{1}{6}$ that of Earth.

Metric				
1 kg	=	1,000 g	=	1,000,000 mg
0.001 kg	=	1 g	=	1,000 mg
0.000001 kg	=	0.001 g	=	1 mg

Customary				
1 T	=	2,000 lb	=	32,000 oz
0.0005 T	=	1 lb	=	16 oz
0.0625 lb	=	1 oz		

1 pound ≈ 0.4536 kilogram
1 kilogram ≈ 2.205 pounds

To convert from one unit to another, first find the 1 for the units you have in the list of equivalents. Then use the conversion factor to calculate the new units.

If you have 32 ounces of peanut butter, how many pounds do you have? 1 oz = 0.0625 lb, so 32 oz = 32 • 0.0625 lb = 2 lb. You have 2 pounds of peanut butter.

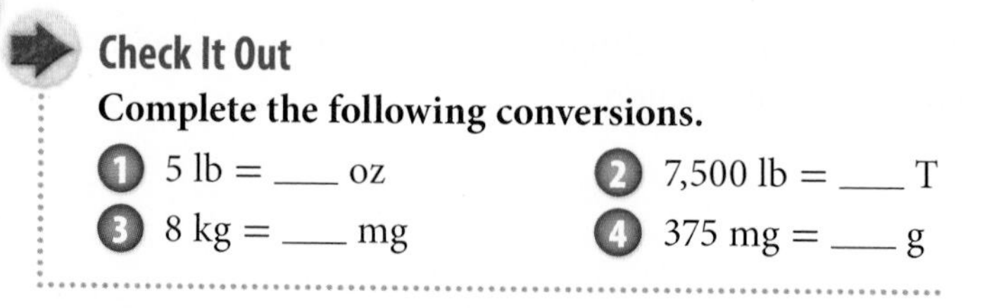

Check It Out

Complete the following conversions.

1. 5 lb = ____ oz
2. 7,500 lb = ____ T
3. 8 kg = ____ mg
4. 375 mg = ____ g

8•4 Exercises

Convert to the indicated units.

1. 1.2 kg = ___ mg
2. 250 mg = ___ g
3. 126,500 lb = ___ T
4. 24 oz = ___ lb
5. 8,000 mg = ___ kg
6. 2.3 T = ___ lb
7. 8 oz = ___ lb
8. 250 g = ___ oz
9. 100 kg = ___ lb
10. 25 lb = ___ kg
11. 200 oz = ___ lb
12. 880 oz = ___ kg
13. 880 g = ___ lb
14. 8 g = ___ oz
15. 16 oz = ___ kg
16. 1.5 T = ___ kg

17. Your cookie recipe calls for 12 ounces of butter to make one batch. For your party, you will make 4 batches of cookies. How many pounds of butter do you need to buy?
18. Two brands of laundry soap are on sale. A 2-pound box of Brand Y is selling for $12.50. A 20-ounce box of Brand Z is on sale for $7.35. Which is the better buy?
19. French chocolates sell for $18.50 per kilogram. A 10-ounce box of domestic chocolates sells for $7.75. Which is the better buy?
20. If an elephant weighs about 3,500 kilograms on Earth, how many pounds would it weigh on the Moon? Could you lift it? Round your answer to the nearest pound.

8•5 Size and Scale

Similar Figures

Similar figures are figures that have exactly the same shape. Figures that are similar may be the same size or different sizes. However, all the corresponding sides of the figure will have the same ratio. Remember that each ratio must be set up in the same order.

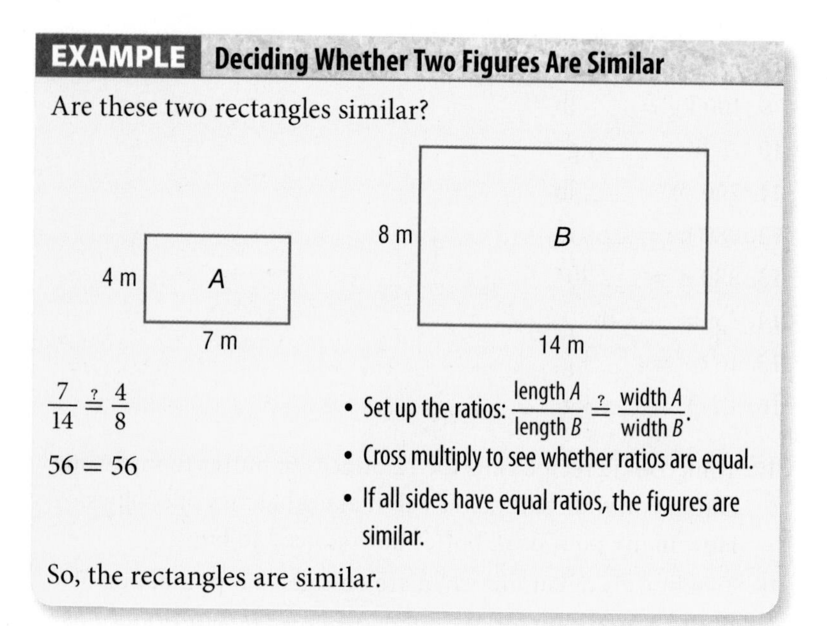

EXAMPLE Deciding Whether Two Figures Are Similar

Are these two rectangles similar?

$\frac{7}{14} \stackrel{?}{=} \frac{4}{8}$

$56 = 56$

- Set up the ratios: $\frac{\text{length } A}{\text{length } B} \stackrel{?}{=} \frac{\text{width } A}{\text{width } B}$.
- Cross multiply to see whether ratios are equal.
- If all sides have equal ratios, the figures are similar.

So, the rectangles are similar.

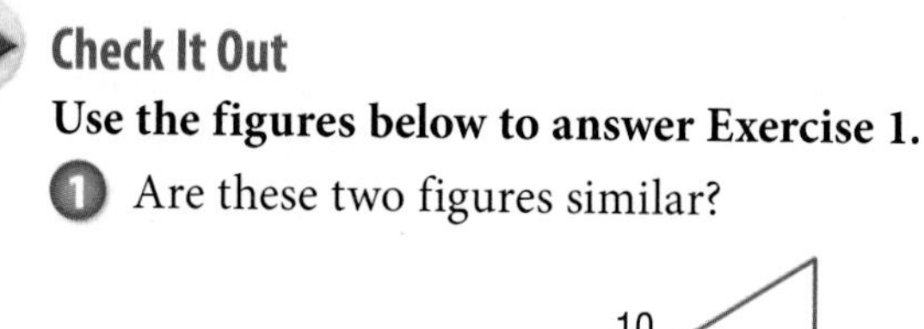

Check It Out

Use the figures below to answer Exercise 1.

1. Are these two figures similar?

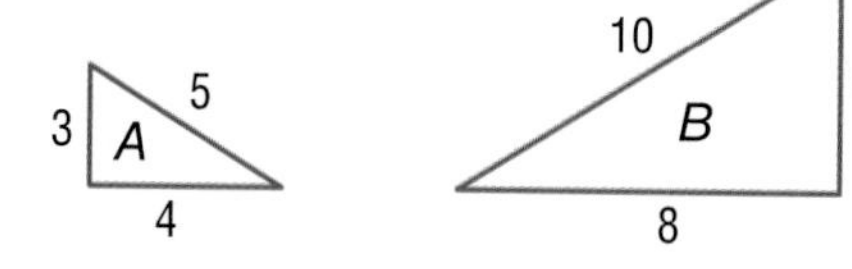

Scale Factors

The **scale factor** of two similar figures is the ratio of the corresponding side lengths.

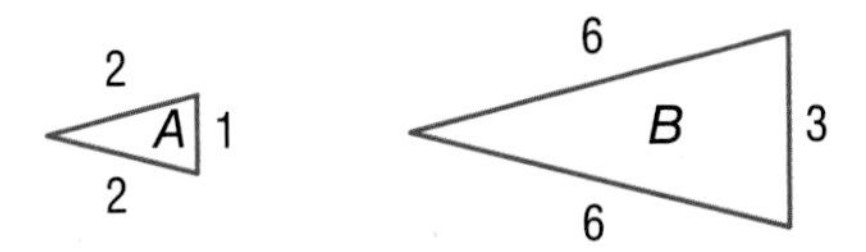

Triangle A is similar to triangle B. ΔB is 3 times larger than ΔA, so the scale factor is 3.

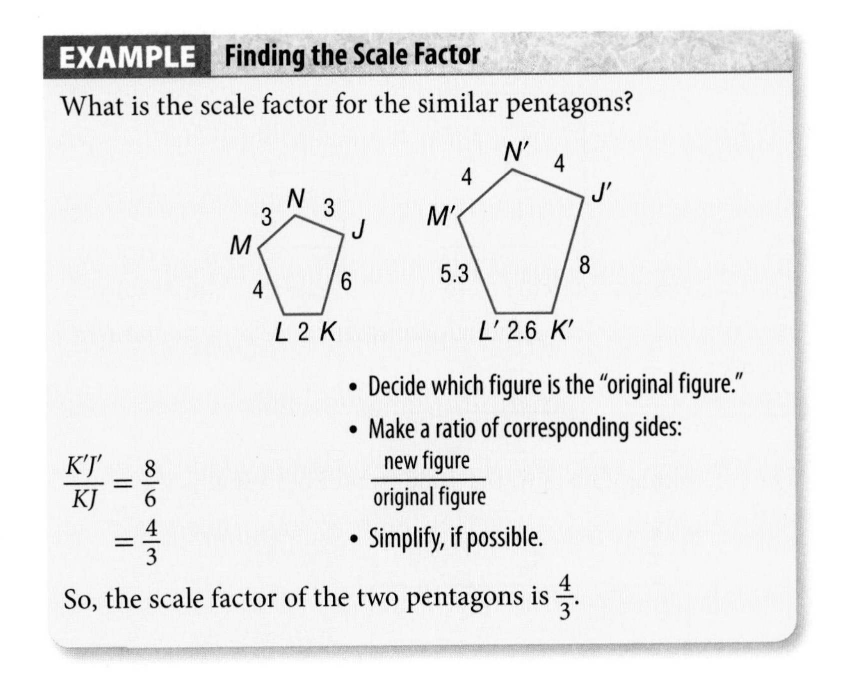

When a figure is enlarged, the scale factor is greater than 1. When similar figures are identical in size, the scale factor is equal to 1. When a figure is reduced, the scale factor is less than 1 but greater than zero.

Check It Out

What is the scale factor?

2

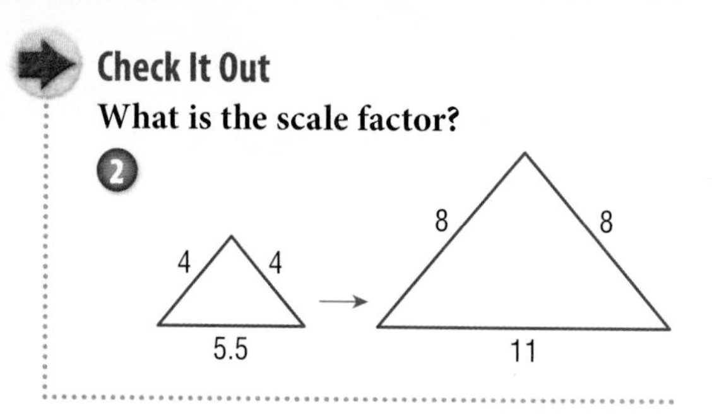

Scale Factors and Area

Scale factor refers to a ratio of lengths only, not of the areas.

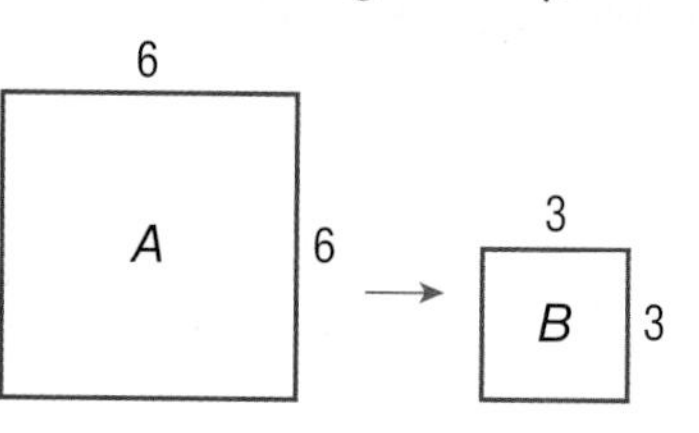

For the squares, the scale factor is $\frac{1}{2}$ because the ratio of sides is $\frac{3}{6} = \frac{1}{2}$. Notice that, although the scale factor is $\frac{1}{2}$, the ratio of the areas is $\frac{1}{4}$.

$$\frac{\text{Area of } B}{\text{Area of } A} = \frac{3^2}{6^2} = \frac{9}{36} = \frac{1}{4}$$

For the rectangles below, the scale factor is $\frac{1}{3}$. What is the ratio of the areas?

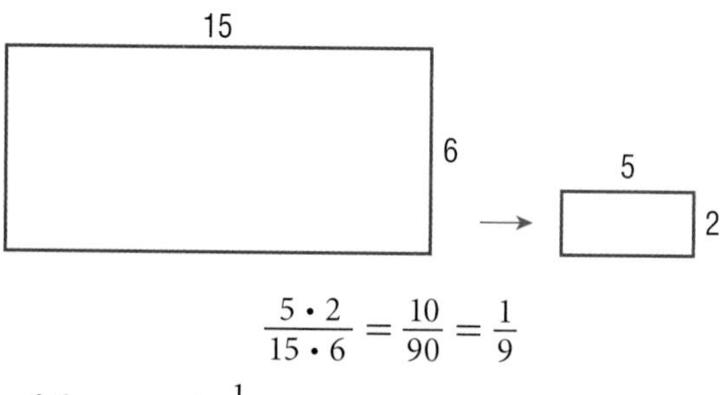

$$\frac{5 \cdot 2}{15 \cdot 6} = \frac{10}{90} = \frac{1}{9}$$

The ratio of the areas is $\frac{1}{9}$.

The ratio of the areas of similar figures is the *square* of the scale factor.

Check It Out

Solve.

3. The scale factor for two similar figures is $\frac{3}{2}$. What is the ratio of the areas?
4. The scale on a blueprint for a garage is 1 foot = 4 feet. An area of 1 square foot on the blueprint represents how much area on the garage floor?
5. Complete the table below.

	Area
Scale Factor **2**	**4** times
Scale Factor **3**	
Scale Factor **4**	
Scale Factor **5**	
Scale Factor ***X***	

Scale Factors and Volume

Remember that scale factor refers to a ratio of lengths.

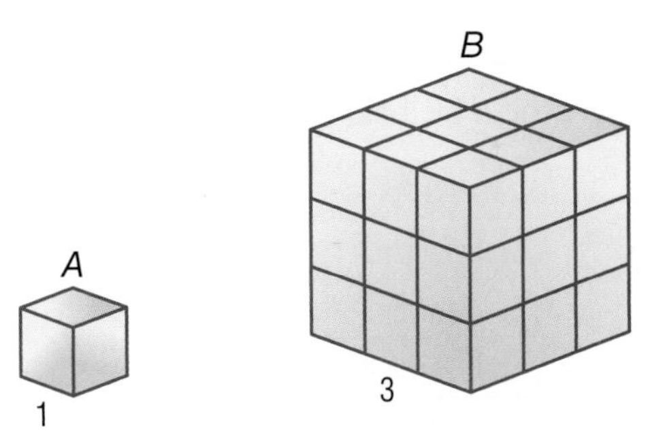

Cube A has a volume of 1 cm^3. Cube B is an enlargement of Cube A by a scale factor 3. The volume changes by a factor of s^3. Multiply $3 \cdot 3 \cdot 3$ to find the volume of the enlarged cube.

$$\frac{\text{Volume of } B}{\text{Volume of } A} = \frac{3 \cdot 3 \cdot 3}{1 \cdot 1 \cdot 1} = \frac{27}{1}$$

The volume of the enlarged cube is 27 cm^3.

If the scale factor is $\frac{2}{3}$, what is the ratio of the volumes?

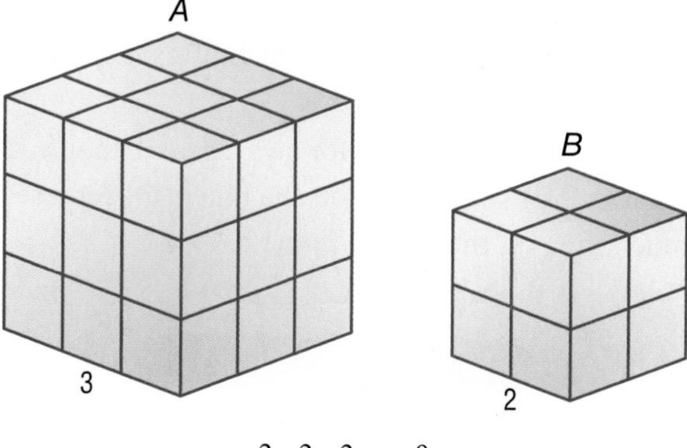

$$\frac{2 \cdot 2 \cdot 2}{3 \cdot 3 \cdot 3} = \frac{8}{27}$$

The ratio of the volumes is $\frac{8}{27}$.

The ratio of the volume of similar figures is the scale factor *cubed.*

Check It Out

Solve.

6. The scale factor for two similar figures is $\frac{3}{4}$. What is the ratio of the volumes?

7. Complete the table below.

	Volume
Scale Factor **2**	**8** times
Scale Factor **3**	
Scale Factor **4**	
Scale Factor **5**	
Scale Factor ***X***	

8•5 Exercises

Give the scale factor.

1.

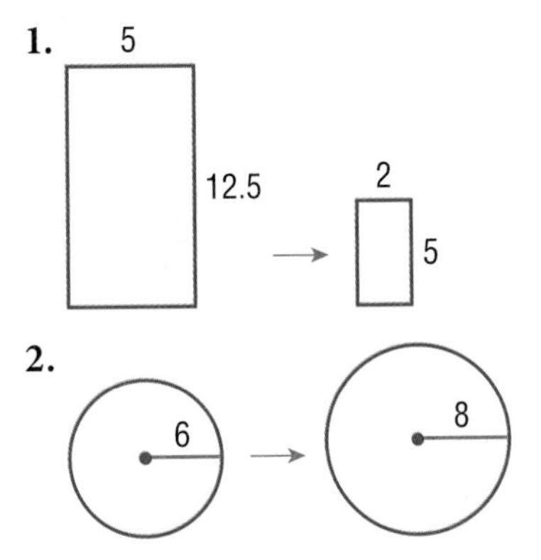

2.

3. A 3-inch by 5-inch photograph is enlarged by a scale factor of 3. What are the dimensions of the enlarged photo?

4. A document 11 inches long and $8\frac{1}{2}$ inches wide is reduced. The reduced document is $5\frac{1}{2}$ inches long. How wide is it?

5. A map shows a scale of 1 centimeter = 20 kilometers. If two towns are about 50 kilometers apart, how far apart will they be on the map?

6. If a map's scale is 1 inch = 5 miles, and the map is a rectangle 12 inches by 15 inches, what is the area shown on the map?

7. A photo is enlarged by a scale factor of 1.5. The area of the larger photo is how many times the area of the smaller photo?

8. On a map, a road appears to be $2\frac{3}{4}$ inches long. The map's scale is $\frac{1}{2}$ inch = 10 miles. About how many miles long is the road?

9. The triangles below are similar. Find the values of x and y.

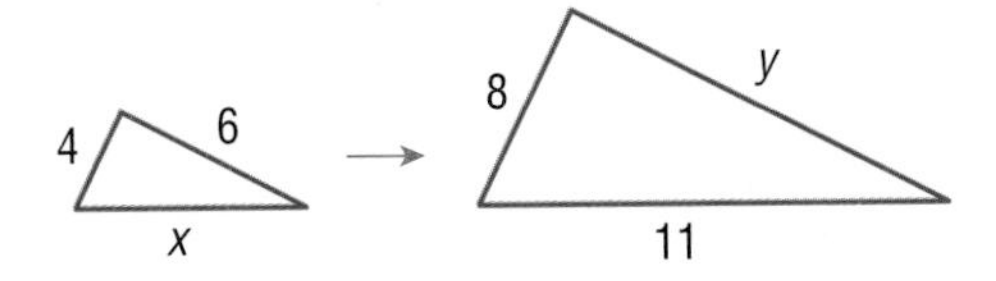

10. The scale factor for two similar figures is $\frac{5}{2}$. What is the ratio of the volumes?

Measurement

What have you learned?

You can use the problems and the list of words that follow to see what you learned in this chapter. You can find out more about a particular problem or word by referring to the topic number (*for example,* Lesson 8•2).

Problem Set

Give the meaning for each metric system prefix. (Lesson 8•1)

1. centi- **2.** kilo- **3.** milli-

Complete each of the following conversions. Round to the nearest hundredth. (Lesson 8•2)

4. 600 mm = ____ m
5. 367 m = ____ km
6. 2.5 mi = ____ ft
7. 288 in. = ____ yd

Exercises 8–13 refer to the rectangle. Round to the nearest whole unit.

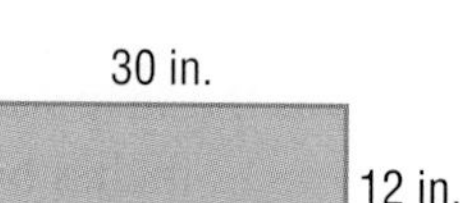

8. What is the perimeter in inches? (Lesson 8•2)
9. What is the perimeter in yards? (Lesson 8•2)
10. What is the perimeter in centimeters? (Lesson 8•2)
11. What is the approximate perimeter in meters? (Lesson 8•2)
12. What is the area in square inches? (Lesson 8•3)
13. What is the area in square centimeters? (Lesson 8•3)

Convert the following area and volume measurements as indicated. (Lesson 8•3)

14. $42.5 \text{ m}^2 =$ ____ cm^2
15. $7 \text{ yd}^2 =$ ____ ft^2
16. $10 \text{ ft}^3 =$ ____ in^3
17. $6.5 \text{ cm}^3 =$ ____ mm^3

18. You pour 3 pints of water into a gallon jar. What fraction of the jar is filled? (Lesson 8•3)
19. A perfume bottle holds $\frac{1}{2}$ fluid ounce. How many bottles do you need to fill $\frac{1}{2}$ cup? (Lesson 8•3)
20. A can of juice holds 1,250 milliliters. How many cans will it take to fill a 15-liter container? (Lesson 8•3)
21. About how many kilograms are in 17 pounds? (Lesson 8•4)
22. How many ounces are in 8 pounds? (Lesson 8•4)

A photograph 5 inches high and 3 inches wide was enlarged to make a poster. The width of the poster is 1 foot.

23. What is the ratio of the width of the poster to the width of the original photo? (Lesson 8•5)
24. What is the scale factor? (Lesson 8•5)
25. A cube has a volume of 216 cubic feet. Suppose that the cube was enlarged by a ratio of $\frac{2}{1}$. What is the volume of the original cube?
26. What is the scale factor of the similar rectangles below?

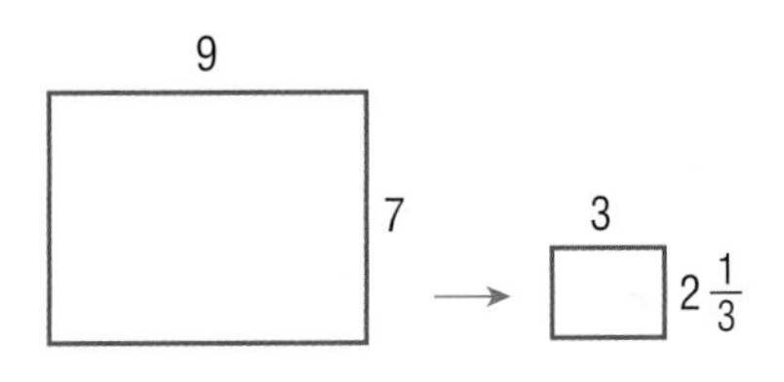

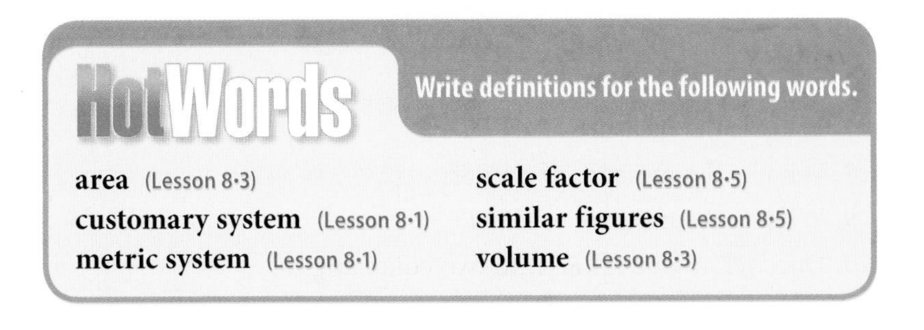

Write definitions for the following words.

area (Lesson 8•3)
customary system (Lesson 8•1)
metric system (Lesson 8•1)
scale factor (Lesson 8•5)
similar figures (Lesson 8•5)
volume (Lesson 8•3)

HotTopic 9

Tools

You can use the problems and the list of words that follow to see what you already know about this chapter. The answers to the problems are in **HotSolutions** at the back of the book, and the definitions of the words are in **HotWords** at the front of the book. You can find out more about a particular problem or word by referring to the topic number (*for example,* Lesson 9·2).

Problem Set

Use a scientific calculator for Exercises 1–6. Round decimal answers to the nearest hundredth. (Lesson 9·1)

1. 8.9^5

2. Find the reciprocal of 3.4.

3. Find the square of 4.5.

4. Find the square root of 4.5.

5. $(8 \cdot 10^4) \cdot (4 \cdot 10^8)$

6. $0.7 \cdot (4.6 + 37)$

Use the protractor to find the measure of each angle. (Lesson 9·2)

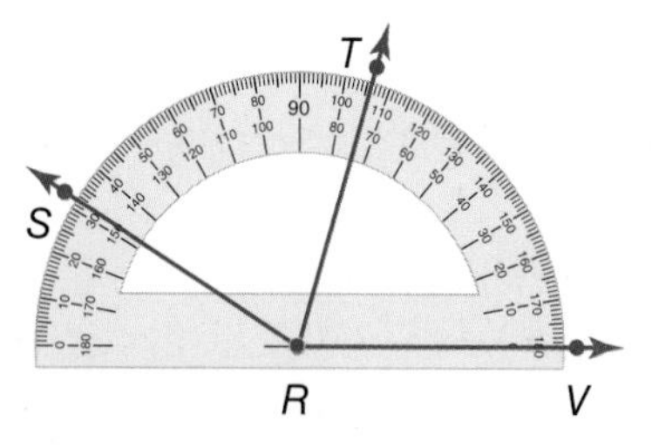

7. What is the measure of $\angle VRT$?

8. What is the measure of $\angle VRS$?

9. What is the measure of $\angle SRT$?

10. Does $\overrightarrow{RT}$ divide $\angle VRS$ into two equal angles?

11. What are the basic construction tools in geometry? (Lesson 9•2)

Using a ruler, protractor, and compass, copy the figures below. (Lesson 9•2)

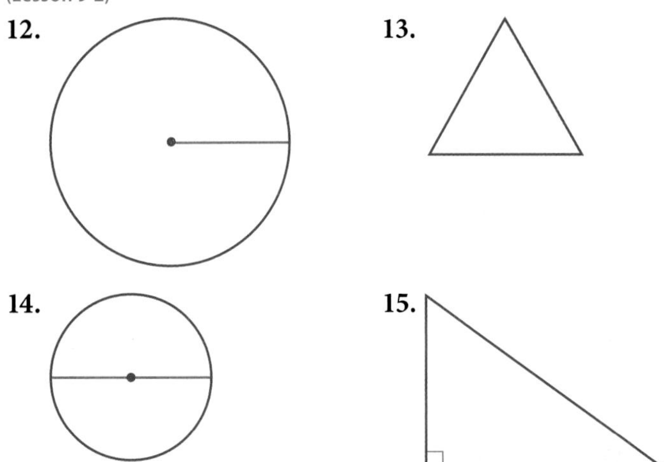

For Exercises 16–18, refer to the spreadsheet below. (Lesson 9•3)

File Edit

Sample.xls

◇	A	B	C	D
1	34	68	100	66
2	14	28	200	
3	20	40	300	

Sheet 1 Sheet 2

16. Name the cell holding 14.

17. A formula for cell C3 is 3 * C1. Name another formula for cell C3.

18. Cell D1 contains the number 66 and no formula. After using the command fill down, what number will be in cell D10?

9•1 Scientific Calculator

Mathematicians and scientists use scientific calculators to help quickly and accurately solve problems. Scientific calculators vary widely; some perform a few functions and others perform many functions. Some calculators can be programmed with functions you choose. The calculator below shows functions you might find on your scientific calculator.

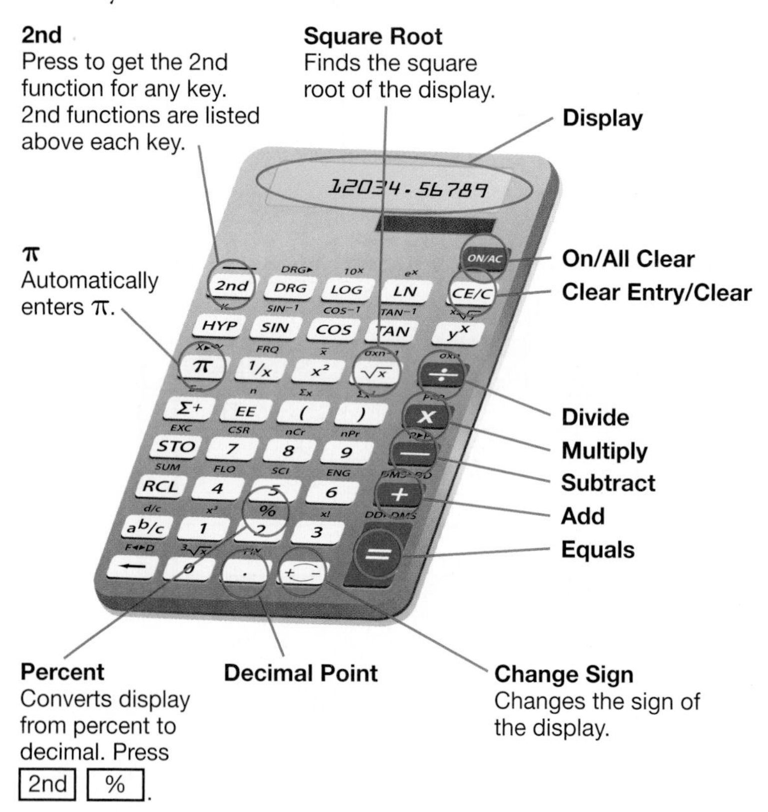

Frequently Used Functions

Because each scientific calculator is set up differently, your calculator may not work exactly as below. These keystrokes work with the calculator illustrated on page 396. Use the reference material that came with your calculator to perform similar functions. See the index for more information about the mathematics mentioned here.

Function	Problem	Keystrokes
Cube Root [$\sqrt[3]{x}$] Finds the cube root of the display.	$\sqrt[3]{343}$	343 [2nd] [$\sqrt[3]{x}$] [7.]
Cube [x^3] Finds the cube of the display.	17^3	17 [2nd] [x^3] [4913.]
Factorial [$x!$] Finds the factorial of the display.	7!	7 [2nd] [$x!$] [5040.]
Fix number of decimal places. [FIX] Rounds display to number of places you determine.	Round 3.046 to the tenths place.	3.046 [2nd] [FIX] 2 [3.05]
Parentheses [(] [)] Groups calculations.	$12 \cdot (7 + 8)$	12 [×] [(] 7 [+] 8 [)] [=] [180.]
Powers [y^x] Finds the x power of the display.	56^5	56 [y^x] 5 [=] [550731776.]
Powers of Ten [10^x] Raises ten to the power displayed.	10^5	5 [2nd] [10^x] [100000.]

Function	Problem	Keystrokes
Reciprocal [1/x] Finds the reciprocal of the display.	Find the reciprocal of 8.	8 [1/x] [0.125]
Roots [$\sqrt[x]{y}$] Finds the *x* root of the display.	$\sqrt[4]{852}$	852 [2nd] [$\sqrt[x]{y}$] 4 [=] [5.402688131]
Square [x^2] Finds the square of the display.	17^2	17 [x^2] [289.]

Some calculators have keys with special functions.

Key	Function
[$\sqrt{x}$]	Finds the square root of the display.
[π]	Automatically enters pi to as many places as your calculator holds.

The [π] key saves you time by reducing the number of keystrokes. The [$\sqrt{x}$] key allows you to find square roots precisely, something difficult to do with pencil and paper.

See how these two keys are used in the examples below.

Problem: $7 + \sqrt{21}$
Keystrokes: 7 [+] 21 [$\sqrt{x}$] [=]
Final display: [11.582575]

If you try to take a square root of a negative number, your calculator will display an error message. For example, if you enter 9 [+/−] [$\sqrt{x}$], the resulting display is [E 3.]. There is no square root of −9, because no number multiplied by itself can result in a negative number.

Problem: Find the area of a circle with radius 3. (Use formula $A = \pi r^2$.)
Keystrokes: [π] [×] 3 [×] 3 [=]
Final display: [28.274333]

If your calculator does not have the [π] key, you can use 3.14 or 3.1416 as an approximation for π.

Check It Out

Use your calculator to find the following.

1. 12!
2. 14^4

Use your calculator to find the following to the nearest thousandth.

3. the reciprocal of 27
4. $(10^3 + 56^5 - \sqrt[3]{512}) \div 7!$
5. the square root of 7,225

APPLICATION Magic Numbers

On a calculator, press the same number three times to display a three-digit number, for example 333. Then divide the number by the sum of the three digits and press the [=] key. Do you get 37? Try this with other three-digit numbers. Write an algebraic expression that shows why the answer is always the same. See **HotSolutions** for the answer.

9·1 Exercises

Use a scientific calculator to find the following. Round decimal answers to the nearest hundredth.

1. 69^2

2. 44^2

3. 13^3

4. 0.1^5

5. $\frac{60}{\pi}$

6. $9(\pi)$

7. $\frac{1}{9}$

8. $\frac{1}{\pi}$

9. $(15 - 4.4)^3 + 6$

10. $25 + (8 \div 6.2)$

11. $5! \cdot 4!$

12. $9! \div 4!$

13. $11! + 6!$

14. 5^{-3}

15. $\sqrt[4]{1{,}336{,}336}$

16. reciprocal of 0.0625

17. reciprocal of 25

Find the value of each expression, using your calculator.

18. $\sqrt{804} \div 17.35 + 620$

19. $\sqrt{68} \cdot 7 + 4$

20. Find the perimeter of the rectangle if $x = 11.9$ cm.

21. Find the area of the rectangle if $x = 9.68$ cm.

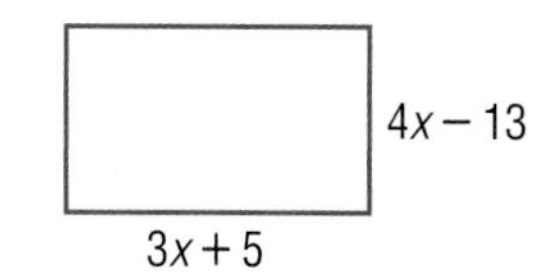

22. Find the circumference of the circle if $a = 3.7$ in.

23. Find the area of the circle if $a = 2$ in.

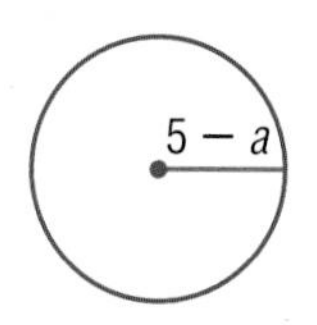

9•2 Geometry Tools

Protractor

Angles are measured with a *protractor.* There are many different kinds of protractors. The key is to find the center point of the protractor to which you align the vertex of the angle.

EXAMPLE Measuring Angles with a Protractor

Find $m\angle CDE$ and $m\angle FDC$.

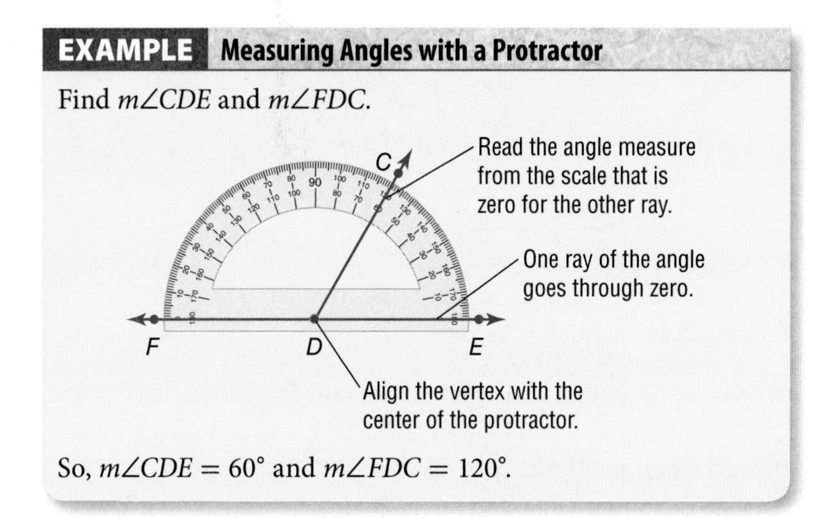

So, $m\angle CDE = 60°$ and $m\angle FDC = 120°$.

To draw an angle, draw one ray first, and position the center of the protractor at the endpoint. (The endpoint will be the vertex of the angle.) Then make a dot at the desired measure, and draw a ray through that dot from the endpoint.

EXAMPLE Drawing Angles

Use a protractor to draw a 45° angle.

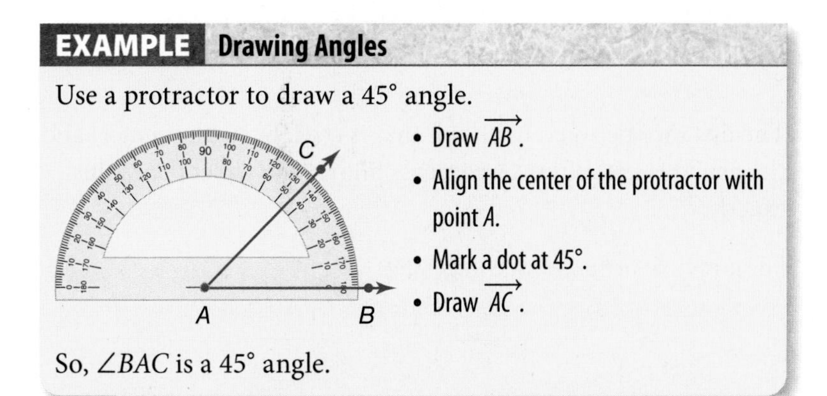

- Draw $\overrightarrow{AB}$.
- Align the center of the protractor with point A.
- Mark a dot at 45°.
- Draw $\overrightarrow{AC}$.

So, $\angle BAC$ is a 45° angle.

Check It Out

Measure each angle to the nearest degree, using your protractor.

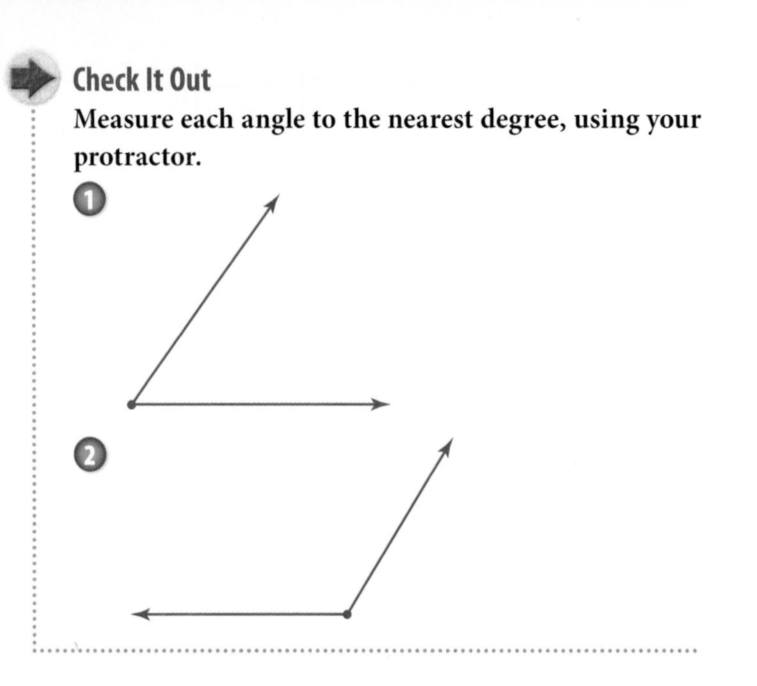

Compass

A *compass* is used to construct circles and arcs. To construct a circle or an arc, place one point at the center and hold it there. Pivot the point with the pencil attached to draw the arc, or circle.

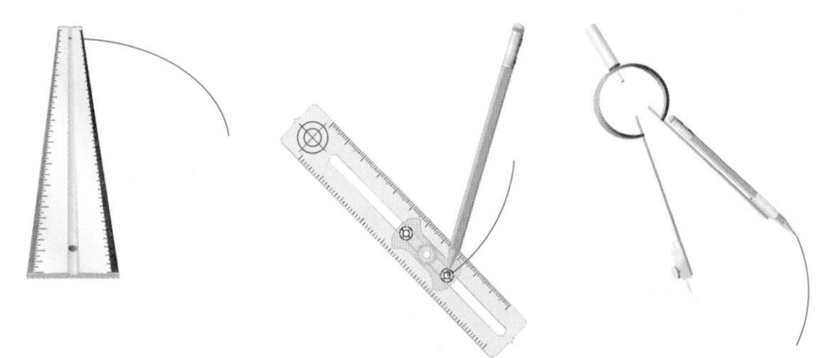

The distance between the point that is stationary (the center) and the pencil is the radius. A compass allows you to set the radius exactly.

For a review of *circles,* see page 359.

To draw a circle with a radius of $1\frac{1}{2}$ inches, set the distance between the stationary point of your compass and the pencil at $1\frac{1}{2}$ inches. Draw a circle.

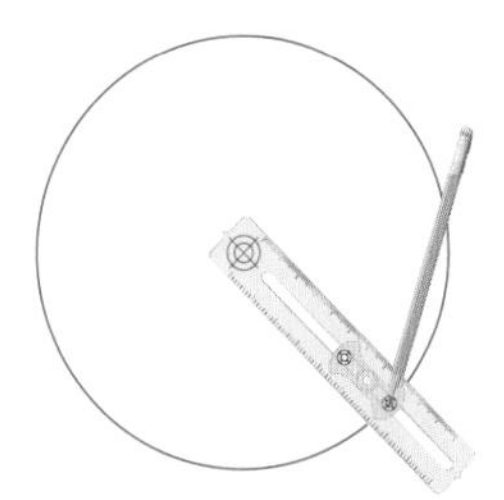

Check It Out

Use a compass to draw these circles.

3. Draw a circle with a radius of 2.5 centimeters.
4. Draw a circle with a radius of 2 inches.

Construction Problem

A construction is a drawing in geometry that permits the use of only the straightedge and the compass. When you make a construction by using a straightedge and compass, you have to use what you know about geometry.

Follow the step-by-step directions below to inscribe an equilateral triangle in a circle.

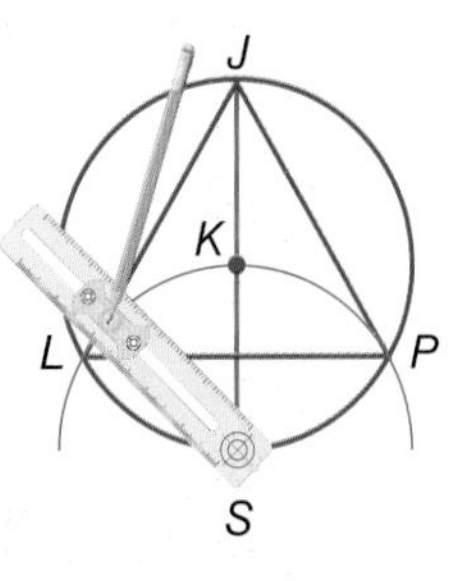

- Draw a circle with center K.
- Draw a diameter ($\overline{SJ}$).
- Using S as a center and $\overline{SK}$ as a radius, draw an arc that intersects the circle. Label the points of intersection L and P.
- Connect L, P, and J to form the triangle.

You can create a more complex design by inscribing another triangle in your circle by using J as a center for drawing another intersecting arc.

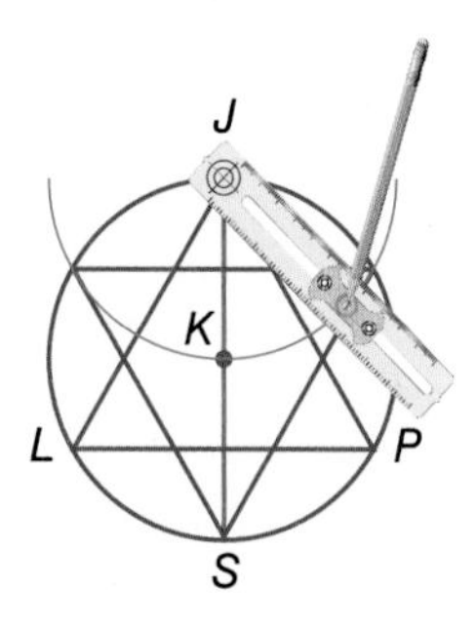

Once you have the framework, you can fill in different sections to create a variety of designs based on this construction.

Check It Out

Construct.

5. Draw the framework based on four triangles inscribed in two concentric circles. Fill in sections to copy the design below.

6. Create your own design based on one or two triangles inscribed in a circle.

9•2 Exercises

Using a protractor, measure each angle in $\triangle ABC$.

1. $\angle A$
2. $\angle B$
3. $\angle C$

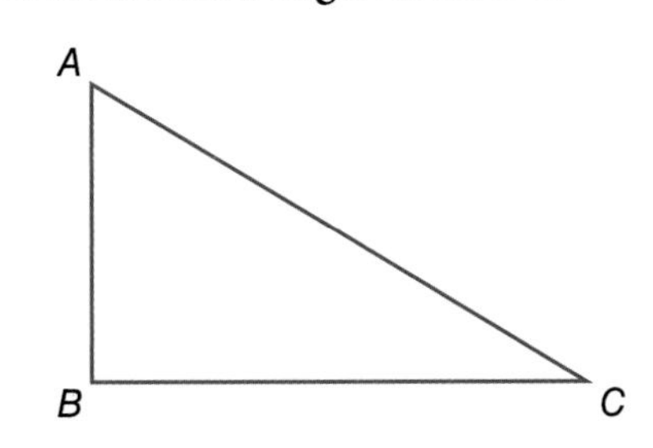

4. When you use a protractor to measure an angle, how do you know which of the two scales to read?

Write the measure of each angle.

5. $\angle GFH$
6. $\angle GFI$
7. $\angle HFI$
8. $\angle JFH$
9. $\angle HFJ$

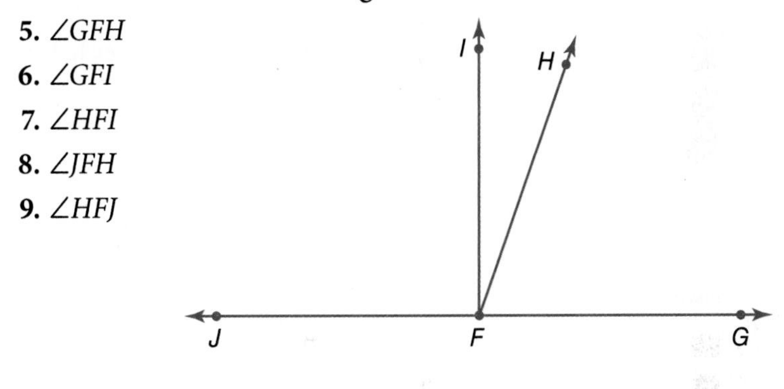

Match each tool with the function.

Tool	Function
10. compass	**A.** measure distance
11. protractor	**B.** measure angles
12. ruler	**C.** draw circles or arcs

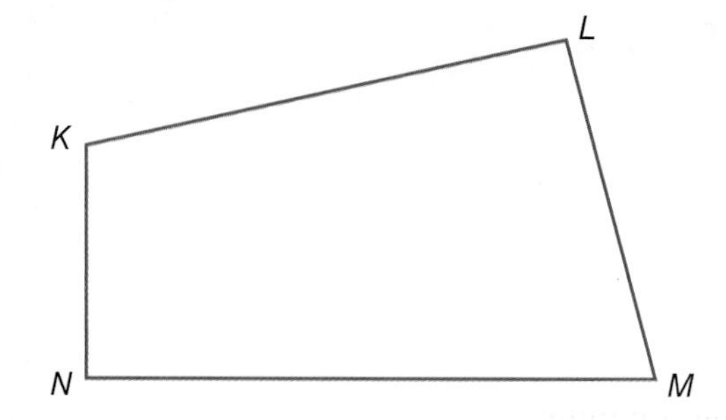

Find the measure of each angle.

13. $\angle KLM$

14. $\angle MNK$

15. $\angle LMN$

16. $\angle NKL$

17. Use a protractor to copy $\angle LMN$.

Using a ruler, protractor, and compass, copy the figures below.

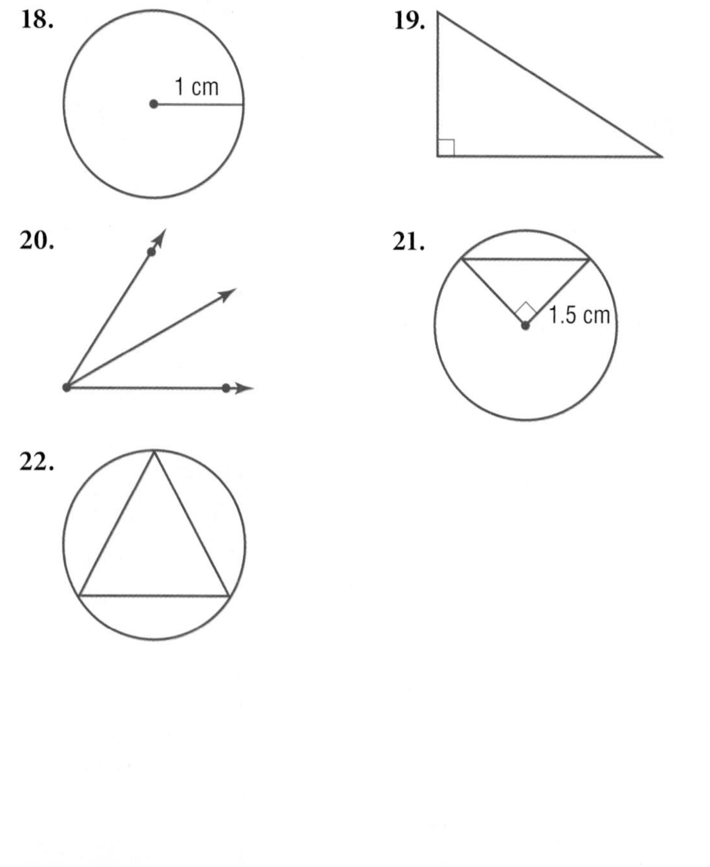

9•3 Spreadsheets

What Is a Spreadsheet?

People use **spreadsheets** as a tool to keep track of information, such as finances. Spreadsheets were paper-and-pencil math tools before being computerized. You may be familiar with computer spreadsheet programs.

A computer program spreadsheet calculates and organizes information in **cells** within a grid. When the value of a cell is changed, all other cells dependent on that value automatically update.

Spreadsheets are organized in rows and columns. Rows are horizontal and are numbered. Columns are vertical and are named by capital letters. The cells are named according to the column and row in which they are located.

File Edit

Sample.xls

◇	A	B	C	D
1	1	3	1	
2	2	6	4	
3	3	9	9	
4	4	12	16	
5	5	15	25	

Sheet 1 Sheet 2 Sheet 3

The cell A3 is in Column A, Row 3. In this spreadsheet, there is a 3 in cell A3.

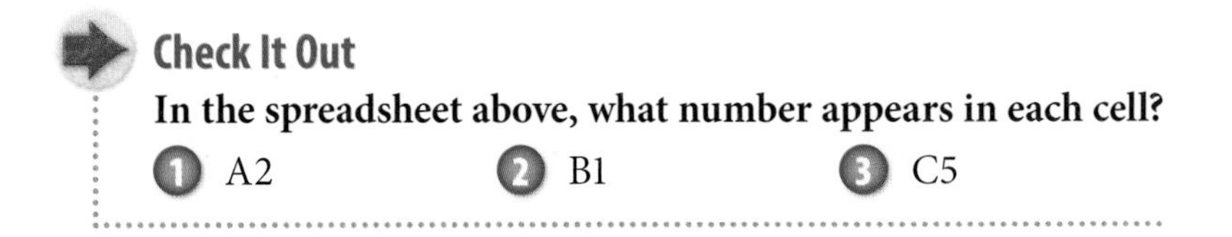

Check It Out

In the spreadsheet above, what number appears in each cell?

1. A2
2. B1
3. C5

Spreadsheet Formulas

A cell can contain a number or a formula. A **formula** generates a value dependent on other cells in the spreadsheet. The way that the formulas are written depends on the particular spreadsheet computer software you are using. Although you enter a formula in the cell, the value generated by the formula appears in the cell. The formula is stored behind the cell.

EXAMPLE Creating a Spreadsheet Formula

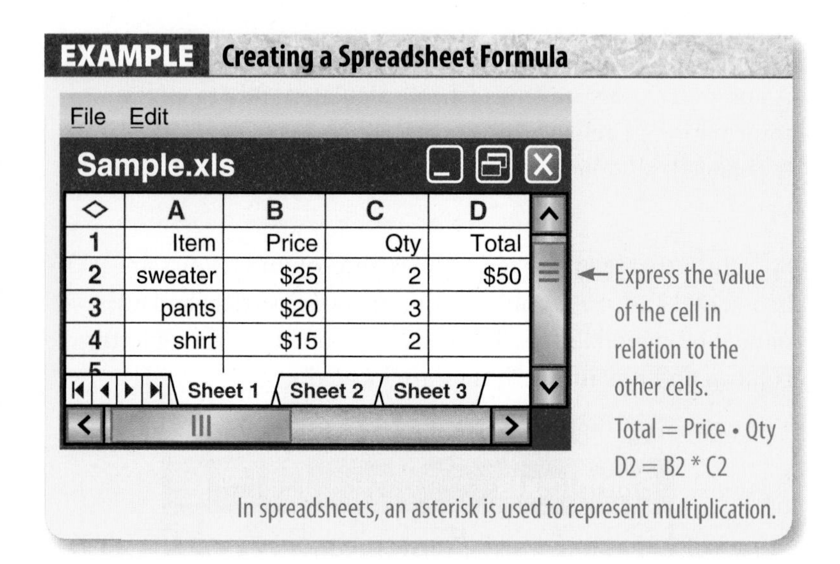

	A	B	C	D
1	Item	Price	Qty	Total
2	sweater	\$25	2	\$50
3	pants	\$20	3	
4	shirt	\$15	2	
5				

In spreadsheets, an asterisk is used to represent multiplication.

If you change the value of a cell and a formula depends on that value, the result of the formula will change.

In the spreadsheet above, if you entered 3 sweaters instead of 2 (C2 = 3), the total column would automatically change to \$75.

Check It Out

Use the spreadsheet above. If the total is always figured the same way, write the formula for:

4. D3
5. D4
6. If D5 is to be the total of column D, write the formula for D5.

Fill Down and Fill Right

Spreadsheet programs are designed to perform other tasks as well. *Fill down* and *fill right* are two useful spreadsheet commands.

To use *fill down,* select a portion of a column. *Fill down* will take the top cell that has been selected and copy it into the rest of the highlighted cells. If the top cell in the selected range contains a number, such as 5, *fill down* will generate a column of 5s.

If the top cell of the selected range contains a formula, the *fill down* feature will automatically update the formula as you go from cell to cell.

The selected column is highlighted.

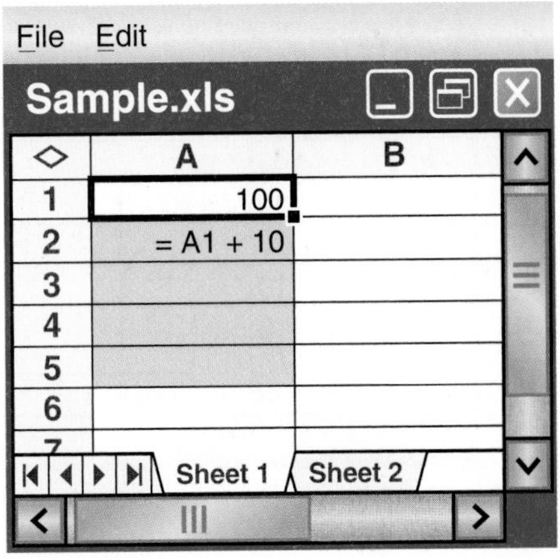

The spreadsheet fills the column and adjusts the formula.

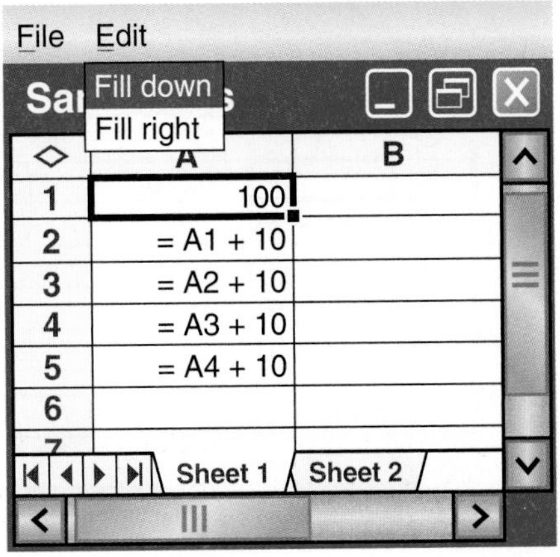

These are the values that actually appear.

File Edit

Fill down
Fill right

	A	B
1	100	
2	110	
3	120	
4	130	
5	140	

Sheet 1 Sheet 2

Fill right works in a similar way. It copies the contents of the leftmost cell of the selected range into each selected cell within a row.

File Edit

Sample.xls

	A	B	C	D	E
1	100				
2	= A1 + 10				
3	= A2 + 10				
4	= A3 + 10				
5	= A4 + 10				

Sheet 1 Sheet 2

Row 1 is selected.

File Edit

Fill down
Fill right

	A	B	C	D	E
1	100	100	100	100	100
2	= A1 + 10				
3	= A2 + 10				
4	= A3 + 10				
5	= A4 + 10				

Sheet 1 Sheet 2

The 100 fills to the right.

If you select A1 to E1 and *fill right,* you will get all 100s. If you select A2 to E2 and *fill right,* you will "copy" the formula A1 + 10 as shown.

File Edit

Fill down
Fill right

◇	A	B	C	D	E
1	100	100	100	100	100
2	= A1 + 10	= B1 + 10	= C1 + 10	= D1 + 10	= E1 + 10
3	= A2 + 10				
4	= A3 + 10				
5	= A4 + 10				

Sheet 1 Sheet 2

File Edit

Sample.xls

◇	A	B	C	D	E
1	100	100	100	100	100
2	110	110	110	110	110
3					
4					
5					
6					
7					

Sheet 1 Sheet 2

The spreadsheet fills the row and adjusts the formulas.

Check It Out

Use the spreadsheet above.

7. "Select" A2 to A8 and *fill down.* What formula will be in A7? What number?
8. "Select" A3 to E3 and *fill right.* What formula will be in D3? What number?

Spreadsheet Graphs

You can also generate a graph from a spreadsheet. As an example, use a spreadsheet that compares the perimeter of a square to the length of its side.

File Edit

Sample.xls

◇	A	B	C	D	E
1	side	perimeter			
2	1	4			
3	2	8			
4	3	12			
5	4	16			
6	5	20			
7					

Sheet 1 Sheet 2

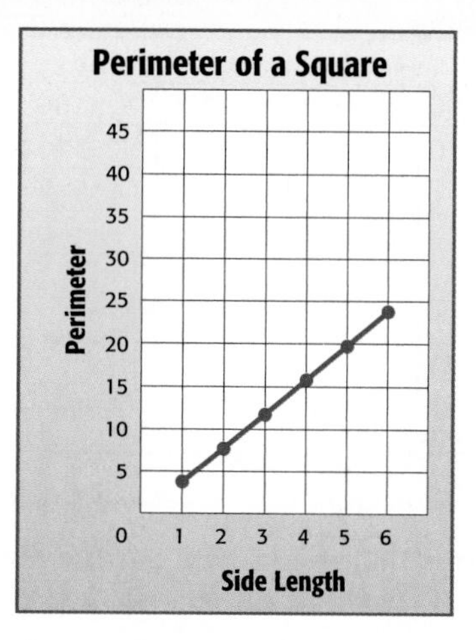

Most spreadsheets have a function that displays tables as graphs. See your spreadsheet reference for more information.

Check It Out

Use the spreadsheet above.

9. What cells give the point (1, 4)?
10. What cells give the point (4, 16)?
11. What cells give the point (2, 8)?

9•3 Exercises

File Edit

Sample.xls

◇	A	B	C	D
1	1	1	200	1
2	2	3	500	6
3	3	5	800	15
4	4	7	1100	28

Sheet 1 Sheet 2

In which cell does each number appear?

1. 15 **2.** 7 **3.** 800

4. If the formula behind cell A2 is A1 + 1 and is copied down, what formula is behind cell A3?

5. The formula behind cell D2 is A2 * B2. What formula is behind cell D4?

6. If row 5 were included in the spreadsheet, what numbers would be in that row?

Use the spreadsheet below to answer Exercises 7–9.

File Edit

Fill down
Fill right

◇	A	B
1	5	10
2	A1 + 6	B1 × 2

Sheet 1 Sheet 2

7. If you select A2 to A5 and fill down, what formula will appear in A3?

8. If you select A3 to A5 and fill down, what numbers will appear in A3 to A5?

9. If you select B1 to E1 and fill right, what will appear in C1, D1, and E1?

Tools

What have you learned?

You can use the problems and the list of words that follow to see what you learned in this chapter. You can find out more about a particular problem or word by referring to the topic number (*for example,* Lesson 9·2).

Problem Set

Use a scientific calculator for Exercises 1–6. Round decimal answers to the nearest hundredth. (Lesson 9·1)

1. 2.027^5
2. Find the reciprocal of 4.5.
3. Find the square of 4.5.
4. Find the square root of 5.4.
5. $(4 \cdot 10^3) \cdot (7 \cdot 10^6)$
6. $0.6 \cdot (3.6 + 13)$

Use the protractor to find the measure of each angle. (Lesson 9·2)

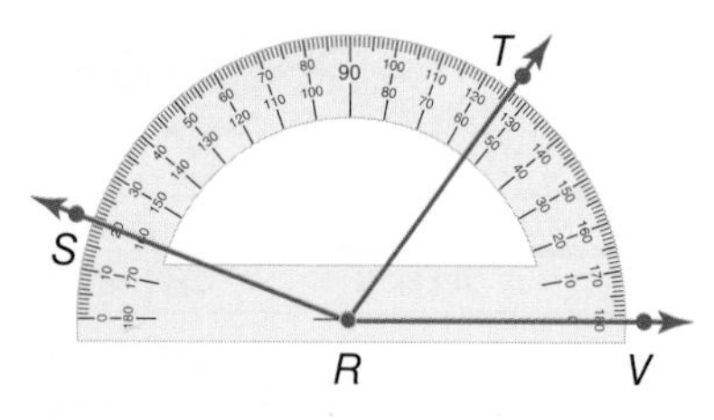

7. What is the measure of $\angle VRT$?
8. What is the measure of $\angle VRS$?
9. What is the measure of $\angle SRT$?
10. Does $\overrightarrow{RT}$ divide $\angle VRS$ into 2 equal angles?

WHAT HAVE YOU LEARNED? 9

Use a compass. (Lesson 9•2)

11. Draw a circle with a radius of 3 centimeters.

12. Draw a circle with a radius of 1 inch.

For Exercises 13–15, refer to the spreadsheet below. (Lesson 9•3)

File Edit

Fill down
Fill right

◇	A	B	C	D
1	10	15	25	77
2	47	28	75	
3	64	36	100	

Sheet 1 Sheet 2

13. Name the cell holding 28.

14. A formula for cell C3 is C1 + C2. Name another formula for cell C3.

15. Cell D1 contains the number 77 and no formula. After using the command *fill down,* what number will be in cell D10?

congruent
outlier
polygon

Part Three 3

Hot Solutions

Chapter 1 Numbers and Computation

p. 72 **1.** $(4 + 7) \cdot 3 = 33$ **2.** $(30 + 15) \div 5 + 5 = 14$
3. no **4.** no **5.** yes **6.** no **7.** $2^3 \cdot 5$ **8.** $2 \cdot 5 \cdot 11$
9. $2 \cdot 5 \cdot 23$ **10.** 4 **11.** 5 **12.** 9 **13.** 60 **14.** 120
15. 90 **16.** 60

p. 73 **17.** 7, 7 **18.** 15, −15 **19.** 12, 12 **20.** 10, −10
21. >; (number line −3 to 5, points at −1 and 3)
−3 −2 −1 0 1 2 3 4 5
22. <; (number line −8 to 4, points at −8 and 4)
−8 −7 −6 −5 −4 −3 −2 −1 0 1 2 3 4
23. >; (number line −6 to 1, points at −4 and −2)
−6 −5 −4 −3 −2 −1 0 1
24. >; (number line −8 to −1, points at −7 and −3)
−8 −7 −6 −5 −4 −3 −2 −1
25. 2 **26.** −4 **27.** −11 **28.** 16 **29.** 0 **30.** 6
31. 42 **32.** −4 **33.** 7 **34.** 24 **35.** −36 **36.** −50
37. It will be a negative integer.
38. It will be a positive integer.

1•1 Order of Operations

p. 74 **1.** 12 **2.** 87

1•2 Factors and Multiples

p. 76 **1.** 1, 2, 4, 8 **2.** 1, 2, 3, 4, 6, 8, 12, 16, 24, 48

p. 77 **3.** 1, 2 **4.** 1, 5 **5.** 2 **6.** 6 **7.** 14 **8.** 12

p. 78 **9.** yes **10.** no **11.** yes **12.** yes

p. 79 **13.** yes **14.** no **15.** yes **16.** no
17. Sample answers: (17, 19); (29, 31); (41, 43)

p. 80 **18.** $2^4 \cdot 5$ **19.** $2^3 \cdot 3 \cdot 5$

p. 81 **20.** 6 **21.** 8 **22.** 14 **23.** 12

p. 82 **24.** 18 **25.** 140 **26.** 36 **27.** 75

1•3 Integer Operations

p. 84 **1.** −6 **2.** +200 **3.** 12, 12 **4.** 5, −5 **5.** 0, 0

p. 85 **6.** >; (number line −6 to 2 with points at −5 and 2)

−6 −5 −4 −3 −2 −1 0 1 2

7. <; (number line −5 to 0 with points at −4 and −2)

−5 −4 −3 −2 −1 0

8. −4, −1, 3, 6 **9.** −5, −2, 4, 7

p. 87 **10.** −2 **11.** 0 **12.** −5 **13.** −3

p. 88 **14.** 10 **15.** −3 **16.** 3

Chapter 2 Rational Numbers

p. 92 **1.** about 9 days **2.** 10 servings **3.** second game

4. 85% **5.** C **6.** B. $\frac{4}{3} = 1\frac{1}{3}$

p. 93 **7.** $1\frac{1}{6}$ **8.** $1\frac{3}{4}$ **9.** $3\frac{1}{4}$ **10.** $8\frac{3}{10}$ **11.** $\frac{2}{5}$ **12.** $\frac{1}{2}$

13. $\frac{3}{4}$ **14.** 3

15. 16.154 **16.** 1.32 **17.** 30.855 **18.** 7.02

19. $\frac{33}{200}$; 1.065; 1.605; $1\frac{13}{20}$ **20.** irrational

21. rational **22.** 30% **23.** 27.8 **24.** 55

2•1 Fractions

p. 95 **1.** Sample answers: $\frac{2}{8}$ and $\frac{3}{12}$
2. Sample answers: $\frac{1}{2}$ and $\frac{5}{10}$
3. Sample answers: $\frac{8}{10}$ and $\frac{40}{50}$
4. Sample answers: $\frac{2}{2}$, $\frac{5}{5}$, and $\frac{50}{50}$

p. 95 **5.** $\neq$ **6.** $=$ **7.** $\neq$

p. 96 **8.** $\frac{4}{5}$ **9.** $\frac{3}{4}$ **10.** $\frac{2}{5}$

p. 97 **11.** $7\frac{1}{6}$ **12.** $11\frac{1}{3}$ **13.** $6\frac{2}{5}$ **14.** $9\frac{1}{4}$

p. 98 **15.** $\frac{37}{8}$ **16.** $\frac{77}{6}$ **17.** $\frac{49}{2}$ **18.** $\frac{98}{3}$

2•2 Operations with Fractions

p. 100 **1.** $1\frac{1}{5}$ **2.** $1\frac{3}{34}$ **3.** $\frac{1}{2}$ **4.** $\frac{1}{2}$

p. 101 **5.** $1\frac{2}{5}$ **6.** $1\frac{3}{14}$ **7.** $\frac{1}{20}$ **8.** $\frac{11}{24}$

p. 102 **9.** $9\frac{5}{6}$ **10.** $34\frac{5}{8}$ **11.** 61

p. 103 **12.** $23\frac{39}{40}$ **13.** $20\frac{1}{24}$ **14.** $22\frac{7}{15}$ **15.** $10\frac{5}{9}$

p. 104 **16.** $3\frac{4}{11}$ **17.** $5\frac{1}{2}$ **18.** $-7\frac{3}{4}$

p. 105 **19.** $7\frac{1}{2}$ **20.** $3\frac{37}{70}$ **21.** $11\frac{1}{8}$

p. 106 **22.** $\frac{1}{3}$ **23.** $\frac{1}{12}$ **24.** 2 **25.** $\frac{77}{5}$ or $15\frac{2}{5}$

p. 107 **26.** $\frac{1}{10}$ **27.** $\frac{8}{15}$ **28.** 2 **29.** $\frac{7}{3}$ **30.** $\frac{1}{3}$ **31.** $\frac{5}{22}$

p. 108 **32.** $1\frac{1}{2}$ **33.** $\frac{1}{14}$ **34.** $\frac{1}{4}$

2•3 Operations with Decimals

p. 110 **1.** 7.1814 **2.** 96.674 **3.** 38.54 **4.** 802.0556

p. 111 **5.** 59.481 **6.** 80.42615 **7.** 62.95383

p. 112 **8.** 900 **9.** 4 **10.** 50 **11.** 21.6 **12.** 5.23
13. 92 **14.** 25.8

p. 113 **15.** 10.06 **16.** 24.8 **17.** 307.625

p. 114 **18.** 0.07 **19.** 0.65 **20.** 1.64 **21.** 0.78

2•4 Fractions and Decimals

p. 117 **1.** 0.8 **2.** 0.55 **3.** 0.875 **4.** $0.41\overline{6}$ **5.** −0.75
6. −0.625 **7.** 3.875 **8.** 2.3125

p. 118 **9.** $2\frac{2}{5}$ or $\frac{12}{5}$ **10.** $\frac{7}{125}$ **11.** $-\frac{3}{5}$ **12.** $-1\frac{3}{8}$ **13.** $\frac{4}{9}$
14. $-3\frac{2}{11}$ **15.** $-\frac{5}{6}$ **16.** $7\frac{8}{25}$

2•5 The Real Number System

p. 121 **1.** integer, rational **2.** rational **3.** irrational
4. irrational **5.** rational **6.** irrational

p. 122 **7.** 3.3 **8.** −5.4 **9.** 4.2

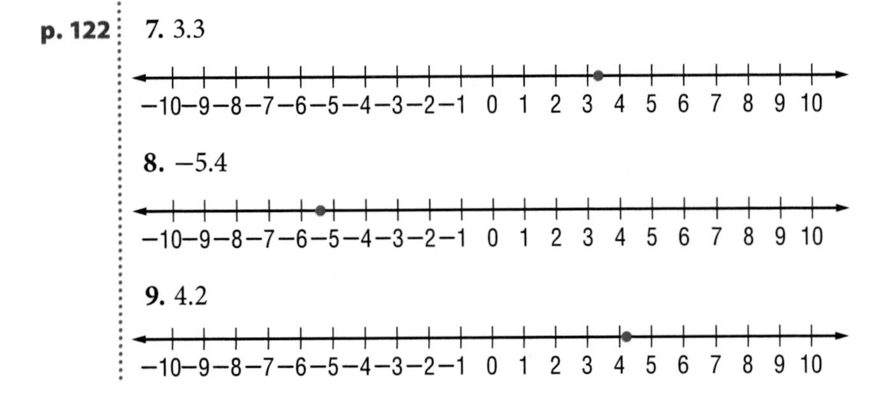

2•6 Percents

p. 123 **1.** 150 **2.** 100 **3.** 150 **4.** 250

p. 124 **5.** $1 **6.** $6 **7.** $9.50 **8.** $10

p. 125 **9.** 80% **10.** 65% **11.** 45% **12.** 38%

p. 126 **13.** $\frac{11}{20}$ **14.** $\frac{29}{100}$ **15.** $\frac{17}{20}$ **16.** $\frac{23}{25}$ **17.** $\frac{89}{200}$ **18.** $\frac{43}{125}$

p. 127 **19.** 8% **20.** 66% **21.** 39.8% **22.** 74%

p. 128 **23.** 0.145 **24.** 0.0001 **25.** 0.23 **26.** 0.35

2•7 Using and Finding Percents

p. 130 **1.** 60 **2.** 665 **3.** 11.34 **4.** 27

p. 132 **5.** 665 **6.** 72 **7.** 130 **8.** 340

p. 133 **9.** $33\frac{1}{3}\%$ **10.** 450% **11.** 400% **12.** 60%

p. 134 **13.** 104 **14.** 20 **15.** 25 **16.** 1,200

p. 135 **17.** 25% **18.** 95% **19.** 120% **20.** 20%

p. 136 **21.** 11% **22.** 50% **23.** 16% **24.** 30%

p. 137 **25.** discount: \$162.65; sale price: \$650.60
26. discount: \$5.67; sale price: \$13.23
27. discount: \$12; sale price: \$67.99

p. 138 **28.** 100 **29.** 2 **30.** 15 **31.** 30

p. 139 **32.** I = \$1,800; total amount = \$6,600
33. I = \$131.25; total amount = \$2,631.25

Chapter 3 Powers and Roots

p. 144 **1.** 5^7 **2.** a^5 **3.** 4 **4.** 81 **5.** 36 **6.** 8 **7.** 125
8. 343 **9.** 1,296 **10.** 2,187 **11.** 512 **12.** 1,000
13. 10,000,000 **14.** 100,000,000,000 **15.** 4 **16.** 7
17. 11 **18.** 5 and 6 **19.** 3 and 4 **20.** 8 and 9

p. 145 **21.** 3.873 **22.** 6.164 **23.** 2 **24.** 4 **25.** 7
26. very small **27.** very large **28.** $7.8 \cdot 10^7$
29. $2.0 \cdot 10^5$ **30.** $2.8 \cdot 10^{-3}$ **31.** $3.02 \cdot 10^{-5}$
32. 8,100,000 **33.** 200,700,000 **34.** 4,000
35. 0.00085 **36.** 0.00000906 **37.** 0.0000007
38. 12 **39.** 13 **40.** 18

3•1 Power and Exponents

p. 146 **1.** 4^3 **2.** 6^9 **3.** x^4 **4.** y^6

p. 147 **5.** 25 **6.** 100 **7.** 9 **8.** $\frac{1}{16}$

p. 148 **Squaring Triangles** 21, 28; It is the sequence of squares.

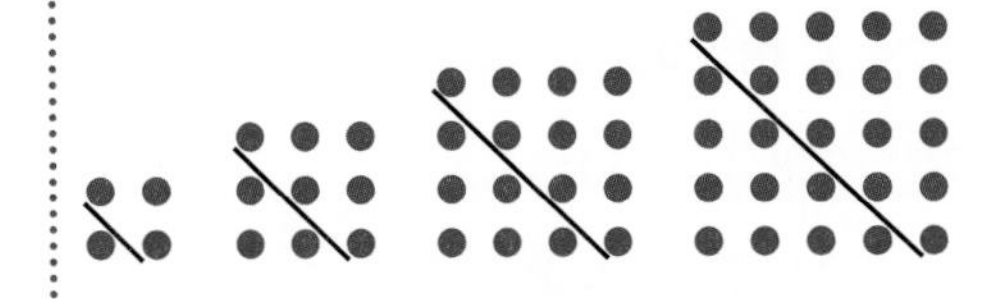

p. 149 **9.** 64 **10.** −216 **11.** 27 **12.** −512

p. 150 **13.** −128 **14.** 59,049 **15.** 81 **16.** 390,625

p. 152 **17.** 1 **18.** $\frac{1}{216}$ **19.** 1 **20.** $\frac{1}{81}$

p. 153 **21.** 0.0001 **22.** 1,000,000 **23.** 1,000,000,000 **24.** 0.00000001

p. 154 **25.** 324 **26.** 9,765,625 **27.** 33,554,432 **28.** 20,511,149

3•2 Square and Cube Roots

p. 156 **1.** 4 **2.** 7 **3.** 10 **4.** 12

p. 157 **5.** between 7 and 8 **6.** between 4 and 5 **7.** between 2 and 3 **8.** between 9 and 10

p. 159 **9.** 1.414 **10.** 7.071 **11.** 8.660 **12.** 9.950 **13.** 4 **14.** 7 **15.** 10 **16.** 5

3•3 Scientific Notation

p. 161 **1.** very small **2.** very large **3.** very small

p. 162 **4.** $6.8 \cdot 10^4$ **5.** $7.0 \cdot 10^6$ **6.** $7.328 \cdot 10^7$ **7.** $3.05 \cdot 10^{10}$

Bugs $1.2 \cdot 10^{18}$

p. 163 **8.** $3.8 \cdot 10^{-3}$ **9.** $4.0 \cdot 10^{-7}$ **10.** $6.03 \cdot 10^{-11}$ **11.** $7.124 \cdot 10^{-4}$

p. 164 **12.** 53,000 **13.** 924,000,000 **14.** 120,500 **15.** 8,840,730,000,000

p. 165 **16.** 0.00071 **17.** 0.000005704 **18.** 0.0865 **19.** 0.000000000030904

HOTSOLUTIONS

3•4 Laws of Exponents

p. 167 **1.** 23 **2.** 17 **3.** 18 **4.** 19

p. 168 **5.** 3^9 **6.** 2^{18} **7.** 8^3 **8.** 30^2

p. 169 **9.** 2^1 **10.** 5^3 **11.** 2^4 **12.** 3^2

p. 170 **13.** 3^8 **14.** 3^{18} **15.** $64a^3b^3$ **16.** $27x^3y^{18}$

Chapter 4 Data, Statistics, and Probability

p. 174 **1.** late morning **2.** seventh **3.** no **4.** bar graph

p. 175 **5.** positive **6.** 34 **7.** mode **8.** 21 **9.** $\frac{2}{15}$

4•1 Collecting Data

p. 177 **1.** adults over the age of 45; 150,000 **2.** elk in Roosevelt National Forest; 200 **3.** motor vehicle drivers in California in 2007; 500

p. 178 **4.** No, it is limited to people who are friends of her parents, and they may have similar beliefs.

5. Yes, if the population is the class. Each student has the same chance of being picked.

p. 179 **6.** Yes, the sample is biased because it favored a part of the population and was only given to the customers that were listening to the country music radio station.

7. The sample is not biased. All students participated in the survey.

p. 180 **8.** It assumes you like pizza. **9.** It does not assume that you watch TV after school. **10.** Do you recycle newspapers? **11.** 6 **12.** bagels **13.** pizza; Students chose pizza more than any other food.

4•2 Displaying Data

p. 182

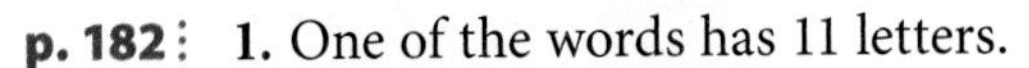

1. One of the words has 11 letters.

2. 1944 Winter Olympics

No. of Gold Medals	0	1	2	3	4	5	6	7	8	9	10	11
No. of Countries	8	4	2	2	1	0	1	1	0	1	1	1

p. 183 **3.** 25 g **4.** 11.5 g **5.** 50%

p. 185

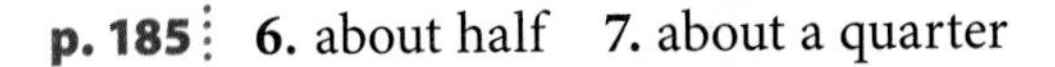

6. about half **7.** about a quarter

8.

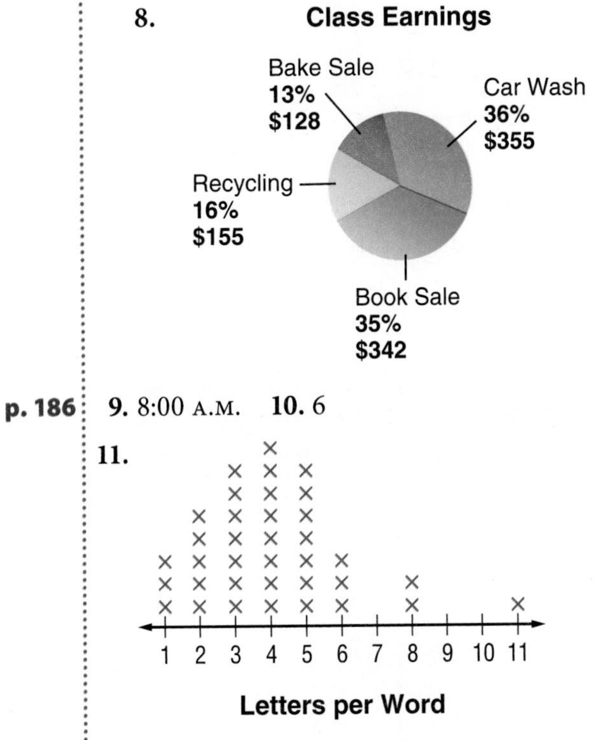

p. 186 **9.** 8:00 A.M. **10.** 6

11.

p. 187 **12.** Gabe **13.** Gabe **14.** 7 **15.** 37.1; 27.2

p. 188 **16.** September **17.** Sample answer: Kirti's earnings increased from May to July.

18.

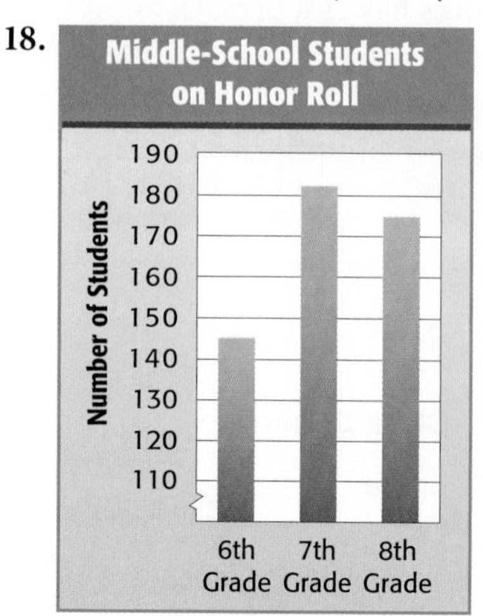

p. 189 **19.** $120,000,000,000 **20.** Sample answer: It increased slightly for the first three years and then remained constant.

p. 191 **21.** 16

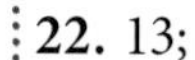
22. 13;

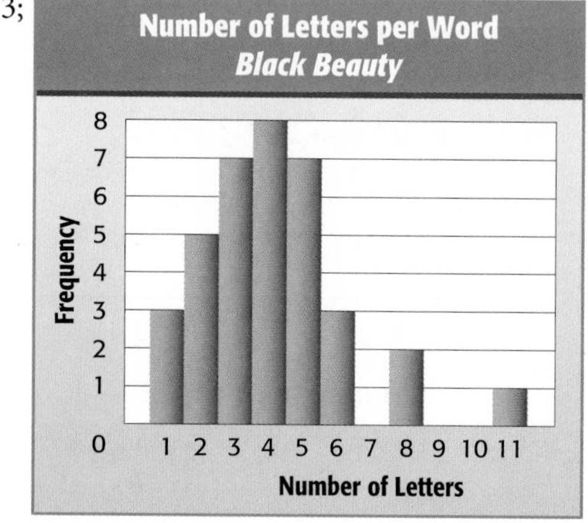

4•3 Analyzing Data

p. 194 **1.** Yes, as the price increases, the number sold decreases.

2.

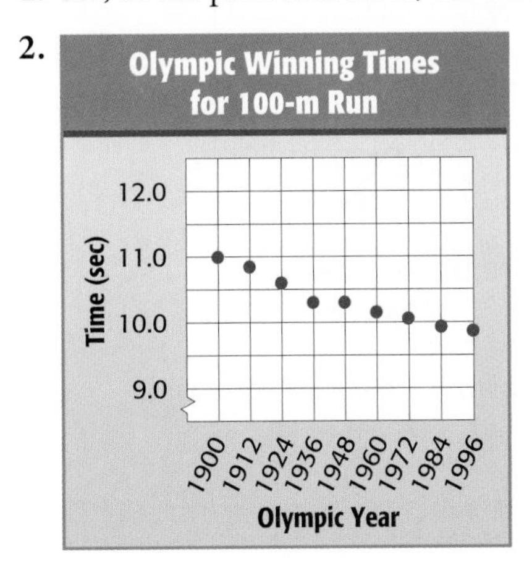

p. 195 **3.** Age and Letters in Name **4.** negative

5. Miles cycled and Hours

p. 196 **How Risky Is It?**

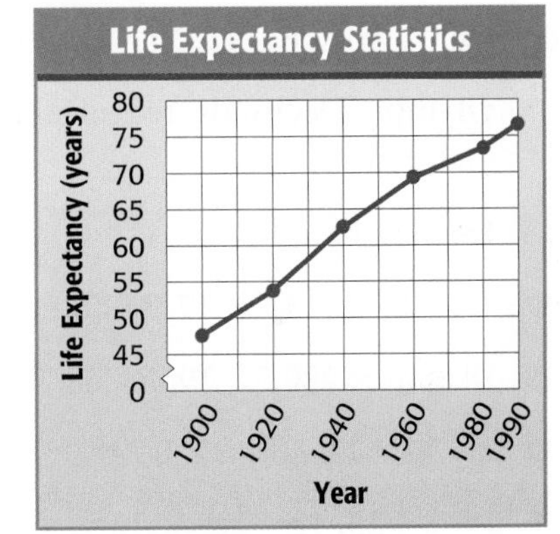

Life expectancy has been steadily increasing during the past century.; Sample answers: better nutrition and healthcare, safer transportation, increased awareness of safe working conditions, better tools to work with

p. 197 **6.**

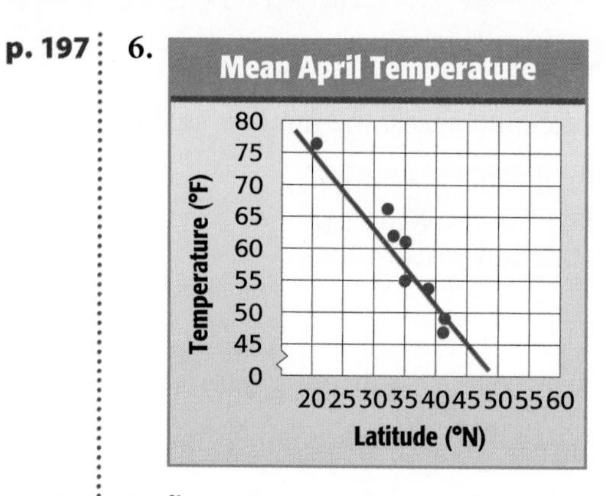

7. about 65°F

p. 199 **8.** flat **9.** normal **10.** skewed to right **11.** bimodal **12.** skewed to left

4•4 Statistics

p. 202 **1.** 8 **2.** 84 **3.** 27° **4.** 92 points

p. 203 **5.** 11 **6.** 2.1 **7.** 18 **8.** 23,916

p. 205 **9.** 7 **10.** 1.6 **11.** 10 **12.** 49

Olympic Decimals technical merit 9.52; composition and style 9.7

p. 206 **13.** 37 **14.** 6.8

p. 207 **15.** 5.1 **16.** 52° **17.** 34 points

p. 209 **18.** 246.5; 290.5 **19.** 58; 68 **20.** 3.10; 3.34

p. 210 **21.** 44 **22.** 10 **23.** 0.24

p. 211 **24.** 2 **25.** 42 **26.** 36.1

4•5 Combinations and Permutations

p. 215 **1.** 216 three-digit numbers **2.** 36 routes **3.** 12;

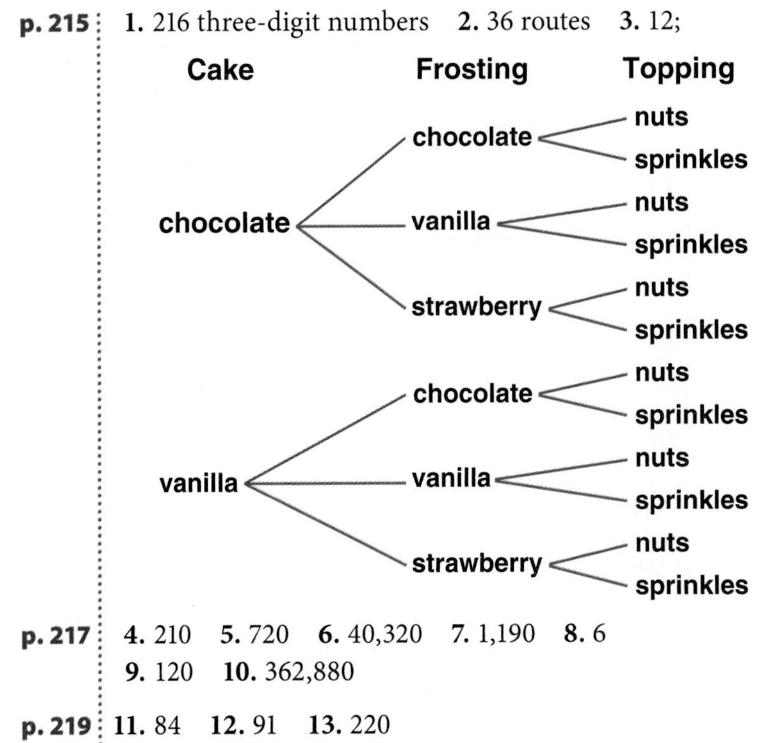

p. 217 **4.** 210 **5.** 720 **6.** 40,320 **7.** 1,190 **8.** 6
9. 120 **10.** 362,880

p. 219 **11.** 84 **12.** 91 **13.** 220
14. There are twice as many permutations as combinations.

4•6 Probability

p. 222 **1.** $\frac{1}{2}$ **2.** $\frac{1}{20}$ **3.** Sample answer: $\frac{2}{50}$ or $\frac{1}{25}$

p. 223 **4.** $\frac{3}{4}$ **5.** 0 **6.** $\frac{1}{6}$ **7.** $\frac{4}{11}$

p. 224 **8.** $\frac{1}{4}$, 0.25, 1:4, 25% **9.** $\frac{1}{8}$, 0.125, 1:8, 12.5%
10. $\frac{1}{8}$, 0.125, 1:8, 12.5% **11.** $\frac{1}{25}$, 0.04, 1:25, 4%

p. 225 **Lottery Fever** struck by lightning; $\frac{260}{260{,}000{,}000}$ is about 1 in 1 million, compared to the 1-in-16-million chance of winning a 6-out-of-50 lottery.

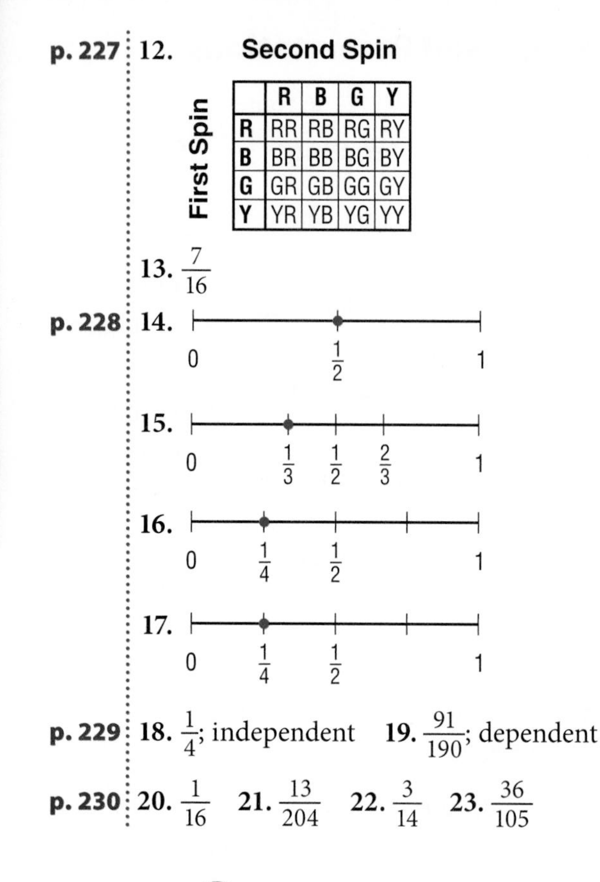

p. 227 **12.** Second Spin

First Spin	R	B	G	Y
R	RR	RB	RG	RY
B	BR	BB	BG	BY
G	GR	GB	GG	GY
Y	YR	YB	YG	YY

13. $\frac{7}{16}$

p. 228 **14.**

15.

16.

17.

p. 229 **18.** $\frac{1}{4}$; independent **19.** $\frac{91}{190}$; dependent

p. 230 **20.** $\frac{1}{16}$ **21.** $\frac{13}{204}$ **22.** $\frac{3}{14}$ **23.** $\frac{36}{105}$

Chapter 5 Logic

p. 234 **1.** false **2.** false **3.** true **4.** true **5.** true **6.** true
7. If it is Tuesday, then the jet flies to Belgium.
8. If it is Sunday, then the bank is closed. **9.** If $x^2 = 49$, then $x = 7$. **10.** If an angle is acute, then it has a measure less than 90°.

p. 235 **11.** The playground will not close at sundown.
12. These two lines do not form an angle.
13. If two lines do not intersect, then they do not form four angles. **14.** If a pentagon is not equilateral, then it does not have five equal sides.
15. Thursday **16.** any nonisosceles trapezoid
17. {a, c, d, e, 3, 4} **18.** {e, m, 2, 4, 5}
19. {a, c, d, e, m, 2, 3, 4, 5} **20.** {e, 4}

5•1 If/Then Statements

p. 237 **1.** If lines are perpendicular, then they meet to form right angles. **2.** If an integer ends in 0 or 5, then it is a multiple of 5. **3.** If you are a runner, then you participate in marathons. **4.** If an integer is odd, then it ends in 1, 3, 5, 7, or 9. **5.** If Jacy is too young to vote, then he is 15 years old. **6.** If you see a cumulus cloud, then it is raining.

p. 238 **7.** A rectangle does not have four sides. **8.** The donuts were not eaten before noon. **9.** If an integer does not end with 0 to 5, then it is not a multiple of 5. **10.** If I am not in Seattle, then I am not in the state of Washington.

p. 239 **11.** true **12.** If an angle is not a right angle, then it does not have a measure of 90°. **13.** If $2x = 6$, then $x = 3$. **14.** If school is not canceled, then it will not snow. **15.** If you do not buy an adult ticket, then you are not over 12 years old. **16.** If you did not pay less for your tickets, then you did not buy them in advance.

5•2 Counterexamples

p. 241 **1.** true; false; counterexample: skew lines **2.** true; true

p. 242 **150,000 . . . . , But Who's Counting** If your town has 150,002 or more people, there are two people with the same number of hairs on their head. If your town has 150,001 or fewer people, you cannot prove whether two have the same number of hairs. This is the same logic as on page 242.

5•3 Sets

p. 244 **1.** false **2.** true **3.** true **4.** {1}; {4}; {1, 4}; ∅ **5.** {m}; ∅ **6.** {a}; {b}; {c}; {a, b}; {b, c}; {a, c}; {a, b, c}; ∅

p. 245 **7.** {1, 2, 9, 10} **8.** {m, a, p, t, h} **9.** {∞, %, $, #, ▲, ♪, ★} **10.** {9} **11.** ∅ **12.** {∞, %, $}

p. 246 **13.** {1, 2, 3, 4, 5, 6} **14.** {1, 2, 3, 4, 5, 6, 9, 12, 15} **15.** {6, 12} **16.** {6}

Chapter 6 Algebra

p. 250 **1.** $4(n + 2) = 2n - 4$ **2.** $a + 3b$ **3.** $11n - 10$
4. 3 hrs. **5.** $y = 54$ **6.** $n = 4$ **7.** 16 girls

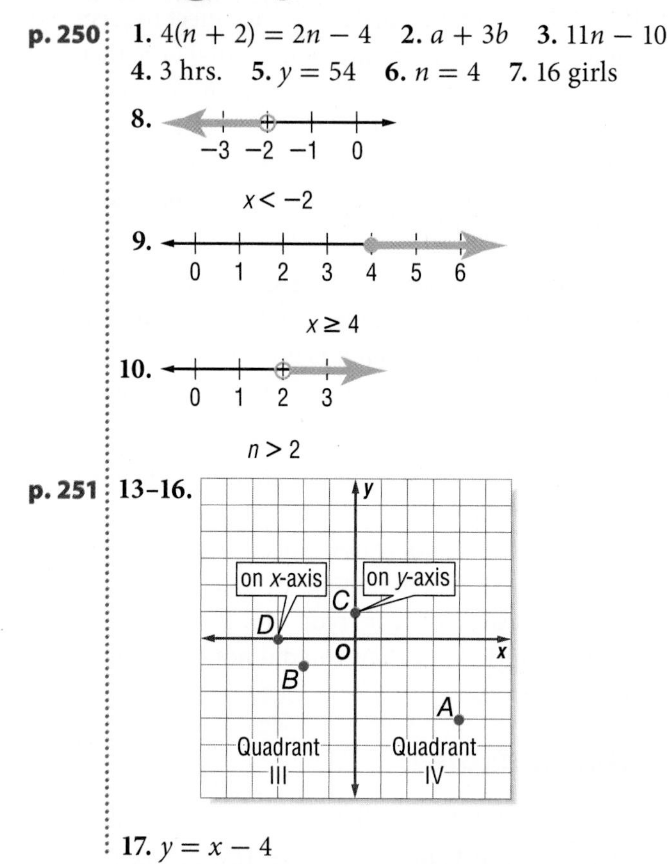

17. $y = x - 4$

6•1 Writing Expressions and Equations

p. 252 **1.** 2 **2.** 1 **3.** 3 **4.** 2

p. 253 **5.** $7 + x$ **6.** $n + 10$ **7.** $y + 3$ **8.** $n + 1$

p. 254 **9.** $14 - x$ **10.** $n - 2$ **11.** $y - 8$ **12.** $n - 9$

p. 255 **13.** $3x$ **14.** $7n$ **15.** $0.35y$ **16.** $12n$

p. 256 **17.** $\frac{x}{7}$ **18.** $\frac{16}{n}$ **19.** $\frac{40}{y}$ **20.** $\frac{a}{11}$

p. 257 **21.** $8n - 12$ **22.** $\frac{4}{x} - 1$ **23.** $2(n - 6)$ **24.** $x - 8 = 5x$
25. $4n - 5 = 4 + 2n$ **26.** $\frac{x}{6} + 1 = x - 9$

6•2 Simplifying Expressions

p. 259 **1.** no **2.** yes **3.** no **4.** yes

p. 260 **5.** $5 + 2x$ **6.** $7n$ **7.** $4y + 9$ **8.** $6 \cdot 5$ **9.** $4 + (8 + 11)$
10. $5 \cdot (2 \cdot 9)$ **11.** $2x + (5y + 4)$ **12.** $(7 \cdot 8)n$

p. 261 **13.** $5(100 - 4) = 480$ **14.** $4(100 + 3) = 412$
15. $9(200 - 1) = 1{,}791$ **16.** $4(300 + 10 + 8) = 1{,}272$
17. Identity Property of Multiplication
18. Identity Property of Addition
19. Zero Property of Multiplication

p. 262 **20.** $14x + 8$ **21.** $24n - 16$ **22.** $-7y + 4$ **23.** $9x - 15$

p. 263 **24.** $7(x + 5)$ **25.** $3(6n - 5)$ **26.** $15(c + 4)$
27. $20(2a - 5)$

p.265 **28.** $11y$ **29.** $5x$ **30.** $14a$ **31.** $-3n$ **32.** $3y + 8z$
33. $13x - 20$ **34.** $9a + 4$ **35.** $9n - 4$

6•3 Evaluating Expressions and Formulas

p. 267 **1.** 22 **2.** 1 **3.** 23 **4.** 20

p. 268 **5.** \$3,000 **6.** \$253,125 **7.** \$7,134.40

p. 269 **8.** 36 mi **9.** 1,875 km **10.** 440 mi **11.** 9.2 ft

Maglev $1\frac{1}{4}$ hr; $2\frac{1}{4}$ hr; $3\frac{3}{4}$ hr

6•4 Solving Linear Equations

p. 271 **1.** -4 **2.** x **3.** 35 **4.** $-10y$

p. 272 **5.** $x = 9$ **6.** $n = 16$ **7.** $y = -7$

p. 273 **8.** $x = 7$ **9.** $y = 32$ **10.** $n = -3$

Prime Time 162,037,037

p. 275 **11.** $x = 3$ **12.** $y = 50$ **13.** $n = -7$ **14.** $a = -6$
15. $m = -9$ **16.** $n = 4$ **17.** $x = -2$ **18.** $a = 6$

p. 277 **19.** $n = 5$ **20.** $t = -2$ **21.** $x = -6$ **22.** $w = \frac{A}{\ell}$
23. $y = \frac{3x + 8}{2}$ **24.** $b = \frac{9 - 3a}{6}$ or $\frac{3 - a}{2}$

6•5 Ratio and Proportion

p. 279 **1.** $\frac{3}{9} = \frac{1}{3}$ **2.** $\frac{9}{12} = \frac{3}{4}$ **3.** $\frac{12}{3} = \frac{4}{1} = 4$

p. 280 **4.** yes **5.** no **6.** yes

p. 281 **7.** 5.5 gal **8.** \$450 **9.** 970,000 **10.** 22,601,000

6•6 Inequalities

p. 284 **1.** −3 −2 −1 0 1 2 3

2. −5 −4 −3 −2 −1 0

3. −4 −3 −2 −1 0 1

4. −2 −1 0 1 2 3

5. let a = driving age; $a \geq 16$ **6.** let c = cost of cell phone; $c > \$19.99$ **7.** let t = age; $t \leq 57$
8. let n = a number; $n - 9 \leq 3$

p. 285 **9.** $x > -3$ **10.** $n \leq 20$

Oops! If a can be positive, negative, or zero, then $2 + a$ can be greater than, equal to, or less than 2.

p. 286 **11.** $x < 4$ **12.** $x \geq 7$ **13.** $x < -4$ **14.** $x \geq -3$

6•7 Graphing on the Coordinate Plane

p. 288 **1.** y-axis **2.** Quadrant II **3.** Quadrant IV **4.** x-axis

p. 289 **5.** (−2, 4) **6.** (1, −3) **7.** (−4, 0) **8.** (0, 1)

p. 290 **9–12.**

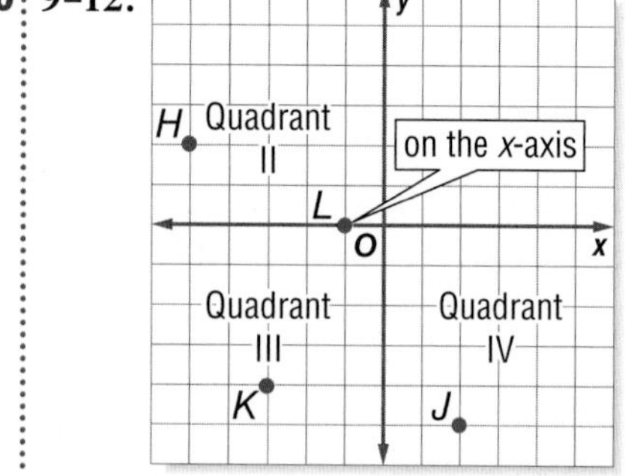

p. 291 **13.** $(n + 1)(-6)$; $-36, -42, -48$ **14.** $\frac{1}{n^2}$; $\frac{1}{25}, \frac{1}{36}, \frac{1}{49}$

15. $3n + 1$; 16, 19, 22

p. 293 **16.** **17.** **18.** **19.**

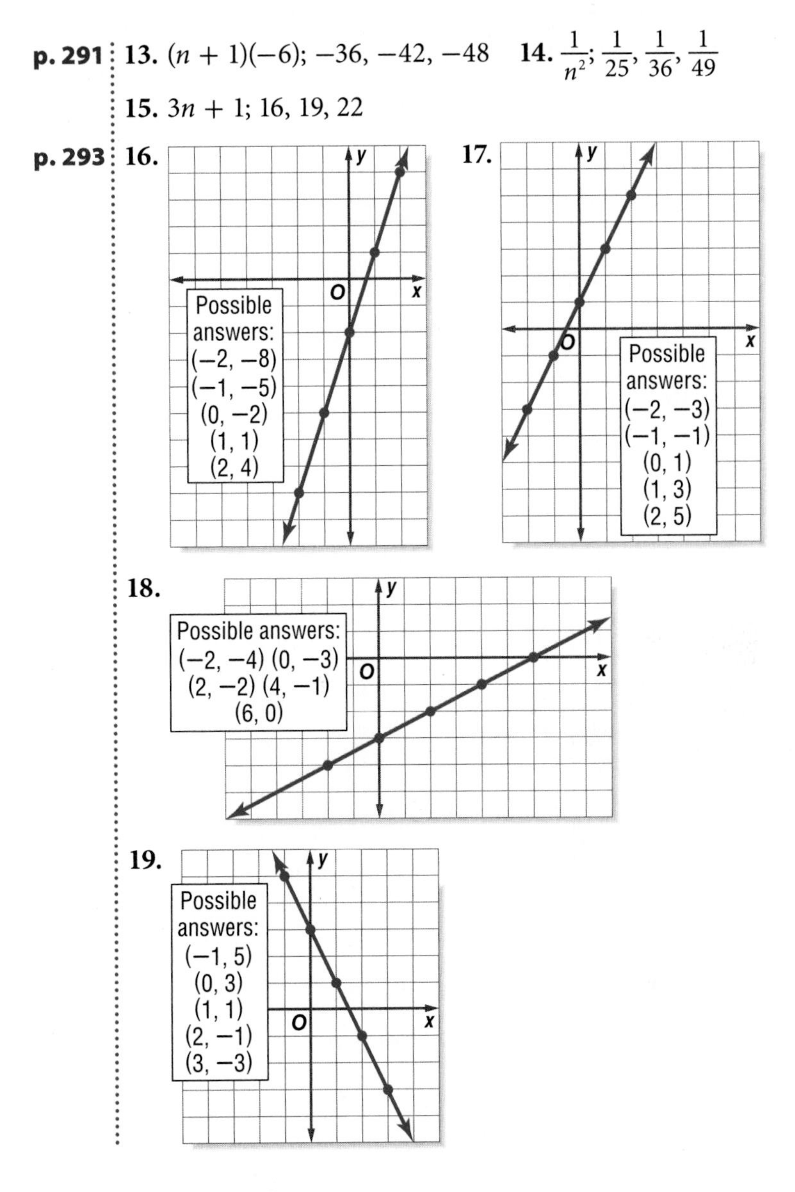

6•8 Slope and Intercept

p. 296 **1.** $\frac{2}{3}$ **2.** $-\frac{5}{1} = -5$

p. 297 **3.** -1 **4.** $\frac{3}{2}$ **5.** $-\frac{1}{2}$ **6.** 5

p. 298 7. 0 **8.** no slope **9.** no slope **10.** 0

11. 0 **12.** $-\frac{1}{3}$ **13.** 1 **14.** no *y*-intercept

p. 299 **15.**

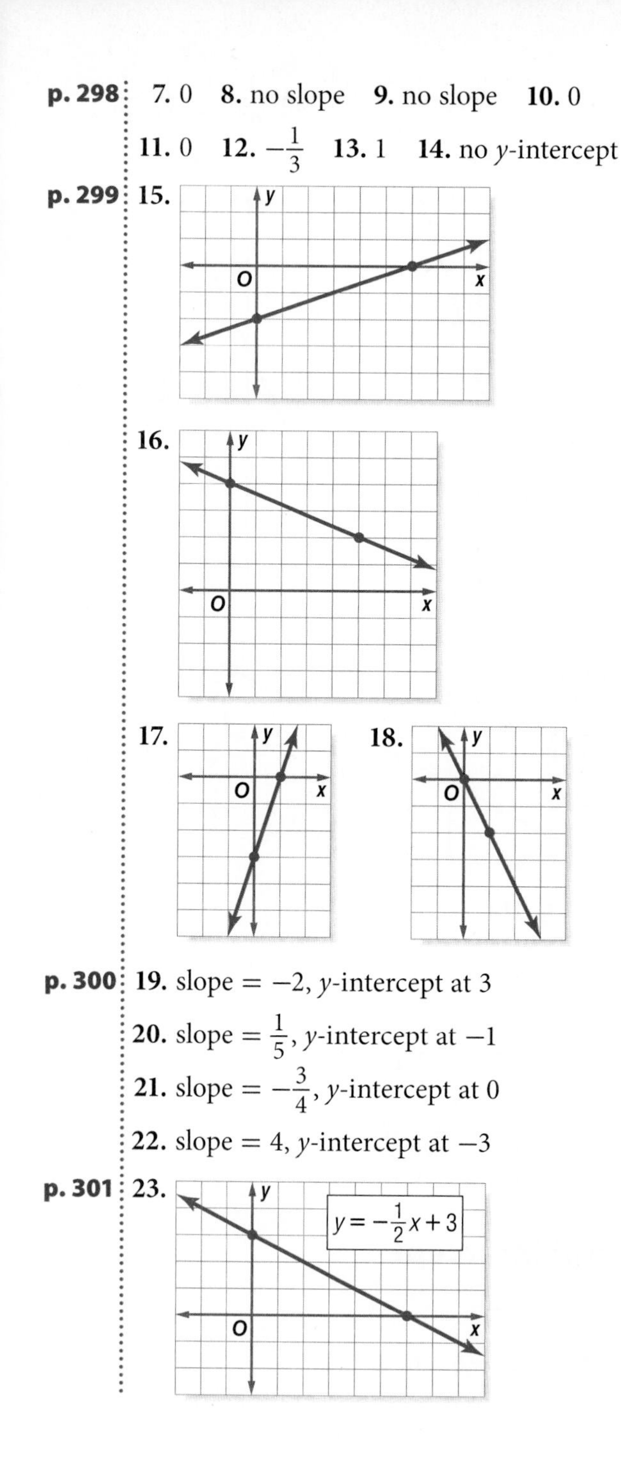

16.

17. **18.**

p. 300 **19.** slope = −2, *y*-intercept at 3

20. slope = $\frac{1}{5}$, *y*-intercept at −1

21. slope = $-\frac{3}{4}$, *y*-intercept at 0

22. slope = 4, *y*-intercept at −3

p. 301 **23.**

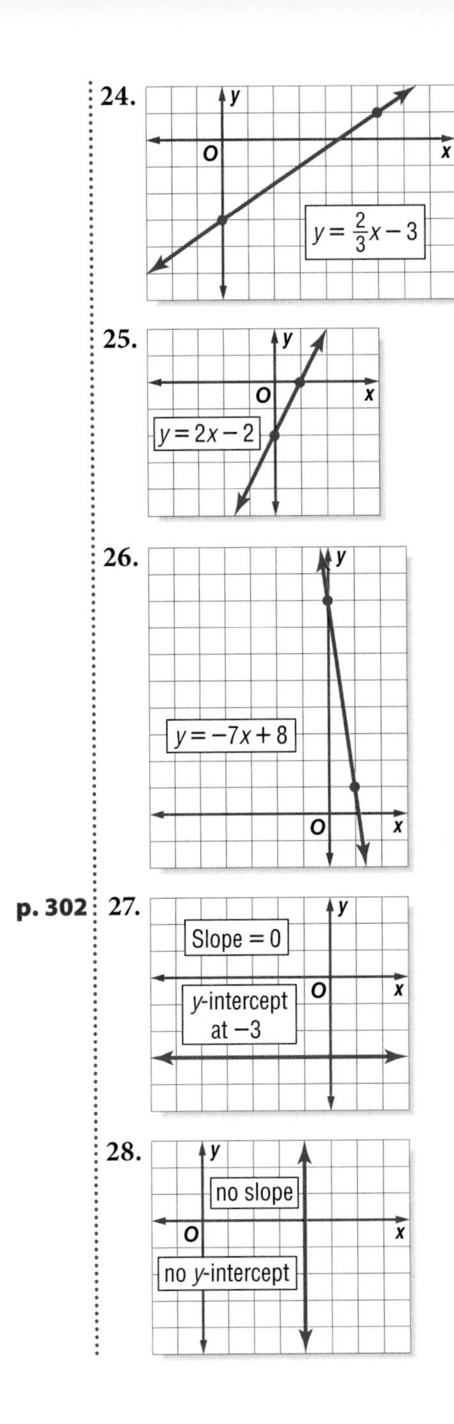
24.
y
O
x
$y = \frac{2}{3}x - 3$
25.
y
O
x
$y = 2x - 2$
26.
y
$y = -7x + 8$
O
x
p. 302
27.
y
Slope = 0
O
x
y-intercept at −3
28.
y
no slope
O
x
no y-intercept

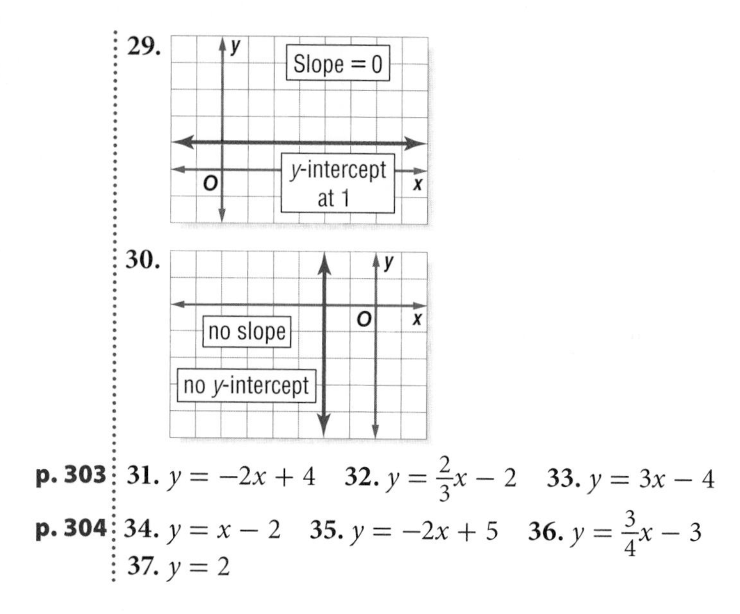

p. 303 **31.** $y = -2x + 4$ **32.** $y = \frac{2}{3}x - 2$ **33.** $y = 3x - 4$

p. 304 **34.** $y = x - 2$ **35.** $y = -2x + 5$ **36.** $y = \frac{3}{4}x - 3$
37. $y = 2$

6•9 Direct Variation

p. 307 **1.** 413.1 kph **2.** \$3.94 **3.** The ratios are not the same, so the function is not a direct variation.

6•10 Systems of Equations

p. 309 **1.** $(-4, -1)$ **2.** $(2, -2)$

p. 310 **3.** no solution

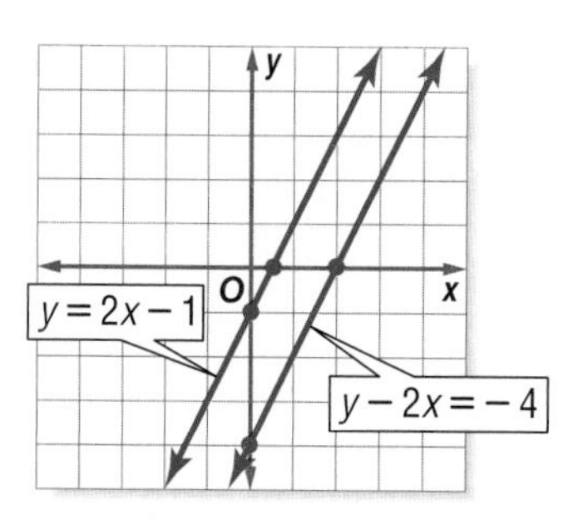

4. (1, 0)

5. no solution

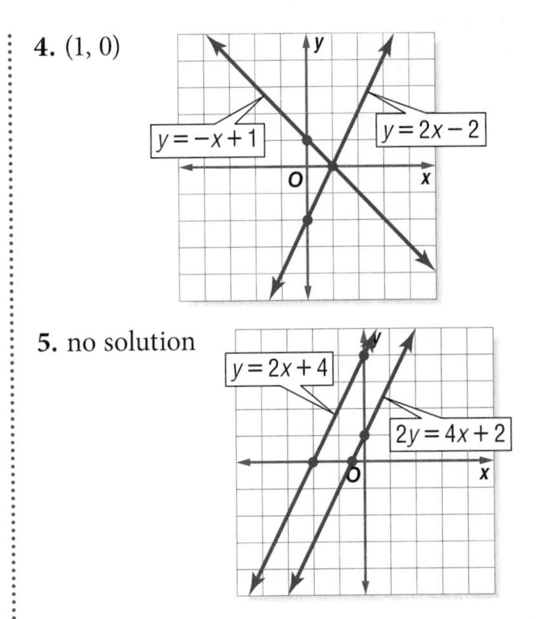

p. 311

6. all ordered pairs of the points on the line $y = 3x - 2$

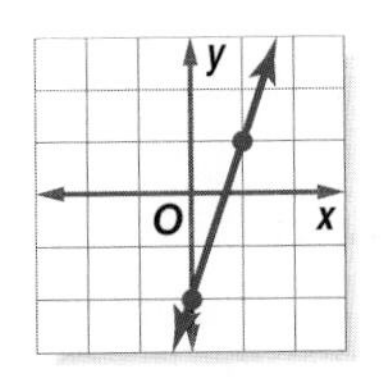

7. all ordered pairs of the points on the line $y = 4x + 6$

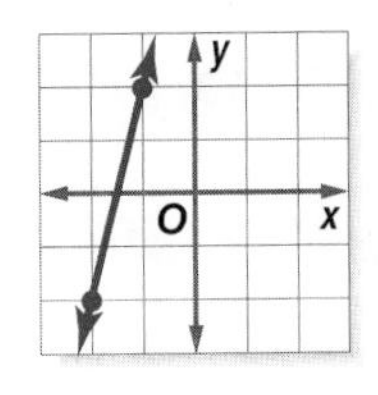

p. 312

8. $x = y + 3$; $x + y = 9$; (6, 3)

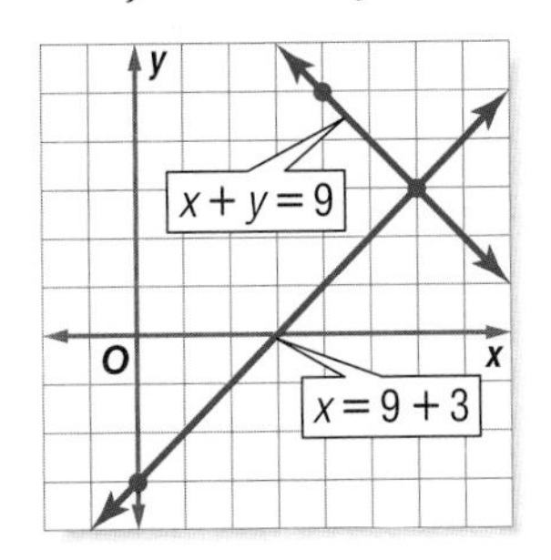

Chapter 7 Geometry

p. 316 **1.** corresponding angles; 131° **2.** 108° **3.** 30 cm **4.** 65 ft^2 **5.** 386.6 cm^2

p. 317 **6.** 118 in^3 **7.** 157 ft and 1,963 ft^2

8. 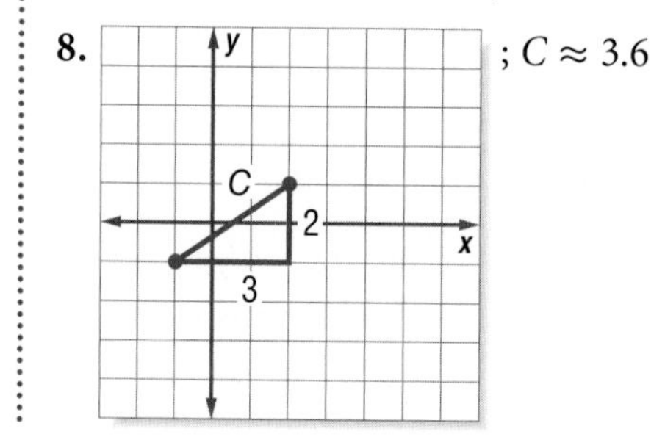

; $C \approx 3.6$

7•1 Classifying Angles and Triangles

p. 318 **1.** reflex angle **2.** straight angle **3.** acute angle **4.** obtuse angle

p. 320 **5.** 30° **6.** 100°

p. 322 **7.** corresponding angles **8.** alternate interior angles **9.** $\angle 1$ and $\angle 7$, or $\angle 2$ and $\angle 8$ **10.** Possible pairs include any two obtuse angles: $\angle 1$, $\angle 3$, $\angle 5$, or $\angle 7$, or any two acute angles: $\angle 2$, $\angle 4$, $\angle 6$, or $\angle 8$.

p. 323 **11.** $m\angle Z = 90°$ **12.** $m\angle M = 60°$ **13.** 39° **14.** 97°

7•2 Naming and Classifying Polygons and Polyhedrons

p. 326 **1.** Sample answers: *RSPQ; QPSR; SPQR; RQPS; PQRS; QRSP; PSRQ; SRQP* **2.** 360° **3.** 105°

p. 327 **4.** no; yes; no; yes; no **5.** yes, because it has four sides that are the same length and opposite sides are parallel **6.** no, because all sides of the rectangle may not be of equal length **7.** no, because it may not have 4 right angles

p. 329 **8.** yes; quadrilateral **9.** no **10.** yes; hexagon

p. 330 **11.** 1,440° **12.** 108°

p. 331 **13.** triangular prism
14. triangular pyramid or tetrahedron

7•3 Symmetry and Transformations

p. 335 **1.**

2.

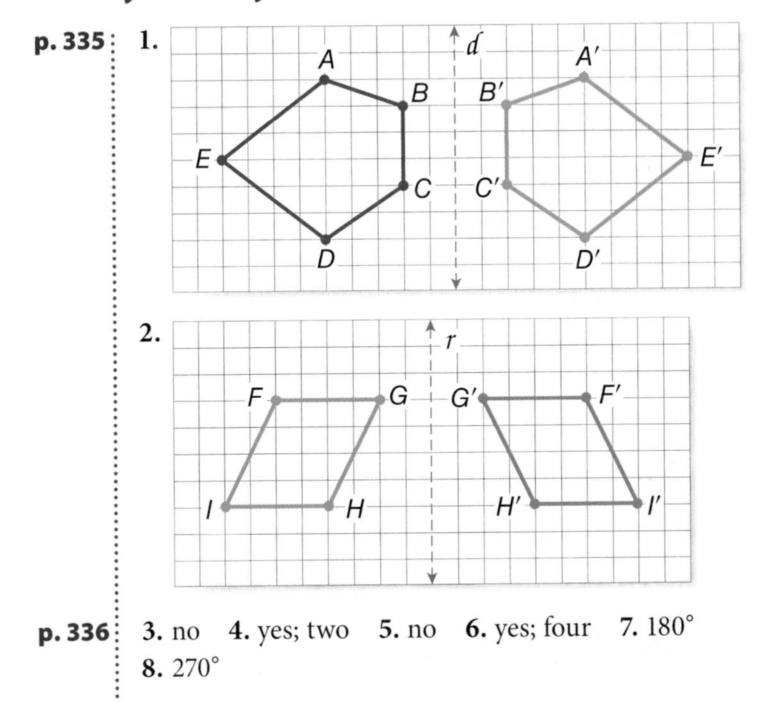

p. 336 **3.** no **4.** yes; two **5.** no **6.** yes; four **7.** 180°
8. 270°

p. 337 **9.** yes **10.** no **11.** no

7•4 Perimeter

p. 339 **1.** 29 cm **2.** 39 in. **3.** 6 m **4.** 20 ft

p. 340 **5.** 60 cm **6.** 48 cm **7.** 5.2 m

p. 341 **8.** 35.69 in. **9.** 38.58 m

7•5 Area

p. 345 **1.** $6\frac{2}{3}$ ft^2 or 960 in^2 **2.** 36 cm^2

3. 54 m^2 **4.** 8 m

p. 346 **5.** 60 in^2 **6.** 540 cm^2

p. 347 **7.** 12 ft^2 **8.** 30 ft^2

7•6 Surface Area

p. 350 **1.** 126 m^2 **2.** 88 cm^2

p. 351 **3.** 560 cm^2 **4.** A **5.** 1,632.8 cm^2

7•7 Volume

p. 353 **1.** 3 cm^3 **2.** 6 ft^3

p. 354 **3.** 896 in^3 **4.** 27 cm^3

p. 355 **5.** 113.04 in^3 **6.** 56.52 cm^3

p. 357 **7.** 9.4 m^3 **8.** 2,406.7 in^3

Good Night, T.Rex about 1,176,000 mi^3

7•8 Circles

p. 360 **1.** 9 in. **2.** 1.5 m **3.** $\frac{x}{2}$ **4.** 12 cm **5.** 32 m **6.** $2y$

p. 361 **7.** 5π in. **8.** 20.1 cm **9.** 7.96 m **10.** $5\frac{1}{2}$ in.

p. 362 **11.** $\angle ABC$ **12.** 90° **13.** 270° **14.** 120° **15.** 240°

p. 363 **16.** 132.7 m^2 **17.** 20.25π in^2; 63.6 in^2 **18.** 177 cm^2

7•9 Pythagorean Theorem

p. 365 **1.** 9, 16, 25 **2.** yes

p. 367 **3.** 14 cm **4.** 55 in.

p. 368 **5.** $c \approx 3.6$ **6.** $c \approx 7.2$

Chapter 8 Measurement

p. 372 **1.** one hundredth **2.** one thousand
3. one thousandth **4.** 0.8 **5.** 5.5 **6.** 15,840
7. 13 **8.** 108 in. **9.** 3 yd **10.** 274 cm **11.** 3 m

p. 373 **12.** 684 in^2 **13.** 4,181 cm^2 **14.** 50,000 **15.** 90
16. 5,184 **17.** 4,000 **18.** $\frac{6}{8}$ or $\frac{3}{4}$
19. 16 bottles **20.** about 13 cans **21.** about 4.4 lb
22. 3 lb **23.** 9:4 **24.** 2.25 or $\frac{9}{4}$

8•1 Systems of Measurement

p. 374 **1.** metric **2.** customary **3.** metric

8•2 Length and Distance

p. 377 **1.** 800 **2.** 3.5 **3.** 4 **4.** 10,560 **5.** 71.1 cm
6. 89.7 yd **7.** B

8•3 Area, Volume, and Capacity

p. 380 **1.** 1,600 mm^2 **2.** 288 in^2 **3.** 14,520 yd^2

p. 381 **4.** 324 ft^3 = 12 yd^3 **5.** 512 cm^3 = 512,000 mm^3
6. 25,920 in^3 **7.** 0.25 cm^3

p. 382 **8.** the juice

In the Soup! 1,792 cartons; 912.6 ft^3

8•4 Mass and Weight

p. 384 **1.** 80 **2.** 3.75 **3.** 8,000,000 **4.** 0.375

8•5 Size and Scale

p. 386 **1.** yes

p. 388 **2.** 2

p. 389 **3.** $\frac{9}{4}$ **4.** 16 ft^2 **5.** Complete the table below.

	Area
Scale Factor 2	4 times
Scale Factor 3	9 times
Scale Factor 4	16 times
Scale Factor 5	25 times
Scale Factor *X*	x^2 times

p. 390 **6.** $\frac{27}{64}$ **7.** Complete the table below.

	Volume
Scale Factor 2	8 times
Scale Factor 3	27 times
Scale Factor 4	64 times
Scale Factor 5	125 times
Scale Factor *X*	x^3 times

Chapter 9 Tools

p. 394 **1.** 55,840.59 **2.** 0.29 **3.** 20.25 **4.** 2.12 **5.** $3.2 \cdot 10^{13}$ **6.** 29.12 **7.** 74° **8.** 148° **9.** 74° **10.** yes

p. 395 **11.** straightedge and compass
12–15. Check that students' drawings match original figures. **16.** A2 **17.** C1 + C2 **18.** 66

9•1 Scientific Calculator

p. 399 **1.** 479,001,600 **2.** 38,416 **3.** 0.037 **4.** 109,272.375 **5.** 85

Magic Numbers $\frac{100a + 10a + 1a}{a + a + a} = \frac{111a}{3a} = 37$

9•2 Geometry Tools

p. 402 **1.** 54° **2.** 122°

p. 403 **3.**

4.

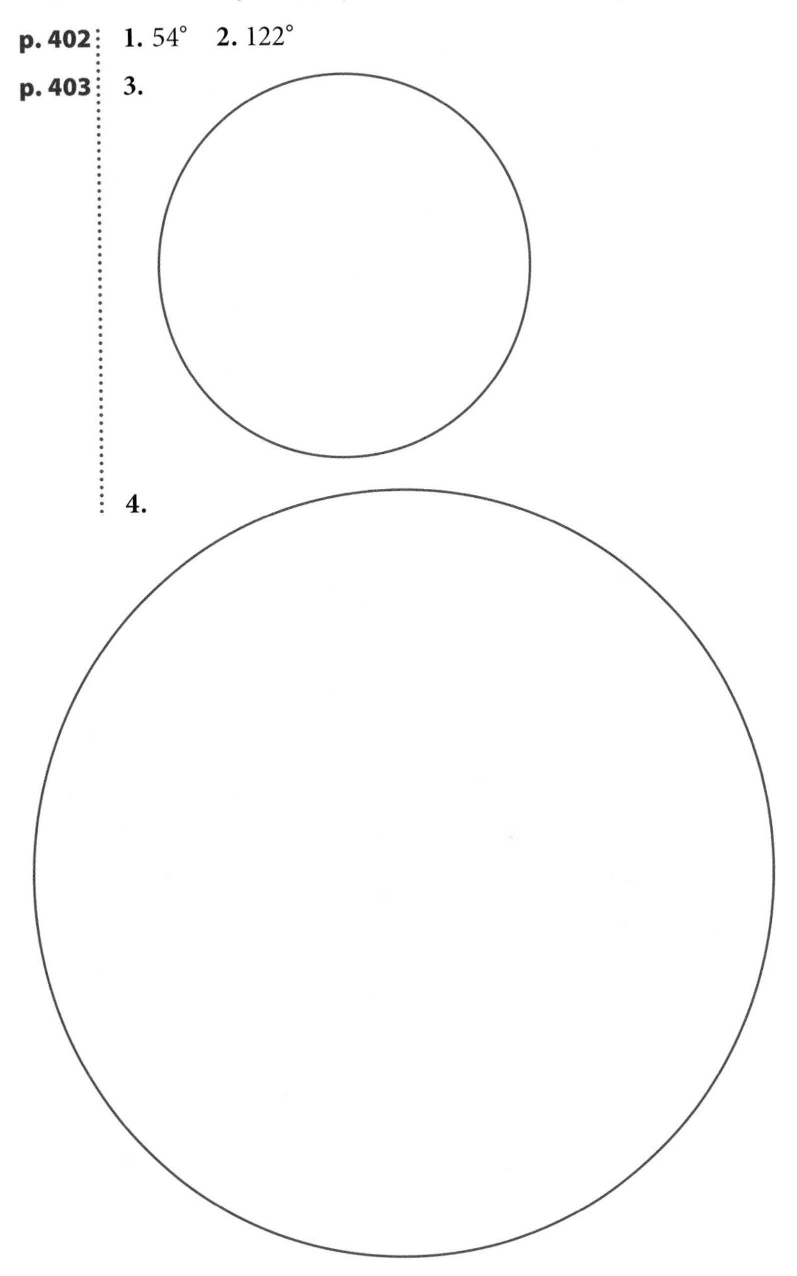

p. 404 **5.**

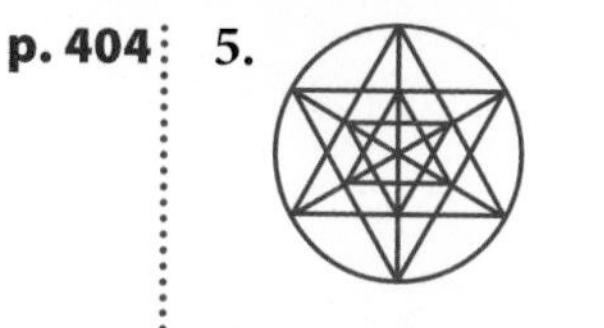

6. Sample answer:

9•3 Spreadsheets

p. 407 **1.** 2 **2.** 3 **3.** 25

p. 408 **4.** B3 * C3 **5.** B4 * C4 **6.** D2 + D3 + D4

p. 411 **7.** A6 + 10; 160 **8.** D2 + 10; 120

p. 412 **9.** A2, B2 **10.** A5, B5 **11.** A3, B3

Index

A

B

I

L

M

S

Photo Credits

All coins photographed by United States Mint.
002–003 CORBIS; **070–071** Jupiterimages; **131** Larry Dale Gordon/Getty Images; **140** Janis Christie/Getty Images; **162** Getty Images; **196** Dale O'Dell/Alamy; **205** CORBIS; **225** Alamy; **242** David R. Frazier/Alamy; **269** Paul Souders/Getty Images; **273** Yasuhide Fumoto/Getty Images; **285** PictureQuest; **357** U.S. Geological Survey; **382** Alamy; **399** SuperStock; **416–417** Alamy.